Competitiveness Strategy in Developing Countries

The ongoing process of globalisation has prompted an extensive debate on how to enhance industrial competitiveness of economies all over the world. This book focuses on the way developing economies, best-practice policies and institutions for improving industrial competitiveness in an integrated world economy are introduced. Leading international contributors cover such important themes as:

- benchmarking competitiveness across countries
- management of competitiveness strategy
- international trade policies
- privatisation, regulation and domestic competition policy
- foreign direct investment policies and promotion
- science, technology and innovation policies
- business development services and financial policies for small enterprises

This comprehensive collection is a vital contribution to an important debate. It will be essential reading for students in the field of development economics, international business and innovation. Not only this, but it should find itself a place on development policy-makers' bookshelves.

Ganeshan Wignaraja is Managing Economist, Maxwell Stamp PLC. He is also a Visiting Fellow at the United Nations University Institute for New Technologies and a Research Associate at the Overseas Development Institute.

Routledge studies in development economics

1 **Economic Development in the Middle East**
   Rodney Wilson

2 **Monetary and Financial Policies in Developing Countries**
   Growth and stabilization
   Akhtar Hossain and Anis Chowdhury

3 **New Directions in Development Economics**
   Growth, environmental concerns and government in the 1990s
   Edited by Mats Lundahl and Benno J. Ndulu

4 **Financial Liberalization and Investment**
   Kanhaya L. Gupta and Robert Lensink

5 **Liberalization in the Developing World**
   Institutional and economic changes in Latin America, Africa and Asia
   Edited by Alex E. Fernández Jilberto and André Mommen

6 **Financial Development and Economic Growth**
   Theory and experiences from developing countries
   Edited by Niels Hermes and Robert Lensink

7 **The South African Economy**
   Macroeconomic prospects for the medium term
   Finn Tarp and Peter Brixen

8 **Public Sector Pay and Adjustment**
   Lessons from five countries
   Edited by Christopher Colclough

9 **Europe and Economic Reform in Africa**
   Structural adjustment and economic diplomacy
   Obed O. Mailafia

10 **Post-apartheid Southern Africa**
    Economic challenges and policies for the future
    Edited by Lennart Petersson

11 **Financial Integration and Development**
    Liberalization and reform in Sub-Saharan Africa
    Ernest Aryeetey and Machiko Nissanke

12 **Regionalization and Globalization in the Modern World Economy**
    Perspectives on the Third World and transitional economies
    Edited by Alex E. Fernández Jilberto and André Mommen

13 **The African Economy**
Policy, institutions and the future
Steve Kayizzi-Mugerwa

14 **Recovery from Armed Conflict in Developing Countries**
Edited by Geoff Harris

15 **Small Enterprises and Economic Development**
The dynamics of micro and small enterprises
Carl Liedholm and Donald C. Mead

16 **The World Bank**
New agendas in a changing world
Michelle Miller-Adams

17 **Development Policy in the Twenty-First Century**
Beyond the post-Washington consensus
Edited by Ben Fine, Costas Lapavitsas and Jonathan Pincus

18 **State-Owned Enterprises in the Middle East and North Africa**
Privatization, performance and reform
Edited by Merih Celasun

19 **Finance and Competitiveness in Developing Countries**
Edited by José María Fanelli and Rohinton Medhora

20 **Contemporary Issues in Development Economics**
Edited by B.N. Ghosh

21 **Mexico Beyond NAFTA**
Edited by Martín Puchet Anyul and Lionello F. Punzo

22 **Economies in Transition**
A guide to China, Cuba, Mongolia, North Korea and Vietnam at the turn of the twenty-first century
Ian Jeffries

23 **Population, Economic Growth and Agriculture in Less Developed Countries**
Nadia Cuffaro

24 **From Crisis to Growth in Africa?**
Edited by Mats Lundal

25 **The Macroeconomics of Monetary Union**
An analysis of the CFA franc zone
David Fielding

26 **Endogenous Development**
Networking, innovation, institutions and cities
Antonio Vasquez-Barquero

27 **Labour Relations in Development**
Edited by Alex E. Fernández Jilberto and Marieke Riethof

28 **Globalization, Marginalization and Development**
Edited by S. Mansoob Murshed

29 **Programme Aid and Development**
Beyond conditionality
Howard White and Geske Dijkstra

30 **Competitiveness Strategy in Developing Countries**
Edited by Ganeshan Wignaraja

Competitiveness Strategy in Developing Countries

Edited by
Ganeshan Wignaraja

LONDON AND NEW YORK

First published 2003
by Routledge
11 New Fetter Lane, London EC4P 4EE

Simultaneously published in the USA and Canada
by Routledge
29 West 35th Street, New York, NY 10001

Routledge is an imprint of the Taylor & Francis Group

© 2003 Ganeshan Wignaraja for selection and editorial
matter; individual contributors their chapters

Transferred to Digital Printing 2003

Typeset in Goudy by
Newgen Imaging Systems (P) Ltd, Chennai, India
Printed and bound in Great Britain by
Biddles Short Run Books, King's Lynn

All rights reserved. No part of this book may be reprinted or
reproduced or utilised in any form or by any electronic,
mechanical, or other means, now known or hereafter
invented, including photocopying and recording, or in any
information storage or retrieval system, without permission in
writing from the publishers.

British Library Cataloguing in Publication Data
A catalogue record for this book is available from the British Library

Library of Congress Cataloging in Publication Data
A catalog record for this book has been requested

ISBN 0-415-22836-0

Contents

List of illustrations ix
List of contributors xi
Editor's note xii

PART I
Introduction, concepts and benchmarking 1

1 Introduction 3
GANESHAN WIGNARAJA

2 Competitiveness analysis and strategy 15
GANESHAN WIGNARAJA

3 Benchmarking competitiveness: a first look at the MECI 61
GANESHAN WIGNARAJA AND ASHLEY TAYLOR

PART II
Supply-side issues and policies for competitiveness 93

4 Science, technology and innovation policy 95
STAN METCALFE

5 Industrial clusters and business development services for small and medium-sized enterprises 131
EILEEN FISCHER AND REBECCA REUBER

6 Government policies towards foreign direct investment 166
DIRK WILLEM TE VELDE

7 Financial sector policies for enterprise development 198
ANDY W. MULLINEUX AND VICTOR MURINDE

PART III
Incentive policies for competitiveness 237

8 Privatisation, regulation and domestic competition policy 239
   CHRISTOS N. PITELIS

9 International trade policies 274
   SHEILA PAGE

Index 298

Illustrations

Figures

2.1	Enterprise-level learning process	23
2.2	National innovation system (NIS)	27
8.1	A decision tree for SOE reform	266

Tables

2.1	Government actions for competitiveness in a developing economy	32
3.1	Construction of MECI for Thailand	70
3.2	MECI ranking by country	71
3.3	Performance by region	74
3.4	Performance by income grouping	74
3.5	Sample composition of MECI, WEF (2001) and IMD (2001) indices	75
3.6	Rankings of MECI, WEF current competitiveness index (2001) and IMD index (2001)	76
3.7	t-Tests to compare the means of high performing countries and low performing countries	81
3.8	t-Tests to compare the means within the low performing sample	83
3.9	High performing and low performing samples	87
3.10	Low performing sub-samples	87
4.1	Breakdown of innovation expenditure	103
4.2	Government support for industrial technology by type, 1995	114
5.1	Examples of clusters in less developed countries	136
5.2	Four kinds of performance criteria for BDS	157
5.3	Contingent development trajectories for industrial clusters	162
6.1	Selected empirical studies on spillovers from FDI	169
6.2	Policies and factors affecting inward FDI	174
6.3	Selected econometric evidence of policy effects on attraction and effects of FDI	183

x *Illustrations*

7.1	Financial sector problems and possible causes	201
7.2	Loan guarantee schemes	213
7.3	Financial sector problems and possible solutions	228
8.1	A decision-making framework for privatisation	262

Boxes

1.1	Knowledge and the new manufacturing context for developing countries	4
2.1	Technological learning in Tatung and Astek fruit processing	24
2.2	Gaps in NIS in developing countries	29
2.3	Denmark's Industrial Network Programme	38
2.4	Public–private consultation mechanisms and partnerships in Côte d'Ivoire	42
2.5	Elements of a competitiveness assessment	44
2.6	Measuring capabilities in manufacturing enterprises	45
3.1	Constructing a simple composite performance index: the HDI	69
4.1	STI policies	97
4.2	Invention, innovation and diffusion	101
4.3	Innovation indicators	113
4.4	UK innovation policies administered by the DTI	120
8.1	Measures of concentration	243
8.2	Mergers, acquisitions and restrictive business practices	244
8.3	Scale, clusters and competition	269
9.1	Countertrade	283
9.2	Industrialisation and development	286
9.3	Unexpected links	287
9.4	Border zones	295

Contributors

Eileen Fischer is Associate Professor at Schulich School of Business, University of York, Canada.

Stan Metcalfe is Professor of Economics at the University of Manchester, UK.

Andy W. Mullineux is Professor of Global Finance at Birmingham Business School, University of Birmingham, UK.

Victor Murinde is Professor of Development Finance at Birmingham Business School, University of Birmingham, UK.

Sheila Page is Research Fellow at the Overseas Development Institute and Co-ordinator of its International Economics Development Group, UK.

Christos N. Pitelis is Director of the Centre for International Business and Management at the Judge Institute of Management in the University of Cambridge and Fellow of Queen's College, UK.

Rebecca Reuber is Associate Professor at Rotman School of Management, University of Toronto, Canada.

Ashley Taylor was a Consultant to the Commonwealth Secretariat but since writing has become an Economist in the International Finance Division of the Bank of England, UK.

Ganeshan Wignaraja is Managing Economist at Maxwell Stamp PLC and a Research Associate at the Overseas Development Institute, UK.

Dirk Willem te Velde is Research Fellow at the Overseas Development Institute, UK.

Editor's note

This volume had its origins in a programme of advisory work by the Commonwealth Secretariat, which sought to assist member governments to formulate public policies to enhance industrial competitiveness in an integrated world economy. This work consisted of country studies of industrial competitiveness for economic ministers as well as executive programmes for senior government officials and the private sector representatives. The first executive programme was held in Barbados in late 1999 in partnership with its government and dealt with the restructuring challenges facing Caribbean enterprises and possible policy options. Subsequent programmes, on a Pan-Commonwealth basis, took place in Singapore in collaboration with Singapore's International Trade Institute of Singapore and its Ministry of Foreign Affairs. These initiatives highlighted the need for a volume on industrial competitiveness in developing countries that would benchmark competitiveness performance, provide a framework for analysis and highlight examples of best-practice policies and institutions. This volume attempts to deal with these and other pressing issues.

The contributors are grateful to the Commonwealth Fund for Technical Co-operation for its support for the volume. They are also grateful to the Barbados Ministry of Industry and International Business and the Singapore's International Trade Institute of Singapore that co-organised the executive programmes as well as participants at these events. Special thanks are due to Richard Gold (Director, Special Advisory Services Division of the Commonwealth Secretariat) for strategic guidance and to Elvis Gannon (its Programme Officer) for his constant support for the volume and efficient management of the details of Barbados and Singapore executive programmes; to Lionel Weekes (Permanent Secretary, Ministry of International Business, Barbados), Tan Song Chuan (President, International Trade Institute of Singapore), Grace Ng (its Vice-President) for their warm hospitality; and to Robert Langham (Senior Economics Editor, Routledge), Terry Clague (Editorial Assistant, Routledge), Prabir Bhambal (General Manager, India Liaison Office of Routledge) and Rupert Jones-Parry (Publications Manager, Commonwealth Secretariat) for assistance with publication of the volume.

The editor owes a great debt to many individuals with whom he has discussed many of the ideas in the volume. In this vein, he is particularly grateful to the other contributors as well as Jagdish Bhagwati, Robert Cassen, Frederic Richard, Vinod Rege, Sanjaya Lall, Lynn Mytelka, Fred Nixson, Razeen Sally, Hubert Schmitz and Frederic von Kirchbach.

Ganeshan Wignaraja

Part I

Introduction, concepts and benchmarking

1 Introduction

Ganeshan Wignaraja

New manufacturing context

Developing countries today face a new manufacturing context. Progressive globalisation, widely acknowledged to be the defining feature of the late twentieth century world economy, seems set to continue into the early twenty-first century.[1] The process of world economic integration has involved a broadening and deepening of the inter-relationships between international trade and foreign direct investment (FDI) flows. World trade growth consistently outpaced world GDP growth during the last two decades and world foreign investment growth far exceeded both of them. Accordingly, international trade and foreign investment flows reached historically unprecedented levels in the late 1990s. The onset of a global recession in 2001–2002 is likely to bring a temporary slowdown in these flows but the integration of the world economy will carry on, albeit at a reduced pace during this period. The outcome is the creation of an international market place for goods and services that seems indifferent to national borders and state regulation. A combination of factors – falling trade barriers (through the implementation of the Uruguay Round Agreements and economic liberalisation), increasing technological progress (especially the information communication and telecommunications (ICT) revolution), declining communications and transport costs and highly mobile multinational enterprises seeking out new investments – have driven world economic integration. This complex process is irreversible and has revolutionary implications for industrial development.

Knowledge and technological progress have become more important to the realisation of economic prosperity within an integrated world economy. These forces are exerting a profound influence on the behaviour of firms as well as the environment around them by altering production processes, new product introduction, supply–chain relationships between firms, demand conditions and regulations (see Box 1.1). This new manufacturing context, based on knowledge and technological progress, provides unparalleled new opportunities and poses new risks for industrialisation in developing countries. It has the potential to offer developing countries (and enterprises within them) with access to new technologies, skills, capital, markets and hence faster industrial growth and greater economic prosperity than ever before. A lack of resources (including skills and technologies) and small markets at national level will pose less of

Box 1.1 Knowledge and the new manufacturing context for developing countries

Progressive globalisation is fundamentally transforming the nature of industrial development and will affect countries at all levels of development. Knowledge and technological progress have become central to economic prosperity in an integrated world economy. This has led some to coin the term 'knowledge driven economy' to describe an economy in which the generation and exploitation of knowledge have come to play the predominant part in the creation of wealth (UK DTI, 1998). This refers to the exploitation of and use of knowledge in all production and service activities and not just those sometimes classified as high tech-tech or knowledge intensive activities. Knowledge and technological progress have always been important but five mutually reinforcing processing are increasing their importance for prosperity:

- *Revolutionary changes in ICT technologies* are transforming every stage of manufacturing (e.g. finding new technology, management of supply–chain relationships and accessing distant markets) and creating entirely new products (e.g. digital televisions) and new services (e.g. software services). Enterprises need to develop efficient manufacturing capabilities (via search, engineering and R&D) to cope with rapid technological progress.
- Rise of *globally integrated value chains*, driven by multinational corporations, are creating 'first mover' advantages for enterprises that manage to insert themselves early into subcontracting relationships. Over time, such enterprises can learn and improve their competitiveness by accessing new technologies, managerial practices and marketing connections of multinational corporations.
- Increasing *global competition* associated with falling trade barriers and transport costs require enterprises to add more value in production processes to stay ahead in their cycles and to compete against lower cost rivals.
- *New rules of the game* (introduced through the World Trade Organisation and by foreign buyers of output) mean that enterprises have to comply with higher technical, environmental and labour standards in export markets. These include things like ISO9000 and ISO14000, technical barriers to trade (TBT) of different kinds and sanitary and phytosanitary measures (SPS).
- Changing *consumer demand* (associated with rising incomes and changing tastes that come with greater prosperity) for more sophisticated, customised and environmentally friendly products places new demands on enterprises.

Taken together, these processes are dramatically transforming the way in which enterprises and governments operate in the knowledge driven

> economy. This calls for a renewed focus on knowledge as a means of improving industrial performance at firm-level and on coherent policies to support industrial restructuring. Close business–government interactions are also increasingly important.
>
> Sources: Based on UK DTI (1998) and UNIDO (2002).

a constraint to industrial growth and structural transformation as developing countries link up with foreign buyers and multinationals and draw on global resources and markets. At the same time, world economic integration brings about a sudden, marked increase in competition from imports and the entry of new foreign investors for local enterprises in domestic markets. There is also likely to be more intense competition within the developing world for export markets, foreign investment and resources. Many developing countries seem to have underestimated the intensity of this global competition and its effects on their enterprises. Accordingly, adjusting to increased global competition has placed unprecedented demands on the capacities of enterprises, institutions and governments in developing countries. In general, old structures, institutions, behavioural patterns and public policies seem ill-adapted to deal with the challenge of global competition.

Not surprisingly, perhaps, the evidence suggests that developing world has witnessed mixed industrial responses during the last two decades or so.[2] A handful of newly industrialising economies (NIEs) in East Asia (and to a lesser extent in Latin America) have taken advantage of the new opportunities offered by globalisation and expanded their manufactured exports, attracted new foreign investment and harnessed their industrial skills and technologies. The first generation NIEs (South Korea, Taiwan, Singapore and Hong Kong) have had sustained rapid manufactured export growth based on an increasing share of technologically complex activities in total exports for over twenty years. A second generation of NIEs (e.g. Malaysia, Thailand, Philippines, Mexico, Chile and Israel) have witnessed rapid manufacturing export growth but with varying achievements in high technology exports. Per capita incomes in the NIEs have typically risen dramatically since the 1970s and several have become high income countries.

However, other developing countries in Sub-Saharan Africa, South Asia, the Caribbean and the Pacific have been slow to reap benefits from globalisation. With a few exceptions, this second group have had sluggish manufactured export growth and limited industrial upgrading of their relatively simple export sectors (which are dominated by natural resource and labour intensive activities).[3] A concentration on relatively simple export activities has made these developing countries vulnerable to the threat of new entrants, with cheaper labour costs and natural resources, in export markets. Some, particularly in Sub-Saharan Africa, have even experienced a contraction in existing industrial capacity and there is little sign of new capacity emerging.[4] Micro and small and medium enterprises (MSMEs) – which often account for bulk of the number of industrial establishments and employment in African economies – have performed poorly relative to large firms in the more open trading environment.[5] Equally serious is that export-oriented FDI has not flowed in the desired volumes into the manufacturing sectors of African economies. Per capita incomes have shown little upward tendency in this second group and the bulk of them have remained as low-income developing economies.

The developing world is thus becoming increasingly polarised into those that have succeeded in becoming industrially competitive in an open, international economy and those that have not done so to date. The prospect of industrial marginalisation is a persistent worry in policy circles in the developing world. There is an urgent need for change in government and private sector attitudes and strategies. Change is no longer an option but a survival algorithm.

Growing interest in competitiveness strategy

Concerns about the process of industrial restructuring in an integrated world economy have sparked widespread interest in the concept of competitiveness as applied to national economies and enterprises within them. This interest originated in developed world but has recently spilled over into developing countries and economies in transition in Central and Eastern Europe. Political leaders and businessmen constantly refer to it during discussions on international rivalries between enterprises and countries. The media makes frequent mention of it in articles and bulletins on the external performance of countries (e.g. trade balance or market shares). International organisations produce indicators (which range from real effective exchange rate measures published by the IMF and World Bank to absolute measures produced by the Institute for Management Development and the World Economic Forum) showing changes in the ranking of the competitiveness of countries.[6] Governments have established commissions to examine national competitive behaviour and formulate policies to improve national and regional competitiveness.[7] Consultancy firms of different kinds conduct studies of competitiveness strategy in what has become a highly profitable industry. The theoretical foundation of some of this work is questionable and its empirical basis is often weak. Thus, the sheer volume of studies that have been generated to date on competitiveness is impressive but this may have been achieved at the expense of academic quality and rigour.

It is not surprising, therefore, that a lively debate is taking place among economists on the subject. Krugman is dismissive of the idea: 'competitiveness is a meaningless word when applied to national economies. And the obsession with competitiveness is both wrong and dangerous' (Krugman, 1994, p. 44). Krugman's critique was particularly directed at policy makers in the Clinton administration in the United States, a group he regards as pop internationalists. He argues that pop internationalists misleadingly advocate that the economic problems of United States (e.g. eroding real wages, stagnating living standards and rising unemployment) will be redressed only when the United States gains a productivity edge over its rivals. Krugman did not discuss specially whether national competitiveness is relevant to developing countries but the clarity and authority of his critique of the concept in relation to the specific case of the United States has attracted a strong response from development specialists.

Howes and Singh, argue: 'once the severe limitations of Krugman's (neoclassical) model in its application to the real world are recognised, his analytical and empirical critique of national competitiveness loses much of its force' (Howes and Singh, 1999, p. 8). They conclude: 'contrary to Krugman, the notion of competitiveness is analytically meaningful and useful to policymakers' (Howes and Singh, 1999, p. 21).

Reinert admits that the word competitiveness is 'often misused and mostly ill-defined' but goes on to argue that 'in spite of its fairly recent appearance on the scene, the term competitiveness, in my view addresses issues which have been central in public policy at least during the last 500 years, albeit under different headings' (Reinert, 1994, p. 1).

Commenting on available competitiveness studies, Lall concurs that 'the output is of variable quality: under its aegis come serious analyses as well as ideological tracts, low level business-school tracts, banal data churning, applications of impressive but vacuous formulae, and straightforward bashing the foreigner…' and adds 'what ever its quality, the analysis of competitiveness clearly responds to a growing policy need' (Lall, 2001, p. 2).

A central issue in the economics debate on competitiveness is the creation of efficient industrial capacity in developing countries to cope with more intense global competition. This has drawn attention to mechanisms of technological advance in developing economies, which lie well behind international technological frontiers. The acquisition and progressive mastering of technologies that are new to them, if not the world, has been a central aspect of the NIEs (particularly those in East Asia) that have witnessed dramatic industrial success during rapid globalisation.[8] In order to put imported technologies into productive use, enterprises in the NIEs had to undergo a lengthy learning process and were supported by coherent policies and institutions. Such interventions were directed at addressing market and institutional imperfections associated with the process of firm-level learning. Capital market imperfections, risk aversion, externalities and coordination failures are widely cited examples of market failures which influence the pace of technological development while missing technology institutions and weak linkages between firms and technology institutions are increasingly recognised as examples of institutional failures.[9] The experience of the NIEs has turned the attention of some economists to the appropriate policy environment for enterprise-level capability building in other industrial latecomers.

In this vein, the five key questions for developing countries are:

- What is the meaning of competitiveness particularly in relation to capability building in the manufacturing sector?
- What are the respective roles of the state and private sector in enhancing industrial competitiveness?

- What specific policies and institutions work better than others?
- What is the optimal way to design and implement an industrial competitiveness strategy?
- What are the necessary conditions for success?

This book seeks to address these questions in the context of developing countries experiencing rapid economic liberalisation and globalisation. The individual chapters are written by leading specialists in their respective fields and, taken together, are intended as a reader on manufacturing competitiveness and policies in developing countries. The book attempts to apply recent developments in economics and business studies (especially in the areas of international trade, competition, finance, technology, foreign investment and small enterprise development) to the study of manufacturing competitiveness issues in developing countries. The focus is on the design and application of specific types of policies to foster manufacturing competitiveness in developing countries rather than on the theories underlying them. The book divides competitiveness policies into two broad groups: *incentive* and *supply-side* policies. The former consist of measures (such as import liberalisation, privatisation of state enterprises and regulation of domestic competition), which change the nature and intensity of competition affecting firms and stimulate their learning process. The latter provide inputs and support for firm-level learning (including industrial finance, attracting of foreign investment and other sources of imported technology, technology support services and small enterprise development). The NIEs emphasised the interaction between these two types of competitiveness policies during their rapid industrial success. In contrast, many industrial latecomers have emphasised the role of incentive policies during industrial adjustment but often neglected supply-side policies to the detriment of their industrial performance. Where possible, the book also tries to assess empirical evidence on the impact of given policies and incorporate examples of policies and institutions from the NIEs and other developing economies to highlight good practices.

Plan of the book

This book has three parts. Part I presents this introduction, the framework for the volume and an empirical analysis of manufacturing competitiveness in the developing world in three chapters. Chapter 2 by Ganeshan Wignaraja explores the relevance of different approaches to competitiveness analysis for designing public policies in developing countries in a globalising economy. Having examined the traditional macroeconomic and business strategy perspectives, he suggests whilst these provide useful theoretical insights they offer only partial guidance for policy formulation in developing countries. Furthermore, he suggests that the newer technology and innovation perspective (derived from the literature on national innovations systems and technological capabilities) offers a more comprehensive policy framework in developing countries because it incorporates domestic distortions, market imperfections and systems failures.

Using this framework, he proposes a liberalisation plus competitiveness approach (which combines market-friendly incentives with supply-side support) and a strong public–private sector partnership for developing countries. Finally, he discusses some of the components of this strategy and crucial issues for the management of industrial competitiveness in developing countries. The chapter concludes that an appropriate competitiveness strategy, tailor-made to national conditions and development objectives, is a vital ingredient of successful industrial adjustment in the face of rapid globalisation.

In Chapter 3, Ganeshan Wignaraja and Ashley Taylor benchmark industrial competitiveness in the developing world using the concept of a manufactured export competitiveness index (MECI). They examine the widely cited business school approaches to national competitiveness (notably those of the World Economic Forum and the International Institute for Management Development) and highlight several criticisms of such approaches. They go on to construct a simple composite measure of manufactured export competitiveness performance based on three subcomponents (current manufactured exports per capita, long-term growth in manufactured exports and the share of technology-intensive exports) and present the results by country, region and income group. The country coverage of the MECI (eighty developing and transition economies) provides a more comprehensive representation of the developing world than the World Economic Forum and International Institute for Management Development exercises. Finally, they examine potential determinants of manufactured export competitiveness and conclude that sound macroeconomic conditions, trade liberalisation and an emphasis on supply-side factors (such as FDI, technological effort, human capital and communications infrastructure) are closely associated with better competitiveness performance in the developing world.

Part II examines supply-side issues and policies for competitiveness in four chapters. In Chapter 4, Stan Metcalfe considers science, technology and innovation policies in developing countries. Following an evolutionary perspective of technical change, he analyses the absorption and adaptation of imported technology in developing countries and the factors affecting this process. He emphasises that innovation and learning in developing countries is the outcome of interactions between firms, institutions and policies within a national innovation system and that this process follows an unpredictable and cumulative technological trajectory. He goes on to present a framework for making science, technology and innovation policy choices in developing countries and reviews the experience of such policies in South Korea, Columbia, Mauritius as well as the United Kingdom. Among other policy measures, he looks at R&D tax incentives and subsidies; matching grant schemes for consultancy services; public purchasing; initiatives for stimulating collaborative innovation; meterology, testing and standards services; finance for innovation; and technology foresight. He concludes that innovation and learning are the driving force of market-oriented development and what matters is the creation of an effective national innovation system.

In Chapter 5 Eileen Fischer and Rebecca Reuber examine industrial clusters and small and medium enterprise (SME) promotion in developing countries. They argue that SMEs can overcome the disadvantages of firm size by participating in geographically concentrated industrial clusters and investing in collective technological learning processes. Cluster dynamics and inter-firm interactions are illustrated through case studies of shoe making and wood working in Brazil, computers in Taiwan and machinery in Japan. They suggest that upgrading of firm-specific capabilities and SME clusters are facilitated by an effective infrastructure of business development services (BDS). They also examine a range of programmes (relating to enterprise training, subcontracting relations, network development and establishing BDS centres) that constitute best practice in the delivery of BDS to small enterprises in developing countries. They conclude with some general principles for effective BDS service design and delivery in developing countries (e.g. customising services, proving for cost-recovery in service delivery, forming networks among BDS providers and conducting cost–benefit analysis) and ways in which BDS outcomes can be evaluated.

Chapter 6 by Dirk Willem te Velde analyses FDI policies in developing countries. He explores empirical evidence on the costs and benefits of FDI and suggests that there are net benefits from FDI (market access, new technologies and finance) for developing countries but their realisation depends on policies in place and other factors. Emphasising that local technological capability development has to go hand-in-hand with the attraction of FDI, he provides a neat three-fold classification of FDI policies into those needed: (a) to attract FDI; (b) to upgrade FDI; and (c) to encourage linkages between foreign affiliates and local firms. Through country cases of highly successful FDI regimes in Singapore and Ireland as well as cross-country studies, he goes on to explore the details of best practice FDI policies. He finds that whilst appropriate FDI policies depend considerably on specific country characteristics and strategic economic objectives, successful cases indicate common elements of a best practice approach. These elements include: sound macroeconomic management, liberalisation of regulations governing the entry of foreign investment, targeting of specific multinationals via a strong investment agency, encouraging enterprise training, fostering local subcontracting via linkage programmes and cluster promotion, excellent trade facilitation and an effective framework for competition policy. The chapter concludes with suggestions on how to attract FDI into industrial latecomers in the developing world and highlights some implementation issues.

Chapter 7 by Andy Mullineux and Victor Murinde deals with financial policies for enterprise development in developing countries. They assess the strengths and weaknesses of financial systems in developing countries and suggest that many do not efficiently allocate industrial finance because of domination by oligopolistic commercial banks and excessive government action in directing credit and interference with commercial interest rate setting. Accordingly, they argue that strategies for stimulating enterprise finance in developing countries should revolve around establishing effective bank regulation and supervision; setting an appropriate positive real interest rate; reducing unnecessary government intervention and direction; and addressing credit rationing and financial exclusion problems affecting small firms. As SMEs are often the largest providers of employment in developing countries, the complex issue of SMEs finance is explored through the experience of loan guarantee schemes, micro finance, mutual and cooperative banks, development banks and venture capital. They conclude that market-led financial reforms are the key to the evolution of financial sectors in developing countries but that the state has a role in financial system restructuring; in regulating and supervising financial systems; and in ensuring access to finance for SMEs (e.g. through loan guarantees, reforming development banks and tax incentives for venture capital).

Part III analyses incentive policies for competitiveness in two chapters. Chapter 8 by Christos Pitelis analyses privatisation, regulation and domestic competition policies in developing countries. He explores alternative approaches to domestic competition, monopoly and industrial organisation and argues that the differential competences perspective – which emphasises productivity and innovation rather than efficient resource allocation – is increasingly challenging the standard neoclassical perspective. He suggests that the domination of many industries in developing countries by large, inefficient state enterprises (insulated from the forces of competition) underlies the drive for privatisation (e.g. direct sales of shares and voucher participation) and state-owned enterprise reform (e.g. performance and management contracts). After reviewing the evidence on the implementation of these programmes in developing countries, he suggests that on balance they have improved economic efficiency but that country-specific conditions determine the most appropriate method and speed of reform. A framework for making appropriate policy choices about privatisation is also provided. The chapter concludes that establishing an effective regulatory framework and mechanisms for learning from experience can be important to realise the benefits of privatisation.

In Chapter 9, Sheila Page examines international trade and multilateral negotiation strategies for developing countries. She assesses historical evidence and theory on trade and development and suggests that a sequence of policies from import substitution to export promotion may be more conducive to fostering technologically efficient enterprises than a single policy. Several factors – country size, the level of development, administrative capabilities and the external environment – influence the speed of this policy transition. She argues that the WTO rules and regulations have restricted the use of certain types of trade policies in developing countries (e.g. import quotas, local content rules and most export subsidies) but that others are still permissible (e.g. import tariffs, subsidies for reasons of industrial/regional policy, public procurement, duty drawbacks for exports, VAT exemption for exports and real exchange rate policy). Apart from such WTO compatible trade policies, she suggests that developing countries can also encourage trade by creating cost-competitive physical infrastructure (especially telecommunications and transport facilities and services) and reducing bureaucratic impediments (e.g. customs procedures and tax administration). She also argues that developing countries can benefit by improving their capabilities to conduct multilateral trade negotiations and discusses ways of doing this (e.g. better inter-ministerial coordination of negotiating strategies, more involvement of private sector organisations in the negotiations and tapping external technical assistance). She concludes by saying that trade is not an end in itself but merely a tool for facilitating the transformation of developing countries.

The book will be of use to all those interested in issues of economic reform, industrial policy and private sector development in developing and transition economies. It will be of particular use to the following: for students interested in policy analysis in developing and transition economies on advanced undergraduate and post-graduate courses in development and transition economics, industrial economics and business studies; for policy makers concerned about manufacturing competitiveness problems and market-friendly policy instruments drawing on international best practice; and private sector organisations seeking strategies for industrial restructuring in an open economy and public–private sector partnerships to foster competitiveness.

Notes

1 See Rodrik (1997), Crafts (2000) and Bordo et al. (forthcoming).
2 Chapter 3 maps out the differential industrial performance in the developing world since 1980 using a composite manufacturing export competitiveness index (MECI). This measure is based on three indicators: per capita manufactured exports, manufactured export growth rates and the share of high technology exports in total exports. UNIDO (forthcoming 2002) contains another mapping of world industrial performance (called an industrial scoreboard) and also reveals a dualistic pattern of industrial performance in the developing world.
3 See Wignaraja (1998) on Sri Lanka and Lall and Wignaraja (1998) on Mauritius, which both have witnessed rapid manufactured export growth based on textiles and garments but little diversification into more technologically sophisticated exports.
4 See Lall et al. (1994) on Ghana; Wignaraja and Ikiara (1999) on Kenya, Deraniyagala and Semboja (1999) on Tanzania and Latsch and Robinson (1999) on Zimbabwe.
5 See Liedholm and Mead (1999).
6 See IMD (2001), IMF (2001), World Bank (2001), WEF (2001).
7 National competitiveness studies conducted or commissioned by governments seem to fall into two broad categories and a few have been published. (i) Official reports put out by governments (e.g. OTA, 1990 on the United States; DTI, 1998 and 2001 on the United Kingdom; Ministry of Trade and Industry, 1998 on Singapore; Ministry for Economic Services, 1999 on Malta; and TRADENZ, 1997 on New Zealand). (ii) Background studies prepared for governments by international organisations and consultants (e.g. Dunning et al., 1998 on Northern Ireland; Lall and Wignaraja, 1998 on Mauritius).
8 See, for instance, Hobday (1995), Stiglitz (1996), Kim and Nelson (ed. 2000), Mathews and Cho (2000).
9 See Chapter 4 by Stan Metcalfe and my Chapter 2.

References

Bordo, M.D, Taylor, A.M. and Williamson, J. (ed.) (forthcoming), *Globalisation in Historical Perspective*, Chicago: University of Chicago Press.

Crafts, N. (2000), 'Globalisation and Growth in the Twentieth Century', *IMF Working Paper* No. 00/44.

Deraniyagala, S. and Semboja, H. (1999), 'Trade Liberalisation, Firm Performance and Technology Upgrading in Tanzania', in S. Lall (ed.), *The Technological Response to Import Liberalisation in SubSaharan Africa*, London: Macmillan, pp. 112–147.

Dunning, J.H., Bannerman, E. and Lundan, S. (1998), *Competitiveness and Industrial Policy in Northern Ireland*, Belfast: Northern Ireland Economic Council.

Hobday, M. (1995), *Innovation in East Asia: The Challenge to Japan*, Aldershot, England: Edward Elgar.

Howes, C. and Singh, A. (1999), 'National Competitiveness, Dynamics of Adjustment, and Long-Term Economic Growth: Conceptual, Empirical and Policy Issues', *Accounting and Finance Discussion Papers* No. 00-AF43, Department of Applied Economics, University of Cambridge.

IMD (2001), *World Competitiveness Yearbook*, Lausanne: International Institute of Management Development.

IMF (2001), *International Financial Statistics Yearbook*, Washington, DC: International Monetary Fund.

Kim, L. and Nelson, R.R. (eds) (2000), *Technology, Learning and Innovation: Experiences of Newly Industrialising Economies*, Cambridge: Cambridge University Press.

Krugman, P. (1994), 'Competitiveness: A Dangerous Obsession', *Foreign Affairs*, 73(2), 29–44.

Lall, S. (2001), *Competitiveness, Technology and Skills*, Aldershot (UK): Edward Elgar.

Lall, S., Barba-Navaretti, G., Teitel, S. and Wignaraja, G. (1994), *Technology and Enterprise Development: Ghana Under Structural Adjustment*, London: Macmillan.

Lall, S. and Wignaraja, G. (1998), *Mauritius: Dynamising Export Competitiveness*, (Commonwealth Economic Paper No. 33), London: Commonwealth Secretariat.

Latsch, W. and Robinson, P. (1999), 'Technology and the Responses of Firms to Adjustment in Zimbabwe', in S. Lall (ed.), *The Technological Response to Import Liberalisation in SubSaharan Africa*, London: Macmillan, pp. 148–206.

Liedholm, C. and Mead, D. (1999), *Small Enterprises and Economic Development: The Dynamics of Micro and Small Enterprises*, London: Routledge.

Mathews, J.A. and Cho, D. (2000), *Tiger Technology: The Creation of a Semiconductor Industry in East Asia*, Cambridge: Cambridge University Press.

Ministry of Economic Services (1999), *Prosperity in Change: Challenges and Opportunities for Industry*, Malta: Ministry of Economic Services.

Ministry of Trade and Industry (1998), *Committee on Singapore's Competitiveness*, Singapore: Ministry of Trade and Industry.

OTA (1990), *Making Things Better: Competing in Manufacturing*, Washington, DC: Office of Technology Assessment, US Senate.

Reinert, E.S. (1994), 'Competitiveness and its Predecessors: A 500 Year Cross National Perspective', Paper prepared for the Business History Conference, Williamburg, Virginia, USA, March 11–13.

Rodrik, D. (1997), *Has Globalisation Gone Too Far?*, Washington, DC: Institute for International Economics.

Stiglitz, J. (1996), 'Some Lessons from the East Asian Miracle', *World Bank Research Observer*, 11(2), 151–177.

TRADENZ (1997), *Competing in the New Millennium: A Global Perspective for New Zealand Business*, Auckland: New Zealand Trade and Development Board.

UK DTI (1998), *Our Competitive Future: Building the Knowledge Driven Economy*, London: Department of Trade and Industry, Her Majesty's Stationary Office.

UK DTI (2001), *Opportunity for All in a World of Change: A White Paper on Enterprise, Skills and Innovation*, London: Department of Trade and Industry, Her Majesty's Stationary Office.

UNIDO (2002), *World Industrial Development Report 2002: Competing Through Innovation*, Vienna: UNIDO.

WEF (2001), *The Global Competitiveness Report*, Geneva: World Economic Forum.

Wignaraja, G. (1998), *Trade Liberalisation in Sri Lanka: Exports, Technology and Industrial Policy*, London: Macmillan.

Wignaraja, G. and Ikiara, G. (1999), 'Adjustment, Technological Capabilities and Enterprise Dynamics in Kenya', in S. Lall (ed.), *The Technological Response to Import Liberalisation in SubSaharan Africa*, London: Macmillan, pp. 57–111.

World Bank (2001), *World Development Indicators 2001*, Washington, DC: World Bank.

2 Competitiveness analysis and strategy

Ganeshan Wignaraja[1]

Introduction

There is a growing literature on competitiveness in economics and business studies but there is little agreement on what the term means, what affects it and the role for public policies.[2] Three distinct views on competitiveness can be conveniently distinguished as follows:[3]

a   a *macroeconomic* perspective which deals with internal and external balance at country-level and focuses on real exchange rate management as the principal tool for competitiveness;
b   a *business strategy* perspective which is concerned with rivalries between firms and countries and a limited role for public policies in fostering competitiveness;
c   a *technology and innovation* perspective that emphasises innovation and learning at the enterprise and national-levels and active public policies for creating competitiveness.

Given the diversity of thinking on the issue, it is not surprising that the academic debate on competitiveness has become so convoluted and emotional. There is also little sign of a consensus being reached over practical guidelines for policy makers. The inconclusive state of the academic debate means that the concept of competitiveness remains somewhat elusive particularly at the national level. Furthermore, the connection between national and enterprise-level competitiveness still seems vague and there appear to be contradictory views on its policy implications.

Against this background, this chapter assesses the concept of industrial competitiveness and policies to enhance it in developing countries attempting to integrate with the world economy. Our main concern is the competitiveness of the manufacturing sector rather than the broader notion of economic competitiveness (i.e. the competitiveness of the economy as a whole). The chapter begins by examining the application of traditional approaches – notably, perspectives emanating from macroeconomics and business strategy – to analysing issues of industrial competitiveness and public policies in developing countries. Having highlighted the strengths and weaknesses of traditional approaches, it focuses on the application of the newer technology and innovation perspective to industrial competitiveness in developing countries. Then it goes on to discuss appropriate public policies and private sector initiatives to improve industrial competitiveness in the developing world and the management of competitiveness strategy.

Macroeconomic and business strategy perspectives

Macroeconomic perspective

This viewpoint on competitiveness is rooted in macroeconomic theory and policy (see Corden, 1994; Boltho, 1996). One of the objectives of macroeconomic policy is to ensure simultaneous internal and external balance in the short run. In this vein, internal balance is usually described as the lowest possible level of unemployment consistent with a tolerable level of inflation (i.e. full employment) and external balance at a given level of the current account (for convenience, equated with current account equilibrium). International competitiveness in this situation could be thought of as the level of the real exchange rate, which in combination with the requisite domestic economic policies, achieved internal and external balance.

Suppose a country goes into a current account deficit for some reason, this would be linked to a real exchange rate appreciation, the domestic price of the tradeables sector declining relative to those of the non-tradeables sector. As the profitability of tradeables declines and that of non-tradeables rises, resources shift from tradeables to non-tradeables. This case implies a loss of competitiveness at country-level. As Corden puts it '... there are gainers and losers, and no general presumption that overall profitability declines. But one can legitimately say that the country has become internationally less competitive. There has been a general loss in competitiveness of the country's tradeable's sector' (Corden, 1994, p. 270).[4] In due course, the current account deficit could be adjusted and competitiveness could be improved using a mixture of exchange rate depreciation and deflation. Thus, from a macroeconomic perspective, competitiveness policy and exchange rate policy are largely synonymous.

The association of short-run competitiveness with the real exchange rate raises the issue of its measurement. As mentioned above by Corden, theory suggests that the appropriate definition of the real exchange is the relative price of non-tradeables to tradeables. A rise in this ratio indicates an appreciation and a fall denotes depreciation. One of the difficulties with this concept of the real exchange is that data on non-tradeable and tradeable prices are not readily available for developed or developing countries.[5]

In practice, therefore, most analysts have resorted to the real effective exchange rate, which is derived from the concept of purchasing power parity (PPP). The real effective change rate is computed by deflating a trade-weighted average of the nominal exchange rates that apply between trading partners. Different price or cost indicators can be used for the home country and its trading partners including relative consumer prices, relative wholesale prices, relative unit labour costs in the manufacturing sector (labour cost divided by output, expressed in a common currency) and relative value added deflators. Owing to conceptual and data limitations, movements in real effective exchange rates need to be used with caution.[6] Consumer prices are the most easily available as nearly all countries conduct cost of living surveys. However, it includes a large number of non-traded goods and services and hence is a crude measure of changes in a country's international competitiveness. The merit of unit labour costs is that sudden changes in key competitiveness factors (wages, productivity and exchange rates) can be associated with movements in profitability and tradeable prices. The difficulties are two-fold: time series data on labour costs are not available for most developing countries and the measure excludes non-labour costs in manufacturing. Bearing in mind these measurement issues, time series information on real exchange rates are readily available from published and unpublished sources.[7]

The macroeconomic perspective has been extensively used to examine competitiveness issues in both developing and developed countries.[8] This approach highlights the links between the changes in the balance of payments, movements in the real exchange rate, shifts in resource allocation between economic activities and changes in competitiveness. It underlines the fact that the exchange rate is a strategic variable that determines whether a country is able to create the requisite macroeconomic conditions for internationally competitive industries. Large current account deficits are related to a real exchange appreciation and hamper the development of tradeables including manufactured exports. Hence, this approach suggests that it is desirable to avoid an exchange rate level displaying an anti-export bias.

The macroeconomic approach to competitiveness has been criticised on various grounds. A major pitfall is the equation of international competitiveness solely with indicators of relative prices or unit costs. Empirical studies have suggested that this view is too simplified and that non-price factors particularly technological capabilities and the ability to compete on delivery are the main factors influencing differences in international competitiveness and growth across countries (Fagerberg, 1988, 1996; Dosi et al., 1990). These studies also find that cost and price factors do also affect competitiveness but to a lesser extent.

Another closely related pitfall is the narrow scope for public policy within the macroeconomic approach to competitiveness (Porter, 1990). It relies heavily on a single instrument (the exchange rate) to remedy balance of payments imbalances, restore the profitability of tradeables relative to non-tradeables and, hence, improve competitiveness. Real exchange rate adjustment is a necessary but not sufficient condition to improve industrial performance in developing countries. A stable, competitive real exchange rate signals to industrial enterprises that tradeables production is more profitable than non-tradeables and is a vital ingredient in a coherent national competitiveness strategy. By itself, however, a competitive real exchange rate has limited impact on a host of other impediments to enterprise behaviour in developing countries such as backward institutions, inefficient business development services, poor quality infrastructure and a lack of scientific and engineering skills.

Business strategy perspective

This perspective on competitiveness comes from the literature on business studies rather than economics and is concerned with issues of business rivalry and the approaches that enterprises employ in competing with each other (see Porter, 1980; Barney, 1991; Yip, 1992; Hamel and Prahalad, 1994; Ohmae, 1994). This literature is geared for the analysis of decision making by managers responsible for domestic and worldwide businesses. Competition from new entrants and existing firms within industries provides the rationale for business strategy at firm-level (to set goals, forecast the industry environment and plan resource deployments). The business strategy literature has responded to this dynamic need by proposing a variety of generic strategies for enterprises to maintain (or increase) market share in an industry and maximise profits. Following Porter (1980), these can be broadly classified into: cost leadership (involving actions like investment in scale efficient plant, designing products for easy manufacture and R&D), differentiation (emphasising branding, brand advertising, design and service) and focus (stressing market niches). Subsequent business strategists have refined these firm-level tactics and combined them under various slogans – for instance, Yip (1992) talks about 'total global strategy', Ohmae (1994) 'strategy for a borderless world' and Hamel and Prahalad (1994) 'core competence leadership'. Most of this literature is thus concerned with manipulating the nuts and bolts of business activity (e.g. marketing, human resources, finance, organisation and technology) to handle inter-firm competition in industries and has not moved beyond these firm/industry-level issues.

A few, however, have applied micro-level business strategy concepts to studying the international economic relations of nations. In this vein, Porter (1990, 1998) has been particularly influential. Porter takes competitiveness and productivity to be synonymous. He argues that: 'the only meaningful concept of competitiveness at the national level is national productivity. A rising standard of living depends on the capacity of a nation's firms to achieve high levels of productivity and to increase productivity over time. Our task is to understand why this occurs' (Porter, 1990, p. 6). His explanation of national competitiveness is based on micro-level foundations. Porter suggests that competitive advantages of nations arise from firm-level efforts to innovate in a broad sense (i.e. develop new products, improve production processes and introduce new brands). This stems from an idea deeply embedded in business strategy literature that firms which are successful in achieving dominant shares in world markets do so not only because they possess some unique resources but also because of their ability in leveraging on these resources.[9] In turn, Porter suggests that innovations in a broad technological and marketing sense can take place in any industry as a result of four elements of his diamond framework, which are as follows:[10]

a   *factor conditions* (the country's endowment of production factors like natural resources, human resources, physical infrastructure and administrative infrastructure);

b *demand conditions* (the nature of home demand for the industry's product or service);
c *related and supporting industries* (the availability of suppliers and ancillary industries that are internationally competitive);
d *firm strategy, structure and rivalry* (the circumstances in the country determining how firms are created, organised and managed and the intensity of domestic competition).

Having set out a wide-ranging framework for the determinants of competitiveness at national level, Porter takes a distinctive stand on public policies. Public policies (and chance) only indirectly affect national competitiveness in an industry via each of the four domestic attributes of the diamond.[11] Public policies can influence factor conditions by investing in new skills and infrastructure; demand conditions by limited public procurement of goods and services; relating and supporting industries by supplying specialised skills and infrastructure particularly to reinforce existing industry clusters; and firm strategy, structure and rivalry through free trade and domestic competition policies. Compared with the macroeconomic perspective on competitiveness, there is a broader role for public policies in Porter's approach but the emphasis is on indirect actions rather than a more comprehensive and integral role related to firm-level and collective learning processes.

Other business strategists seem to echo Porter's emphasis on an indirect or minimalist role for public policies in influencing competitiveness at national level. In the context of his total global strategy approach, Yip (1992) views the role of government as a facilitator of trade liberalisation, public procurement and privatisation of state-owned enterprises. Ohmae's (1994) strategy for a borderless world approach limits the role of government to providing defence against external threats and deregulation of past controls on international trade and inward investment. To the limited extent that they devote to public policies, Barney (1991) and Hamel and Prahalad (1994) also see government as a liberaliser/deregulator rather than an active participant in the creation of national competitiveness.

Porter (1990) and related business strategy work on national competitiveness issues have attracted some attention in the literature on competitiveness. In one line of argument, Krugman (1994) takes exception to the proposition put forward by business strategists that countries compete with each other like large corporations on world markets and countries' economic fortunes are determined by success on world markets. Krugman adds that 'international trade is not a zero sum game' but one in which specialisation and trade according to comparative advantage brings gains to all nations (Krugman, 1994, p. 34).

Another line of argument is concerned with Porter's definition of national competitiveness in terms of national productivity. Reinert (1994) argues: 'this is hardly an operational definition since there is sometimes little relationship between absolute level of productivity and national wealth' (Reinert, 1994, p. 2). He suggests that it is difficult for nations and firms to be competitive if they are not efficient and have high productivity. At the same time, however, Reinert argues: 'it is by no means obvious that being the most efficient producer of an internationally traded product makes a country competitive – i.e., enables it to raise the standard of living' (Reinert, 1994, p. 3).

A related point is that Porter (1990) does not make clear what national productivity means – whether it is total factor productivity or some indicator of partial productivity – and how it should be measured. In this vein, it should be noted that Porter has been collaborating with the World Economic Forum (WEF) to develop a composite indicator to rank competitiveness across leading developed and developing countries on an annual basis. The WEF's competitiveness index is compiled from rich database of published and survey data but suffers from many methodological weaknesses. This discussion will be taken up in Chapter 3.

Others suggest that the limited and indirect role given to government policies in Porter's diamond framework of competitiveness lacks economic rationale. Lall (2001) argues that enterprise learning in developing countries is affected by various market failures (e.g. capital market imperfections, risk aversion, externalities and coordination failures) which can constrain the development of competitiveness. According to Lall, Porter and other business strategists neglect this important feature of the market system in developing countries.[12] Furthermore, Lall argues that the presence of market failures in technological development provide a significant role for public policies to foster competitiveness in developing countries. In a related line of criticism, Moon et al. (1998) argue that whilst Porter's diamond model suggests some important determinants of a nation's global competitiveness, it is incomplete mainly because it does not incorporate multinational activities. Moon et al. (1998) go on to propose a new approach, a generalised double diamond, and offer insights into policies for attracting inward investment.

Summary

The two traditional approaches provide insights on industrial competitiveness in developing countries. Each is useful depending on the purpose at hand. At one extreme, the established macroeconomic perspective deals with aggregate, economy-level issues and links the balance of payments, real exchange rates and resource allocation. It also emphasises prudent real exchange rate management (and, more generally, macroeconomic stability) to foster industrial competitiveness in the developing world. Few would disagree with the view that competitive real exchange rates matter for industrial competitiveness and may be even a fundamental ingredient of success. However, empirical evidence suggests that real exchange rates alone are not panacea for industrial competitiveness and that micro-level influences such as technological capabilities and the ability to compete on delivery may be more important. It follows that a comprehensive competitiveness strategy should include price-based measures and non-price instruments.

At the other extreme, the popular business strategy perspective focuses on micro-level behaviour and is concerned with rivalry between enterprises and strategies that enterprises employ in competing with each other. Porter's influential diamond framework adds an additional dimension by applying business strategy concepts to the international economic relations of nations. This perspective offers a bottom-up view of industrial competitiveness and give some scope for public policies but has been criticised on different grounds: (a) that it is misleading to suggest nations compete like large corporations on world markets; (b) that the definition of national competitiveness in terms of national productivity is unclear and that national productivity itself is not defined for empirical purposes; and (c) the limited and indirect role for government policies to promote competitiveness lacks economic rationale.

The next section examines the alternative, technology and innovation viewpoint that takes on board some of the criticisms of traditional approaches to competitiveness.

Technology and innovation perspective

Origin and link to industrial competitiveness

Technology has long been regarded as an important determinant of competitive advantage in world markets. This is the main driver of competitive advantage in the Schumpeterian perspective as well as the neo-technology theories of comparative advantage articulated by Posner and Vernon (see Dosi et al., 1990). Following in this tradition, the more recent technological and innovation perspective on competitiveness, stems from significant advances in the microeconomic literature on innovation, learning and economic development. A common perception in neoclassical economics literature is that the successful accumulation of technologies in developing countries can be encouraged by a smooth flow of new information (e.g. via foreign direct investment (FDI)), ensuring stable macroeconomic conditions and increasing expenditures on education. The technology and innovation perspective suggests that these factors have a role to play, but on their own, are insufficient to ensure a continuous process of domestic technological development in developing countries.

One strand of the technology and innovation literature (referred to as the technological capabilities approach) is concerned with the process of absorbing imported technologies in enterprises in developing countries within a system of imperfect markets.[13] Another strand (the national innovations systems perspective) is primarily concerned with the emergence of innovations in developed countries through a complex process of interactions between the firms and institutions of different kinds.[14] The technological capabilities approach emphasises learning behind world technological frontiers in industrial latecomers while the national innovation systems perspective largely deals with the generation of new products and processes at (or beyond) world frontiers in advanced industrial countries. Moreover, the technological capabilities approach suggests a range of policies to deal with market imperfections to technological development while the national systems of innovation perspective gives somewhat less emphasis to normative issues connected with the innovation process. Inspite of differing origins, notions of technology and country focuses, both approaches are underpinned by a common evolutionary framework to technological change and innovation, which is conceptually different from the traditional neoclassical approach.[15] For this reason, the technological capabilities approach and the national innovations systems perspective can be combined into a technology and innovation perspective on industrial competitiveness in developing countries.

This directly leads to the issue of defining the notion of industrial competitiveness and making the link between the enterprise and national levels. A concise definition of micro and macro-level competitiveness from a technology and innovation perspective can be found in OECD (1992):

> In microeconomics, competitiveness refers to the capacity of firms to compete, to increase their profits and to grow. It is based on costs and prices, but more vitally on the capacity of firms to use technology and the quality and performance of products. At the macroeconomic level, competitiveness is the ability of a country to make products that meet the test of international competition while expanding domestic real income.
>
> (Adapted from OECD, 1992, p. 237)

The OECD (1992) definition of industrial competitiveness is simple and internally consistent. It highlights the relevance of price and non-price factors at the micro-level and emphasises that technological and marketing considerations are the paramount drivers of enterprise success. It translates these ideas to the national level by suggesting that industrial outputs (i.e. goods and services) have to meet the price, quality and delivery standards of increasingly open, domestic and international markets. This is particularly pertinent in today's progressively integrated world economy with falling trade barriers, accelerating technological progress and increasing MNC activity. The definition also links the performance of a country's industries to raising living standards thereby adhering to the empirically observed relationship between exports and economic growth. This last point is particularly important for policy purposes and the UK Government White Papers on competitiveness and the US Competitiveness Policy Council have followed the OECD in viewing competitiveness as the ability to raise living standards (see Eltis and Hingham, 1995).

Distinctive features of the approach

The technology and innovation perspective recognises that developing countries have access to a global pool of technologies and are typically users of imported technology rather than producers. The distinctive feature of this perspective is its focus on manufacturing enterprises as the main actors in the process of accumulating technological capabilities. It emphasises the notion that enterprises have to undertake conscious investments to convert imported technologies into productive use. New technologies have a large tacit element (i.e. person embodied information which is difficult to articulate in hardware or written instructions) that can only be acquired through experience and deliberate investments in training, information search, engineering activities and even research and development.[16]

A simple representation of learning process at firm-level in a developing country is depicted in Figure 2.1. The diagram links critical four elements of this process: imported technology, firm-level effort, inputs into enterprise learning and phases of technological development. Starting at the top left of the diagram, enterprises begin by importing technology in embodied forms (FDI, licensing, equipment and copying). Then they invest in building their abilities to master the tacit elements of the technology. They draw upon a variety of internal (human resources, technological effort, management effort and organisational effort) and external inputs (other firms, technology support, skills, finance and infrastructure) to build up their capabilities. The process starts with capabilities needed to master the technology for production purposes, and may deepen over time into improving the technology and creating new technology. These concepts are further illustrated by examples of simple learning in a small and medium enterprise (SME) in the food industry in Ghana and complex learning in a large electronics firm in Taiwan (see Box 2.1).

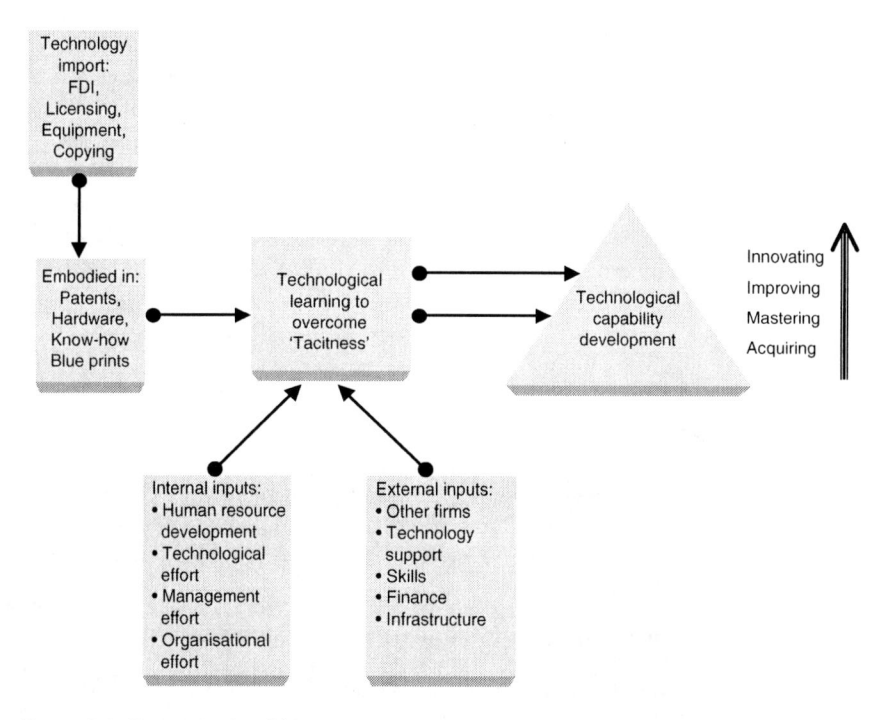

Figure 2.1 Enterprise-level learning process.

Box 2.1 Technological learning in Tatung and Astek fruit processing

Tatung: complex learning in a large Taiwanese electronics firm

Tatung, the largest electronics manufacturer in Taiwan, was one of the country's leading industrial conglomerates in the 1990s. Within thirty years, it advanced from making simple consumer electronics items (e.g. black and white TVs) in the mid-1960s to computers, colour displays and TV monitors. By the mid-1990s, the firm's electronics sales exceeded US$ 1 billion and it had eight overseas subsidiaries (including the United States, Japan, Germany and Ireland).

Tatung assimilated manufacturing know-how initially under technical cooperation arrangements from abroad, then by licensing technology and original equipment manufacture (OEM) deals with foreign multinationals. It began by acquiring relatively mature process technologies for household appliances and consumer electronics from both United States and Japanese companies through technical assistance deals and by forming joint ventures with foreign companies. The company learned many of its technological skills through OEM (a specific form of subcontracting). Under such arrangements, Tatung undertook contract manufacturing of electronics products for foreign multinationals, which in turn sold them under their own international brand names. OEM often involved the foreign multinationals assisting Tatung with equipment selection, training of engineers and advice on production and management. By the late 1980s, about half its colour TVs, PCs and hard disk drives were exported under OEM. Most of this production embodied little original R&D but the company also invested in technological effort to close process and product technology gaps with its competitors. Tatung learned the ability to absorb and adapt advanced foreign technology and modify, re-engineer and re-design consumer goods for different types of customers in regional markets. Its in-house engineering capabilities were used to scale down production processes, adjust capital to labour ratios and implement continuous improvements in its production technology. By the 1990s, the firm had a 500 strong team of engineers and technicians engaged in applied R&D activities.

Tatung began establishing sales offices in developed countries in the 1970s to service those markets. Subsequently, overseas production enabled it to compete in foreign markets, reinforce its brand and acquire advanced technologies and skills. By the mid-1990s it had eight manufacturing operations making TVs, washing machines, refrigerators and other household items.

Astek: simple learning in a Ghanaian food processing SME

With eighty staff, Astek Fruit Processing was one of Ghana's leading SMEs in the early 1990s. Using high quality local pineapples, the firm produced an orange pineapple drink as well fresh fruit juices and concentrates for the domestic market. Its volume of sales grew at 15–20 per cent per year between the late 1980s and the early 1990s. In the same period, its capacity utilisation rate doubled to 40 per cent and was expected to reach 80 per cent by the mid-1990s.

The firm made a good initial choice of technology. New equipment was purchased on a turnkey basis from Italy. The Italian equipment was cheaper and more suited to the smaller scale of production of the local market than rival sources of technology. The Italian equipment supplier sent two engineers to Ghana for two weeks to install the equipment and to train the workers. Prior to this, the Ghanaian production manager spent a month at the equipment suppliers' factory in Italy. The two Italians did the layout and provided the necessary engineering services but the Ghanaian production manager and other local technical staff also participated in designing the layout of the plant and positioning and wiring the equipment. Local technical staff worked alongside foreign engineers in a subsequent investment in a Tetra Pack technology (which sought to substitute paper packing for cans to reduce costs).

Learning about the technology during the start-up and expansion had a significant influence on Astek's acquisition of plant operation capabilities. The firm had a comprehensive quality control system and laboratory (with trained scientists) that performed checks on the fruit, the process and the final products. The equipment was well maintained by a full-time maintenance team headed by a graduate engineer. Moreover, it developed its main product, the orange pineapple drink, through in-house efforts and experimentation with different formulations.

Two factors underlie the strong local market and technological performance of this SME. It is owned and managed by a highly educated scientist (PhD in chemistry from London University who previously worked for the Ghana Standards Board) and his two sons (who have degrees in business studies and mechanical engineering, respectively). Moreover, it developed close relationships with technology centres and banks in Ghana and had ready access to technological services and finance.

Sources: The Tatung case is based on Hobday (1995) and the Astek case on Lall et al. (1994).

Five features of the process of building technological capabilities in developing countries are particularly relevant here (see Lall, 1992; Bell and Pavit, 1993 and Radosevic, 1999):

1   *The process of acquiring technological capabilities is unpredictable.* Investments in technological capabilities, like financial investments, carry considerable

risk and the outcome is uncertain. Firms face technical difficulties and financial uncertainties especially in research activities. Moreover, rarely can firms ensure against failure in capability building. The implications of fundamental uncertainty are clear: the reality cannot be fully modelled and the direction of change never achieves equilibrium.

2 *Capability building is an incremental and cumulative process.* Enterprises cannot instantaneously develop the capabilities needed to handle new technologies; nor can they make jumps into completely new areas of competence. Instead, they proceed in an incremental manner building on past investments in technological capabilities and moving from simple to more complex activities.[17]

3 *Capability building involves close cooperation between organisations.* Firms rarely acquire capabilities in isolation. When attempting to absorb imported technologies, they interact and exchange technical inputs with other firms (e.g. competitors, suppliers and buyers of output) and support institutions (e.g. technology institutions, training bodies and SME service providers) in a national innovation system. Hence, interaction and interdependence between organisations (i.e. collective learning) in a national innovation system is a fundamental characteristic of capability building.

4 *Success in acquiring firm-level technological capabilities can spillover into export success.* The major theories of comparative advantage – notably the factor proportions theory, theories of economies of scale and neo-technology theories – offer powerful explanations on the causes and evolution of international trade.[18] The technology and innovation perspective adds an additional insight to previous theories of comparative advantage. The major theories ignore the basis of comparative advantage as being 'minor learning' in developing countries. By contrast, the technology and innovation perspective holds that comparative advantage may arise from minor learning in developing countries and that this process is costly, unpredictable and cumulative. Hence, differences in the efficiency with which firm-level capabilities are created are themselves a major source of differences in comparative advantage between countries. Developing countries with relatively efficient firm-level learning processes will witness rapid export growth and industrial upgrading while weak learning processes in others will be associated with poor export performance.

5 *Capability building is affected by a host of national policy and institutional factors.* Firm-level learning can be stimulated by the trade, industrial and macroeconomic regime as well as supported by institutions providing industrial finance, training and information and technological support. In general, macroeconomic stability, outward-oriented trade and investment policies, ample supplies of general and technical manpower, ready access to industrial finance and comprehensive support from technology institutions are conducive to rapid capability building. The details of these measures along with some examples will be discussed in the section on 'Core elements of competitveness stratergy'.

Determinants and policy implications

The last point highlights the determinants of capability building at enterprise-level and has a direct bearing on the formulation of policies to enhance industrial competitiveness in developing countries. Perhaps the best way to analyse the influences on capability building is to view them as being a part of a system of inter-connected elements that are all geared to collective learning and, hence, attainment of industrial competitiveness. This scheme, called the national innovation system (NIS) or systems of innovation for development (SID) by systems theorists like Nelson (1993) or Lundvall (1992), is shown in Figure 2.2. The NIS approach emphasises that innovation and learning are processes that involve more than firms, support institutions, governments and other actors because of synergies and systems effects. It also suggests that the innovation and learning processes hinge on the internal interactions between the actors in the system and the external links of the system.

There are three levels in an NIS in a developing country:

- The first level is made up of the industrial clusters within a country. This contains all the firms (producers, buyers and suppliers) engaged in a given industry.[19] The tyre shape in Figure 2.2 indicates national industrial clusters. In turn, national industrial clusters are linked to various players (e.g. foreign buyers of output and multinationals) in global industrial clusters (represented by global knowledge in Figure 2.2). As they provide access to

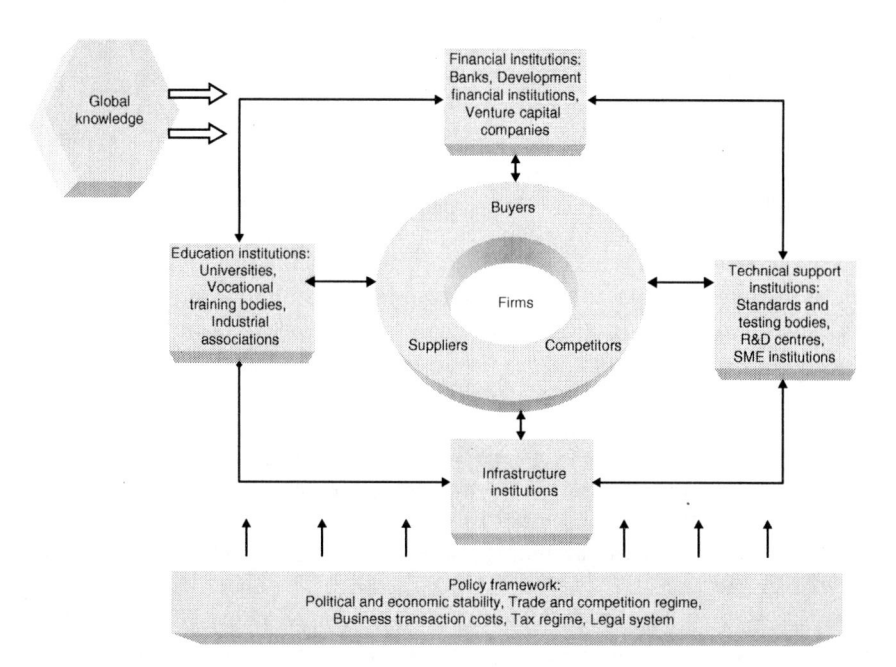

Figure 2.2 National innovation system (NIS).

imported technologies, skills and international markets, these external links are crucial to local technological development and competitiveness.
- The second level is the set of institutions and factor markets which support learning processes in industrial clusters. There is a strong emphasis on processes of interactive learning, that is, the exchange of knowledge and information between organisations involved in the development of capabilities. These institutions and factor markets include: education, finance, technological support and physical infrastructure.
- The third level is the set of policies that stimulate the learning processes between industrial clusters and institutions. A range of policies which influences technological activity falls under this heading including: political and macroeconomic environment, trade and competition regime, business and transactions costs, tax regime and the legal system.

NISs differ markedly in the developing world in the technological strengths of enterprises within them (firms, suppliers, buyers, competitors and service providers), the efficiency of their collective learning processes, systems effects and the intensity of their external links. A few systems (particularly in East Asia) have a good base of technologically competent firms and efficient institutions, display significant collective learning, have strong systems effects and have developed extensive external links with foreign sources of knowledge. Conducive incentive and regulatory frameworks (characterised by exposure to competitive pressures and low transactions costs) have also stimulated collective learning processes in efficient systems. As a result, these innovation systems have witnessed smoother transitions to higher levels of national capability development. This means moving from a stage of collective acquisition of technologies to collective improvement and eventually to collective innovation (see Figure 2.1 for different phases of technological development). Higher levels of national capability development are in turn associated with better industrial growth, export competitiveness and technologically advanced production.

Most NISs in developing countries, however, are deficient in all of these aspects and witness gaps in industrial competitiveness relative to industrial leaders. Weaknesses in NIS in developing countries arising from missing markets, deficiencies in key institutions, poor quality and intensity of internal interactions and weak external links are generally referred to as systems failures.[20] There are different ways of classifying systems failures and Box 2.2 shows a scheme developed by UNIDO for developing countries. Since systems failures directly affect how a developing country copes with globalisation, remedying them is the principal aim of an industrial competitiveness strategy.

Thus, the technology and innovation perspective associates the concept of competitiveness with the accumulation of technological capabilities at enterprise-level and collective learning within the NIS. The major strengths of this approach are that it: (a) provides a realistic and comprehensive portrayal of how enterprises in developing countries (supported by institutions) become competitive by mastering technologies from abroad; (b) translates the influence of this

Box 2.2 Gaps in NIS in developing countries

There are pervasive weaknesses in NIS in developing countries, which inhibit collective learning processes and the development of industrial competitiveness. UNIDO (forthcoming 2002) advocates a linking, leveraging and learning (LLL) industrial learning strategy to transform NIS in developing countries and classifies these weaknesses under three broad headings as follows:

1 *Those that relate to the ability of national industries to link with global industrial clusters:*

   - Firms and clusters might lack strategic intelligence about the organisation and dynamics of global industrial clusters (e.g. evolution of technologies, markets and MNC behaviour) and are unable to diagnose their relative strengths and weaknesses;
   - Business associations and trade promotion organisations may lack information about global industrial clusters and be unable to provide services to forge links with foreign partners;
   - Policy-induced barriers (e.g. high and variable effective protection and controls on MNCs) or geographical isolation may hamper the entry of foreign partners (e.g. foreign buyers and multinationals) and imports.

2 *Those that relate to the ability of national industries – which are already linked to global industrial clusters – to leverage technology from them:*

   - Firms may not be able to devise favourable contracts with foreign partners, which provide for extensive transfers of technology, skills and marketing expertise;
   - Leveraging institutions (such as investment promotion organisations, SME promotion agencies and regional development agencies) may not be able to provide a range of services to firms to leverage resources because they are top-down, bureaucratic and lack technical manpower.

3 *Those that relate to the ability of national industries – which are already linked to global industrial clusters and leveraging resources – to initiate collective learning processes:*

   - Firms may not be aware of the need for learning and lack basic manufacturing capabilities;
   - Producers, suppliers and buyers in an industrial cluster may not connect with each other or form clusters;
   - Firms and other actors may be unable to organise collective learning processes to reap systems effects;

> - Institutions providing training, technological support, industrial finance and physical infrastructure may be weak or fragmented and unable to provide high quality support services to firms;
> - The incentive and regulatory framework may not be conducive to innovation and learning because of residual import protection, overvalued exchange rates, high corporate taxes, poor enforcement of patent laws, excessive rules on MNC operations and weak enforcement of competition laws.
>
> Source: Adapted from UNIDO (forthcoming 2002).

process on national industrial performance via the theories of comparative advantage; and (c) provides scope for a comprehensive public policy agenda to enhance competitiveness in order to remedy systems failures affecting collective learning. Next we turn to the components of this policy agenda.

Core elements of competitiveness strategy

The technology and innovation perspective provides strong theoretical grounds for a different competitiveness agenda than the macroeconomic and business strategy perspectives. In common with previous approaches, the technology and innovation perspective emphasises sound macroeconomic management and a predictable policy framework, a liberal trade regime, a strong and efficient legal framework conducive to private enterprises and a pivotal role for the market mechanism in resource allocation. However, in order to remedy pervasive weaknesses in NIS in developing countries, it also makes the case for a more holistic approach with additional policies to support enterprise-learning processes.

An industrial competitiveness strategy from a technology and innovation perspective can be expressed in terms of three core elements:[21]

a   A national partnership involving complementary actions by government and the private sector for industrial competitiveness.
b   A 'liberalisation plus strategy' involving a mix of policy instruments.
c   Where appropriate on economic grounds, micro-level policies to promote the competitiveness of particular industrial clusters.

The importance of the first element is largely self-explanatory. The successful implementation of any set of economic policies (including competitiveness strategy) requires a national consensus on goals and broad policy direction amongst a developing country's principal social partners (i.e. the government, the private sector, trade unions and NGOs). Whilst noting the importance of trade unions and NGOs, this chapter focuses on the interaction between government and the private sector. As discussed in the section on 'Role of the private sector' the private sector needs to contribute actively, alongside government, to the management of competitiveness strategy.

The second element is somewhat more complicated and can be classified into two sets of policy measures. The first, *incentive policies*, are instruments to remove economic distortions created by past government actions discouraging private sector growth and competitiveness. Import liberalisation, tax reform, removal of redundant bureaucratic procedures on private sector activity and privatisation all fall under this heading. The second, *supply-side policies*, are geared to overcome systems failures which impede collective learning processes and the creation of new competitive advantages by enterprises. These include education and training, technological support, industrial finance and attraction of FDI and small firm promotion. The point to stress is that it is the interaction of incentive and supply-side policies that determines industrial success in developing countries.

Whereas the second element involves largely non-sector specific measures, the third element emphasises detailed actions to improve the competitiveness of specific industrial clusters within a developing country. Intervention at the micro-level is often directed towards acquiring technological capabilities, promoting upgrading and improving links between different parts of the cluster. Public policies might range from industry-specific tax measures to the provision of specialised institutional support facilities for a particular cluster. Joint actions (e.g. setting up a specialised training school) between a business association in the cluster and an aid donor or a government agency are also commonplace.

The remainder of this section explores these three core elements of competitiveness strategy in more detail.

Liberalisation plus approach

Table 2.1 identifies a sample of constraints to improving competitiveness and elements of public policies to enhance competitiveness in a typical developing economy, which has recently adopted a more outward-oriented, market-friendly policy stance. The issues presented in Table 2.1 are illustrative rather than comprehensive. For reason of space, some important headings (e.g. labour market reform and administrative reform) have been omitted. The items under incentive policies are macroeconomic policy, trade policy and competition policy while the rest fall under supply-side policies. There is also a separate item called policy management, which refers to an institutional mechanism for integrating incentive and supply-side policies and for involving the private sector in policy making.

Macroeconomic policy. A stable, predictable macroeconomic environment is a *sine qua non* for improved export competitiveness but the situation is asymmetric: macroeconomic stability cannot ensure competitiveness but macroeconomic instability inevitably hurts competitiveness. Experience suggests that many developing countries embark on a competitiveness strategy (or even for that

Table 2.1 Government actions for competitiveness in a developing economy

Policy area	Constraint	Suggestion
Policy management	Lack of a coordinating vision and mechanism	Establish a national competitiveness council to formulate strategy and monitor implementation
Incentive policies		
Macroeconomic policy	High inflation and large fiscal deficit	Develop a plan to reduce fiscal deficit within a specified time period
	Appreciating real exchange rate	Adopt a more aggressive approach to exchange rate management
	Lack of policy credibility	Implement reforms and involve private sector in pre-budget consultations
Trade policy	High and variable effective protection	Persist with import liberalisation to achieve low, uniform effective protection
	Weak export drive	Revamp trade promotion organisation to become more pro-active and allocate more funds for overseas marketing
	Long delays in refunds on imported inputs	Streamline bureaucratic procedures and introduce computerisation at customs
	Ad hoc participation in the WTO and passive role in international trade negotiations	Develop trade negotiations capabilities within government, co-opt leading trade lawyers into trade delegations and set up an embassy at the WTO
Competition policy and privatisation	Domination of key industries by inefficient SOEs	Conduct a study of SOEs and implement a privatisation programme
	No framework for regulating anti-competitive practices	Pass a competition law and set up an enforcement agency (e.g. a monopolies and mergers commission)

Supply-side policies

Human resources	Skill gaps in potential areas of comparative advantage	Conduct a survey of future skill needs benchmarked against competitors and prioritise future skill needs
	Inefficient public sector training institutions	Introduce partial cost recovery of services for public institutions and assist industry associations to launch training centres
	Limited enterprise training	Introduce an information campaign to educate enterprises about skill gaps and a tax deduction for training investments
Technology support	Weak quality standards in industry	Provide part-grants for SMEs to obtain ISO9000 certification
	Low industrial productivity	Establish a productivity centre to improve industrial productivity to world standards
	Inadequate linkages between technology institutions and industry	Introduce partial cost recovery of service for public institutions and an aggressive marketing campaign
Foreign investment policy	Unfocussed foreign investment promotion strategy	Develop a pro-active foreign investment promotion strategy which targets a few realistic sectors and host countries
	Poor international image/lack of contact with potential investors	Establish overseas investment promotion offices as a joint venture with the private sector
	Uncompetitive EPZ package	Evaluate EPZ incentives against competitors and change offer to attract flagship multinationals
Industrial finance	High interest rates and an oligopolistic banking system	Manage prudent monetary policies and introduce competition into the banking sector
	Anti-SME bias in credit allocation by banks	Promote training for bank staff on assessing SME credit, specialist SME funding windows and micro-finance schemes
Infrastructure	High costs of sea and air freight	Liberalise air and sea cargo entry to foreign operators
	Long delays in accessing utilities connections	Consider commercialisation/privatisation of infrastructure parastatals with an effective regulatory framework

matter an economic liberalisation programme) in the wake of a macroeconomic crisis associated with unsustainable fiscal and balance of payments deficits and high inflation. In such circumstances, an overvalued real exchange rate might also arise from expansionary fiscal and monetary policies and exchange controls. Macroeconomic instability is likely to send confusing signals about the profitability of resource reallocation to enterprises and investors and may lead to a subdued (or mixed) supply response. If a lack of policy credibility leads to a low supply response, then the processing of implementing a new competitiveness strategy may itself become difficult to sustain. Much has been written about what constitutes good macroeconomic policy for competitiveness (see Corden, 1994; Esser et al., 1996). Regardless of the author, in essence the medicine would be the same for our typical developing economy: budget deficits should be kept low, inflation tightly controlled, the real exchange rate kept competitive and debt crises avoided in order to encourage savings, investment and growth. Much emphasis is also placed on consistency, coherence and continuity of macroeconomic policies.

Trade policy. Development experience suggests that a country's trade policy has a great deal of influence on its competitiveness. Economists typically classify countries by their trade policies into inward and outward-oriented economies. The difference between them is in terms of the effective protection granted to production for the home market compared with exports (see Bhagwati, 1988). An inward-oriented trade policy grants significant protection for home market production and is biased against exports where as an outward-oriented policy has limited protection and favours exports. Economists agree that an outward-oriented trade regime is better associated with improved export competitiveness than an inward-oriented one. An outward-oriented trade regime induces better resource allocation according to comparative advantage; the realisation of economies of scale; access to new technologies, imported inputs and markets; investments in new technologies due to competitive pressures; and less rent-seeking behaviour.

But there is an active debate on how fast quantitative restrictions and tariffs should be removed within a policy thrust of outward-orientation (see Amsden, 2001). Gradual import liberalisation may be justified on economic grounds in large developing economies with a long industrial experience, a core of potentially competitive enterprises, well functioning technological and marketing support institutions and a business-friendly government. China might be a case in point. Otherwise, rapid import liberalisation would be optimal for our typical developing economy. Regardless of the speed of liberalisation, quantitative restrictions should be eliminated immediately and be followed by staged reductions in the dispersion of tariffs. This way, price signals would influence resource reallocation and industrial efficiency at an early stage.

Another priority is to ensure that the developing country plays an active part in international trade negotiations, WTO debates and develops the requisite capabilities to do so (see Chapter 9 by Sheila Page in this volume). Policy suggestions might include: developing a specialised trade negotiations capability within government (by recruiting trade lawyers and economists), inviting leading trade lawyers and private sector representative to be part of official international trade delegations and setting up an embassy at the WTO.

Domestic competition policy and privatisation. Like exposure to international competition, more domestic competition can also act as a spur to improvements in resource allocation and industrial efficiency. The industrial sectors of many developing countries are dominated by large state-owned enterprises (SOEs), which are not only inefficient and unprofitable but also crowd out domestic private sector activity (see Chapter 8 by Christos Pitelis in this volume). This type of SOEs can seriously impede the growth of SMEs in key sectors. Furthermore, there is often a lack of an effective policy framework for domestic competition. Policy options in this vein might include: carrying out a detailed assessment of SOEs, formulating a privatisation programme appropriate to the domestic context, passing a strong competition law and setting up an authority to regulate monopolies and mergers.

Human resources. Investments in skills at all levels are a vital pre-condition for improved competitiveness (see Cassen and Wignaraja, 1997; Kim and Nelson, 2000). Investments in primary and secondary education help to develop literate, muti-lingual, numerate production workers, who are the bedrock of successful labour-intensive industrialisation. As industry moves into more complex activities, investments in industry-specific vocational training and tertiary-level technical skills (particularly mathematics, science, engineering and information technology) become necessary to enter new activities. Equally important are investments by local firms (particularly SMEs) in formal employee training to tailor-make skills for industry. There are many ways of approaching human resource development in our typical developing economy but one of the main priorities seems to be a survey of future skill needs benchmarked against competitor countries to identify skills gaps in potential areas of comparative advantage. This would also guide budgetary allocations for education and training. Other ideas might include: partial cost recovery of service approaches for turning around ineffective public sector training institutions, assistance for industry associations to launch training centres, an information campaign to educate firms about skills gaps and a tax deduction scheme for enterprise-level training investments (especially for SMEs).

Technology and SME support. In a world of rapidly changing technologies and long learning cycles, comprehensive technological support is critical for improving competitiveness (Nelson, 1993; Lall and Teubal, 1998; Chapter 4 by Stan Metcalfe and Chapter 5 by Fischer and Reuber in this volume). In our typical developing country, with technological capabilities that lag well behind world frontiers in most industries, technological support should not be directed at R&D activities to create new products and processes at world frontiers. Instead, the first call on scarce resources should be for institutions and schemes to deal with basic production capabilities including productivity improvement, testing and metrology services, introducing ISO9000 quality management and total quality management practices, and technical extension services for SMEs.

As the economy matures and becomes more technologically sophisticated, the emphasis could shift somewhat towards design and R&D capabilities. Some policy suggestions are part-grants for SMEs to obtain ISO9000 certification, a productivity centre to improve productivity to world standards, partial cost recovery of services for public institutions, and an awareness campaign for industry about technology gaps and available services. A scheme to encourage joint projects between firms and technology institutions (e.g. designing new products) to stimulate linkage creation between these organisations would also be useful.

Foreign direct investment (FDI) policy. Attracting export-oriented FDI is viewed by economists as a shortcut method for entering the production of manufactures for export and for technologically upgrading export competitiveness over time in developing economies. For the same reasons that large firms are expected to outperform small firms – including economies of scale in investments in R&D and capital market imperfections which confer an advantage on large firms – subsidiaries of multinationals can be expected to have higher export capabilities than local firms. Moreover, domestic affiliates of multinationals are better placed to acquire export capabilities because of their ready access to the 'ownership advantages' (such as technologies, managerial and technical skills, and marketing-know) of their parent corporations.

Many developing economies are unable to reap the full potential of export-oriented FDI because of weaknesses in their investment environment such as high cost, unproductive and illiterate labour; political instability; pervasive corruption; inefficient physical infrastructure; and an inadequate legal system and accounting standards. A weak foreign investment promotion strategy also contributes to a lack of inward investment. Commonly found problems might include: unfocused investment promotion efforts with limited targeting of individual firms and investor markets; a poor international image as an investment destination; uncompetitive fiscal incentive in export processing zones (EPZ); and cumbersome foreign investment approval processes which lead to long delays in approvals and provide scope for rent-seeking behaviour. In this vein, attracting export-oriented FDI in our typical developing economy should emphasise: a pro-active FDI promotion strategy which targets a few realistic sectors and host countries; creating of overseas investment promotion offices in a few host countries as a joint venture with the private sector to act as a first point of contact for investors and to permit active networking among potential investors; keeping EPZ competitive against international competitor destinations; and a streamlining of approval procedures with provision for their eventual abolishment (see Chapter 6 by te Velde in this volume).

Industrial finance. Access to finance at competitive interest rates (for working capital and capital investments) is an obvious requirement for creating and sustaining export competitiveness. Yet enterprises, particularly SMEs, in most developing economies would complain about high interest rates (in part due to oligopolistic banking practices) and a lack of access to bank credit. Industrial finance is a whole subject in itself (see Chapter 7 by Andy Mullineux and Victor Murinde in this volume) but our initial suggestions for our typical developing economy would be to prudent monetary policy management, competition into the banking sector, training schemes for bank staff on assessing SME credit, specialist 'soft terms' funding windows for SMEs and micro-finance schemes. Once the financial system has deepened, there could be scope for introducing different financing mechanisms including venture capital arrangements.

Infrastructure. An efficient infrastructure is necessary for improved export competitiveness. Persistent infrastructural problems in a given developing economy can significantly raise transaction costs for its enterprises relative to those in competitor economies. For instance, bottlenecks in the availability of air and sea cargo space and high charges feed into uncompetitive pricing, missed foreign buyer deadlines, poor country reputation and cancellation of repeat ordering. Similarly, long delays in accessing utilities like telephone and electricity connections during business start-up or expansion raise production costs and wastage of management time. The overall recommendation for our typical developing economy is for it to invest a given percentage of GDP in new infrastructure like ports and airport facilities and in maintaining the quality of existing infrastructure like roads and electricity supply capacity. Other recommendations might include: liberalisation of air and sea cargo entry to low cost foreign operators; and consider commercialisation/privatisation of infrastructure parastatals and put in place an effective regulatory framework.

Promotion of industrial clusters

Policies to promote industrial clusters have been widely implemented in many developed countries and are being adopted in some developing countries. As defined by a recent OECD study, cluster policies refer to 'the set of policy activities that aim to stimulate and support the emergence of production networks among firms, strengthen the inter-linkages between different parts of the networks and increase the value added of their actions' (Boekholt and Thuriaux, 1999, p. 381). These policies have been applied to particular industrial sectors and geographical locations. As a result, cluster policy is sometimes equated with an old style industrial policy of providing subsidies to inefficient industries in a protected domestic market. However, this is rarely the case in practice. Most cluster policies are implemented in a more market-oriented policy framework characterised by a liberal trade regime and few barriers to entry and exit. Thus, competitive pressures typically form the backdrop for the operation of cluster policy.

Thus, within a market-oriented developing country, the aim of cluster policy is to improve the competitiveness of specific industrial clusters. As discussed in Box 2.2, intervention at the micro-level is justified on the grounds of systems failures, missing markets and policy-induced distortions that hamper collective learning processes. Public policy actions (for acquiring technological capabilities, promoting upgrading and improving links between different parts of the cluster) might include the following:[22]

- Providing strategic information to clusters via benchmarking studies.
- Removing bureaucratic regulations that hamper the emergence of clusters, helping to bring firms together by acting as a broker (see Box 2.3).
- Facilitating joint actions (e.g. setting up a specialised training school or productivity centre) between a business association in the cluster and an aid donor or a government agency to support technological upgrading.
- Attracting foreign investors and buyers to a cluster and supporting the attendance of a group of local firms at an international trade fair.

Box 2.3 Denmark's Industrial Network Programme

> The Danish Industrial Network Programme (DINP) has been the inspiration for many such programmes in OECD and developing countries. It was in operation between 1989 and 1992 and provided over 300 networks with about US$ 25 million in grants. Carried out by the National Agency for Industry and Trade, DINP sought to remedy the lack of large firms in the Danish economy by 'bulking up' SMEs through networks which would promote greater inter-firm cooperation. The programme provided financial grants to help SMEs to implement cooperative projects to improve their manufacturing capabilities. Support was based on a three-phase network development model with support for each stage dependent on success in the previous stage. To be eligible for support, the network had to consist of a minimum of three firms, which could apply for subsidies to cover the external expenses incurred in finding partners and to carry out a feasibility study of cooperation possibilities.
>
> The programme also covered the training of forty network brokers who were to assist the participating companies. Their role was to identify the right companies and to provide the networks with administrative support. Their role in phases 1 and 2 are fundamental as they had to deal with the initial administrative problems of the network. In phase 3, the network broker tends to have a minimal role as the relationships between the firms become established. In addition, the programme financed the services of lawyers to develop standard contracts laying out the obligations of the different parties in the network.
>
> Many companies were encouraged to set up a new legal entity (a new firm). This new entity served as the network centre, controlling the joint resources of the network. The programme enabled SMEs to emulate the behaviour of a larger firm in the following ways:
>
> - Firms were able to purchase or resource joint solutions to common problems, for example, monitoring markets, competitors and technology;

> purchasing advanced equipment; joint R&D; joint finance and credit line guarantees;
> - Firms could specialise within the network to exploit complimentary advantages in much the same way as large firms have divisions specialising in different areas;
> - Firms were able to leverage their subcontracting links both downstream to pooled contractors and upstream through access to markets that had previously been denied to them because of their small size.
>
> Source: Boekholt and Thuriaux (1999).

Whilst cluster policies offer a means to enhance the competitiveness of particular industrial sectors and geographical locations, they should not be applied in an *ad hoc* manner as there is a high risk of government failure. The following principles might guide cluster policy in industrial latecomers:

- The choice of activities should be guided by near future comparative advantage rather than long-term comparative advantage.
- Interventions should be at a broad industry-level rather than firm-level and be justified by systems imperfections.
- Interventions should be strictly time bound.
- Clearly defined performance measures should be used to evaluate the success of interventions.
- The interventions should be bound by WTO rules for subsidies, local content etc.

More applied policy research is needed as to what works and what does not and the conditions for successful cluster policy in the developing world.

Role of the private sector

Competitive markets are widely accepted as the best way to efficiently organise the production of goods and services in an economy. The traditional role for enterprises and business associations in a developing market economy might be articulated as follows:

1 Enterprises exist to produce and distribute their output at the lowest possible cost in order to maximise profits. Competition involves enterprises in a struggle to create new competitive advantages by investing in technological upgrading, marketing capabilities and skill formation. Those that successfully acquire competitive capabilities will survive while the less efficient ones will go under. Over time, new enterprises will be set up in response to market opportunities and incentives.

2   Business associations are set up as self-help bodies by groups of businesses to further the interests of and respond to external events of their members. These consist of federations, chambers of commerce, industry-specific bodies and small business organisations. On behalf of their members, business associations present business viewpoints and interests to government and lobby them on the enabling environment. While lobbying is the main role of business organisations, some also provide direct help to members in the form of information on market opportunities, tax and legal matters as well support services such as organising participation in overseas export promotion missions, helping to forge joint ventures with overseas partners and conducting training for upgrading technical skills.[23]

Accordingly, there are complementary ways in which enterprises and business associations can pull together in a developing market economy. Where this synergy works well, there can be significant gains for the private sector as a whole.

A similar analogy can be applied to the relationship between the private sector as a whole and the government in a developing market economy. If markets work well, and are allowed to, there can be large economic gains. If markets fail, and governments intervene carefully to remedy market failures, then there can be further gains. When markets and government work together, the whole is likely to be greater than the sum as evidenced by the economic success of economies like Singapore and Taiwan and more recently Chile, Mauritius, Costa Rica and Hungary. In an environment of a withering of the state in developing economies (as evidenced by public expenditures cuts and declining public sector responsibility for economic activity) and a rising share of private sector output in GDP, the private sector increasingly needs to contribute actively to long-term economic policy particularly competitiveness strategy.

Successful development experience indicates that the private sector make at least four important contributions to the management of industrial competitiveness strategy:

Help weaker firms to help themselves. Most developing economies are characterised by a dualistic industrial structure with a small base of efficient exporting enterprises and a long tail of under-performers. The challenge is how to raise the efficiency of weaker firms to best practice levels. Many government institutions dealing with quality, productivity and other aspects of technology are actively involved in this process but such efforts are generally inadequate given the scale of the problem. Leading internationally competitive enterprises can also actively assist in upgrading the export capabilities of weaker enterprises and SMEs by establishing industry-specific training centres; carrying out productivity benchmarking exercises and quality awareness training; and providing advice on effective marketing strategies. Actively attempting to develop local subcontracting and supplier relations is another important way of transferring technologies and skills from larger firms to SMEs. Likewise, business associations can play a pivotal role in fostering production relations between large firms and SMEs and directly providing business development services to small firms (training, technology support and marketing assistance).

Help government to plug information gaps. Access to timely and detailed information is a key determinant of national and enterprise-level competitiveness but such information tends to be unevenly distributed within any economy. Owing to its involvement in world markets, the private sector has better access than government to information about a host of competitive threats (e.g. new technologies, external demand conditions, overseas government policies, WTO rules and regulations) and new market opportunities. Active private sector participation in national policy making bodies and international trade negotiations are possible ways of making this knowledge socially effective. Private sector situation-specific knowledge, experience of the enterprise-level impact of policy changes and negotiation skills would be valuable inputs into these exercises. Private sector business associations can further contribute to policy making by carrying out regular surveys of enterprise confidence and obstacles to competitiveness (see Box 2.4 on public–private sector consultation in Côte d'Ivoire).

Augment government capabilities. In many developing economies, the private sector has developed a solid base of modern management, financial, marketing and technical skills while government has become increasingly weak in these areas. In part this may be due to relatively higher compensation packages in the private sector (which attract the best university and technical school graduates) and investments in employee training in private firms. A short-term secondment programme – whereby experienced private sector managers and technicians can work in government departments on specified projects for a given period – may be a useful way to improve government capabilities and develop a better public–private sector dialogue on competitiveness issues. The United Kingdom has such a short-term secondment programme whereby the government pays a private manager the equivalent civil service grade salary and the private firm pays the rest. Another route may be to undertake joint public–private sector overseas investment promotion missions to improve the country's image as a favourable destination for FDI in selected host countries. Singapore and Ireland are well-known for this practice. Private sector presentational and marketing skills would be a useful input into image building and investment generation activities.

Participate in infrastructure projects. Many developing countries are characterised by a wide range of infrastructure gaps ranging power fluctuations to uncompetitive utility costs and are faced with the challenge of investing in new infrastructure and maintaining the quality of existing infrastructure. The problem is that infrastructure investments often involve very high project costs which debt-ridden governments are unable to meet. One solution might be joint-ventures between government and the private sector to develop new infrastructure projects. This may involve arrangements, which call for some private sector funding and private sector management with government funding and financial guarantees. The British Government encourages variants of these arrangements through its Build-Operate-Transfer (BOT) and Design-Build-Finance-Operate (DBFO) schemes (see UK DTI, 1998 and 1999).

Box 2.4 Public–private consultation mechanisms and partnerships in Côte d'Ivoire

> The economic crisis that plagued many African countries in the 1980s aggravated the poor performance of the industrial sector in Côte d'Ivoire. The wisdom of public enterprise was questioned and movements were made to make the state a facilitator of industrial activity rather than an investor. Like several other African countries, Côte d'Ivoire embarked on a reform programme under the advice of the IMF and the World Bank. Between 1991 and 1995, several programmes including the Programme for Adjustment for Sectoral Competitiveness, the Programme for Adjustment for the Financial Sector and the Programme for Development of Human Resources were set up.
>
> Various mechanisms for public–private sector consultations/partnerships emerged, the most important of which were the Committee on Competitiveness and the Committee on Private Sector Development. The Committee on Competitiveness addressed such issues as price control, fiscal regimes and other issues of regulatory impediments for industrial development. The Committee on Private Sector development emerged as a result of a seminar on this subject organised by the Government and the World Bank. The Committee, which is chaired by the Prime Minister and has several prominent businessmen as members, makes recommendations on issues such as the enabling environment, institutional reforms, macroeconomic management and other critical economic and industrial policy issues. The Industrial Partnership Council was established by Presidential Decree in 1999 in the context of the Alliance for Africa's Industrialisation. It is hoped that the Council will play a key role in transforming the country into a newly industrialised economy within a generation.
>
> The experience of Côte d'Ivoire suggests that consultative mechanisms are useful for clear management of the economy and thus for instituting reasonable and realistic reforms. The efficiency of such mechanisms requires that there be representation of all relevant actors and that they have confidence in each other. In addition, partnerships must continue to move forward in an open and frank manner that allows contentious issues to be discussed or else they will become moribund.
>
> Source: UNIDO (2000).

It is worth emphasising that an active public–private sector partnership is a necessary condition for an efficient and practical competitiveness strategy in our typical developing economy. In turn, a successful strategy will fuel future national competitiveness, economic growth, rising per capita incomes, more internal consumption of goods and services and more employment. Thus, a more competitive economy will generate a 'win win situation' for both government and the private sector.

Management of competitiveness strategy

A road map for effective management

The specific context, objectives and policies of an industrial competitiveness strategy will vary between developing countries in order to reflect each developing country's unique economic and political circumstances. Yet, evidence suggests that developed countries and the East Asian NIEs, have broadly followed similar approaches to managing their industrial competitiveness strategies and ensuring that broad goals are translated into industrial results (see Stiglitz, 1996; El-Agraa, 1997; Amsden, 2001 and Teubal, 2001). Evidence also suggests that the management of a coherent competitiveness strategy is an extremely complex, interactive and demanding process (see UK DTI 1998 and 1999; Singapore Ministry of Trade and Industry, 1998). It involves a range of overlapping activities including data collection, assessing industrial performance and existing policies, designing a new policy framework, implementing the new policy framework, coordinating policies, monitoring the results, refining the policy framework and capacity building. East Asian and developed country experience offer insights on these activities, which can be represented as a stylised road map for other industrial latecomers in the developing world.

For simplicity, the process of managing an industrial competitiveness strategy can be represented as a cycle of three inter-related phases:

1   Assessing existing industrial competitiveness performance and the policy framework.
2   Designing a new competitiveness policy framework.
3   Implementing the new policy framework.

Developing countries will differ in the time taken to accomplish each of these phases and the phases may overlap with each other but none of them can be left out if a developing country is to put in place a sustainable industrial competitiveness strategy. The key issues under the three phases will be highlighted in the following sections.

Assessment

A diagnostic assessment is the starting point for developing an industrial competitiveness strategy in a developing country. Its purpose is to collect information and to analyse existing industrial performance, economic policies and the external environment. This type of exercise is increasingly referred to as an industrial competitiveness assessment in policy circles.[24] In essence, it is a benchmarking exercise which compares important aspects of a country's industrialisation experience with international competitors including the strengths and weaknesses in industrial performance (often proxied by manufactured exports), the capabilities of enterprises and industries, support institutions and factor markets, and the regulatory and incentive regime affecting industry. Box 2.5 highlights some of the elements of a typical competitiveness assessment for a developing country and Box 2.6 shows how to benchmark capabilities at enterprise-level within industries.

Box 2.5 Elements of a competitiveness assessment

> Industrial competitiveness assessments on individual developing countries are being increasingly conducted on behalf of national governments by international agencies, consulting firms and research organisations. Some are conducted from an economics viewpoint, others from a business strategy perspective and some from a combined economics/business strategy perspective. Competitiveness assessments seek to provide a detailed picture of the international positioning and performance of a country's industries as well as the policy, institutional and external factors that affect past industrial performance. They also attempt to provide an assessment of future industrial prospects, opportunities and threats on the horizon and suggestions on how to overcome them. Individual competitiveness studies tend to vary in what they cover depending on the purpose at hand and the complexity of the economic challenges faced by the country. Bearing this in mind, a typical competitiveness study would highlight the following elements:
>
> - A review of initial conditions for industrial development including history of economic and liberalisation policies, sources of comparative advantages (e.g. technology, skill base, other labour issues and markets), geographical location and other locational advantages and disadvantages;
> - An examination of industrial growth and structure as well as export dynamism (e.g. export values, structure, growth rates, market shares and competitive positioning in global value chains), revealed comparative advantage;
> - A detailed assessment of firm-level capabilities to manufacture efficiently, the organisation of local value chains and clusters as well as industry positioning in global value chains and foreign actors involved in the process;
> - An assessment of the extent of technology transfer from abroad and the policy regime for technology transfer (e.g. foreign investment regime, investment promotion agencies and other actors involved in technology transfer from abroad);
> - An evaluation of the adequacy of support institutions and factor markets involved in supporting innovation and learning including

education and training, technological support, SME extension services, industrial finance and physical infrastructure;
- An assessment of the internal (firms/local value chains and institutions) and external links (local and global value chains) of industrial innovation and learning systems;
- An examination of the macroeconomic, trade and industrial policy regime, rules and regulations and pertinent actors therein which influence the direction of innovation and learning;
- An assessment of opportunities and threats on the horizon and a detailed evaluation of policy options for the transformation of the industrial sector and the industrial innovation and learning system. This may be a valuable input into a common vision and a strategy.

Competitiveness assessments are not one-off exercises which can be done once every decade or so. A world of globalisation, falling trade barriers and rapid technological change has far-reaching and unpredictable effects on national industrial behaviour in open economies. Thus, competitiveness assessments need to be conducted (and updated on regular intervals) to take account of new statistical information on national industrial performance and changing internal and external circumstances affecting national industrial performance. In this regard, it is relatively cost-effective to develop strong local capabilities within developing countries to conduct and update such assessments (by involving local research organisations, universities and public sector centres of excellence such as central banks).

Box 2.6 Measuring capabilities in manufacturing enterprises

Comparisons of the capabilities of enterprises relative to best practice levels are useful to indicate critical gaps in manufacturing capabilities in developing countries. For instance, a comparison of local enterprises with multinational affiliates in the Thai automobile components industry could reveal weaknesses in the ability to design new products and manage process quality in the former. Similarly, an evaluation of large and small local firms in the food processing industry in Nigeria might show up differences in capabilities to maintain production equipment and test raw materials. More important, such assessments can also show how firms compare with international best practice, a critical element in developing competitive capabilities.

It is not easy to benchmark enterprise capabilities – a proper evaluation can be extremely intensive in information and skills. However, it is possible to devise shortcuts that yield useful results. One is to define the essential technical functions performed by enterprises and give each a subjective ranking indicating levels of competence. The basic assumption is that firms that perform these essential functions well also perform well on the larger spectrum of technological activity. It allows for a 'quick and cheap' method of benchmarking that gives plausible rankings. The average capability score can be aggregated in various ways, for instance by ownership groups (local and foreign firms), size (SMEs and large firms) or market-orientation (domestic market and export market) within an industry. The scores can also be related to performance in terms of growth, profitability or exports. The results are of obvious interest to technology development and policy.

A recent study uses this approach to assess the capabilities of Mauritian garment enterprises (see Wignaraja, 2002). It calculates a 'technology index' (TI) for each firm based on two categories of technical functions: production and linkages with other firms. The larger category, production, is captured by ten technical functions, ranging from process engineering tasks like quality management (measured by internal reject rates and ISO 9000 accreditation) to product engineering (like copying existing products, improving existing products and introducing new products). Linkages are captured by technology transfers through two types of intra-firm relationships – subcontracting and marketing relationships with overseas buyers of output. Each of the twelve technical activities is graded at different levels (0, 1 and 2) to represent different levels of competence within that function. Thus, a firm is ranked out of a total capability score of 24 and the result is normalised to give a value between 0 and 1. The table shows the average TI scores (overall capabilities and separately, for process, product and linkages capabilities) by firm size in the Mauritian sample. It shows that average capabilities in large firms are significantly higher than in SMEs.

Average TI scores for large firms and SMEs in Mauritius

Size category[a]	TI score	Process engineering score	Product engineering score	Linkages score
21 Large firms	0.51	0.53	0.37	0.40
19 SMEs	0.17	0.20	0.23	0.04

Source: Wignaraja (2002).

Note
a SMEs have < 100 employees, large firms have > 100 employees.

Regression analysis confirms the validity and usefulness of the TI measure. For instance, in the Mauritius sample, firm size, the share of engineering and technical manpower in employment, training expenditures as a percentage of sales and the number of times a firm used external technical

> assistance (foreign consultant or technology institution) have positive and significant relations with TI (Wignaraja, 2002). This confirms that investments in human capital and seeking information, both facilitated by size, improve technical performance. This is strengthened by the finding that TI and foreign ownership (the share of foreign equity) have positive and statistically significant effects on export performance by each firm. Simple as this method is, it has great promise as a practical and efficient tool for preliminary benchmarking.

One of the criticisms that is levied against diagnostic work on competitiveness is that they often sit on shelves collecting dust rather than being implemented. Experience suggests the most widely used competitiveness studies have the following common features: (a) they are tailored to an audience of busy policy makers (i.e. they are relatively short and clearly written); (b) they are practically oriented and have a high policy content; and (c) their findings are communicated to key stakeholders who can influence national policy making.

These stakeholders not only include politicians, ministries and government agencies but also the private sector, labour organisations, NGOs and the academic community. Indeed, a regular dialogue involving committed and influential leaders drawn from these different groups of stakeholders can help to ensure commitment to a process of reforming existing policies and institutions (see Esser et al., 1996). In turn, the proceedings of these stakeholder dialogues need to be disseminated to society at large via the media to encourage greater national awareness of the competitiveness challenges facing a country, understanding of the vision for improving future competitiveness and acceptance of requisite policies. Sometimes there is an external event (e.g. a sudden fall in demand for exports, a hike in oil prices or the exit of some major foreign investors) that has a negative impact on a national economy and motivates action to shape an industrial and technological transition. Other times, such a national dialogue takes place against a slowly evolving external threat such as steadily increasing foreign competition which confronts a dominant, local value chain or a segment of that value chain.

Designing the policy framework

Having undertaken a competitiveness assessment and initiated a dialogue between stakeholders, the next major challenge is to define a common vision and an industrial competitiveness strategy to realise the vision. The rationale for working together on a common approach is based on the recognition that there are shared returns to the participating stakeholders from collectively addressing the factors (e.g. various policy-induced distortions, inherent market imperfections and systems failures) that impede enterprise-level learning processes. These shared returns manifest themselves at the national level in improved manufactured export growth and technological upgrading and at the firm-level in improved innovation and learning, competitiveness and profitability.

Arriving at a common vision and strategy is itself quite a difficult process. The need for an over-arching vision to guide firm-level learning arises because developing countries can adopt many possible strategies. The section on 'Core elements of competitiveness strategy' has already outlined some of the elements of international best practice industrial competitiveness strategies in the developing world. These need to be tailor-made to suit individual national goals, economic circumstances and political ideology in particular developing countries.[25] It is hard to generalise beyond this but the following seven issues might be relevant to developing a common vision and strategy:

- What are the *marketing opportunities and external threats* faced by the industrial sector and leading sub-sectors? New opportunities may include new markets and subcontracting possibilities from foreign buyers while threats could encompass rapid technological progress, increased competition from low cost producers, stringent WTO compliance and market access constraints.
- What are the broad *national objectives* for industrial transformation? These can range from diversification of the export base into new areas of comparative advantage (either non-traditional labour-intensive exports or more complex technology-intensive products), deepening within leading export activities (e.g. backward linkages and increasing local content), more domestic technological activity (for instance, promoting design and product-centred R&D) or attraction of foreign investment.
- What is the *time horizon* for achieving these objectives? Several possible time frames are possible for the validity of the vision and the implementation of the strategy including short term (under 3 years), medium term (5–10 years) or long term (more than 10 years). It is not unusual to combine short- and long-term elements within the same strategy.
- What are the desired roles for the *market-mechanism and state intervention* in the allocation of resources? These can range from an entirely market-driven approach to industrial competitiveness to an entirely state-driven approach or a mix between the two. The choice might reflect many things – location, external pressures, economic circumstances and ideology.
- What *policy interventions* are required to promote collective learning processes and competitiveness? The diagnostic assessment of competitiveness in a given developing country (see Box 2.5) would indicate the main competitiveness strengths and weaknesses of the industrial sector and the policy and institutional regime. The relevant remedies (a mixture of incentive/regulatory actions and supply-side policies such as those in the section on 'Core elements of competitiveness strategy') would stem from this.
- Is there a case for selecting *broad industrial sectors or sub-sectors* for special promotion measures? The strategic question is identifying sectors or

sub-sectors that have the greatest potential for dynamic growth or create the most beneficial externalities for other activities. This issue is probably one of the most contentious in industrial competitiveness policy debates and needs to be carefully handled. Two important considerations that can inform sectoral policy making include short vs long-term comparative advantage and the depth of government capabilities to intervene at sectoral level.

- What *resources* are available for implementing the vision and the strategy? This covers financial resources to meet the costs of different policies and perhaps more importantly the technical resources (economic, engineering, legal and management) to design, implement and monitor industrial competitiveness strategy. It goes without saying that strategies have to be tailored to suit financial and technical resource availability. Both are likely to be in short supply locally and developing countries may have to rely on multilateral institutions, bilateral donors and consultancy firms to bridge the gaps.

Some developing countries have separate documents for their vision statement and their industrial competitiveness strategy while others combine them in a single integrated document. The differences are not simply a matter of presentation but reveal attitudes to long-term strategy formulation. Each approach has its own merits and the final choice is a matter of judgement.

In the former approach, the vision sets the long-term perspective (e.g. next twenty-five years) for the economic and industrial transformation while the strategy can be updated at regular events within this timeframe to reflect changing external events, domestic economic circumstances and industrial achievements. The country's future economic and industrial aspirations are well defined but yet there is a degree of flexibility on how to achieve this as the strategy can be altered at particular intervals. Malaysia, with a separate Vision 2020 Statement and two Industrial Master Plans, illustrates this approach.

In the latter approach, the common vision and industrial competitiveness strategy might be expressed with the same timeframe (e.g. the medium term) and can be readily disseminated as one document. The long-term industrial direction is not well known and may cause some uncertainty in private sector circles but there is a closer association between the country's medium-term vision and strategy. Moreover, there is a strong potential role for national stakeholders (including private sector representatives) in shaping the vision and the strategy at regular intervals. Trinidad and Tobago, and Sri Lanka, with a single Industrial Policy Statement that articulates both their medium-term national visions and approaches to improving industrial competitiveness, are examples of the second approach.

Implementation

Implementing a policy framework for enhancing industrial competitiveness often pose particular challenges for developing countries and it is most often at this stage that efforts tend to falter. In this vein, the two main issues that need to be addressed at the implementation stage are:

- Coordination of different parts of the policy agenda and responsible agencies.
- Continuous monitoring of the competitiveness strategy.

The issue of coordination arises because most governments and private sector organisations are not structured for effectively implementing an industrial competitiveness strategy. In the case of governments, for instance, the responsibilities and functions that affect industrial competitiveness are scattered over an array of ministries and institutions: finance, trade, industry, labour, education, science and technology and many others. These often have different objectives and do not communicate with each other on a regular and intimate basis. Turf battles and competition among ministries and agencies are often commonplace. Day-to-day political and macroeconomic issues assume greater prominence in policy discussions than long-term industrial competitiveness strategy. The net result is that little may get accomplished in the way of implementing the different aspects of the strategy. Similarly, private sector organisations might lack the experience and capacity to work with government agencies in implementing the strategy.

An important step in strategy implementation is therefore to set up an agency that can cut across competing interests and coordinate the ministries (and private sector organisations) concerned. There is no one prescribed form for a successful coordination and implementation agency. Such an organisation can be either *ad hoc* (such as a semi-permanent committee on competitiveness) or a formally organised entity (such as council on industrial competitiveness). Singapore follows the former approach while the United Kingdom has a more permanent structure and both have their merits (see UK DTI 1998 and 1999; Singapore Ministry of Trade and Industry, 1998). These bodies may include a combination of business leaders, heads of support institutions and ministries. A leading businessman or minister could chair it. In some cases, chairmanship by the head of state has proved to be very effective and, in other cases, less so. Whatever their form, the basic mission of these organisations is to ensure that the goals of the industrial competitiveness strategy are adhered to, the policies and programmes are implemented according to a strict timetable, the results are monitored and policies and programmes are altered in the light of performance, and new policies and programmes are devised as necessary.[26]

Necessary conditions for success

A well-managed competitiveness strategy is clearly a necessary, but not sufficient, condition for industrial success in the developing world. Even the best managed competitiveness strategies can fail due to a variety of factors and the developing world contains examples of failures as well as successes. Three groups of factors can be readily identified.

One group of factors relates to external shocks (e.g. a sudden fall in world demand, a rise in world interest rates or an international financial crisis) that are outside events that are beyond the control of policy makers. These events are very disruptive to the management of a competitiveness strategy and there is often little one can do to mitigate their effects.

Another group of factors relates to country size, geographical location and resource base. The presence of counter examples, makes it hard to generalise about the relevance of these influences on competitiveness performance and strategy management. However, geographically isolated countries may be at a disadvantage compared with those close to major markets, tiny island countries might lack the stimulus of a large domestic market and be less attractive to FDI, and natural resource and skill-poor countries may find they have fewer options for industrial development.[27] The design of a coherent competitiveness strategy needs to take these influences into account and develop long-term policy remedies.

The final group are internal to an economy and within the scope of national policy making. These consist of political stability, macroeconomic performance, government capabilities, government commitment and relations with the private sector. These issues are briefly discussed below.

Political stability. Civil conflict, domestic political violence and international disputes also reduce the government's capacity to undertake competitiveness strategies. Defence expenditures are often increased at the expense of foreign investment, export promotion and technology support budgets. Key policy makers are sometimes switched from economic management to crisis management. In this vein, negotiations with arms dealers and aid donors assume a higher priority than a dialogue with the private sector over competitiveness strategy or a focus on policy implementation and monitoring. Moreover, the country's reputation suffers as a destination for foreign investment and foreign buyers may seek out more reliable suppliers. Country reputations and international goodwill are 'a scarce national resource' and can take many decades and many millions of dollars in promotion campaigns to rebuild.

Sound macroeconomic performance. Good macroeconomic conditions assist the implementation of national competitiveness strategies while macroeconomic crises are a hindrance. Difficulties in containing inflation, sudden exchange rate devaluation, sharp declines in commodity prices, collapses in external demand and domestic recession often contribute to reversals in certain aspects of competitiveness policies after their implementation including the re-imposition of exchange controls or import controls and cuts in education and training expenditures.

Strong government capabilities. While there is a theoretical case for public action to enhance competitiveness, in practice governments may lack the requisite skills and information to formulate, implement and monitor such strategies. Undertaking detailed national competitiveness strategies (involving carefully designed foreign investment targeting, export contests, training programmes and technology development schemes) demands a host of economic, management engineering and information technology skills that are in short supply in many developing country civil services. In part this may be due to civil service recruitment practices and compensation schemes which typically focus on recruiting generalists and giving them on-the-job training rather than hiring specialists with relevant private sector experience.

Sustained government commitment. Owing in part to their concern with structural issues (like skills, technology and institutional reforms) competitiveness policies can take time to show results. Inadequate commitment by government has often limited the seriousness of policy implementation and backsliding has sometimes affected sustainability. Changing governments and leaders (as well as internal opposition to changes within government) have frequently led to policy reversals.

Good private sector capabilities and relations with government. Countries which have had a long period of inward-looking policies are sometimes characterised by a tiny private sector with limited industrial experience. Such an 'infant private sector' is unlikely to have the requisite technological and marketing capabilities to respond quickly to changes in incentive policies or have the relevant international exposure to advise to government on good competitiveness policies. Similarly, a government with a 'socialist overhang' is likely to regard the private sector with suspicion, particularly multinational affiliates, and may not seek their advice on economic policy matters and implementation issues. A significant national welfare loss would occur if the private sector were not actively involved in the consultative process.

Developing countries that have dealt with these five factors, alongside the effective management of their competitiveness strategies, appear to have the best chance of enhancing their long-term competitiveness. As these issues largely relate to governance, economic theory offers little guidance on how to deal with them (apart from, of course, macroeconomic management). Williamson (1994) provides some interesting thoughts on governance issues underlining the implementation of economic reform programmes and these ideas may also be relevant to competitiveness strategy. His ideas include:

- The presence in government of a team of economists with a common, coherent vision of what needs to be done.
- The presence at the top of a leader with a vision of history rather than a politician unable to lift his sights beyond the next election or a dictator preoccupied with defeating the next coup attempt.
- A comprehensive programme involving radical transformation of the economy, to be implemented rapidly.
- The will and the ability to appeal to the general public through the media, by passing vested interests.

Conclusion

The notion of competitiveness is likely to be shrouded in academic controversy for the foreseeable future. At the same time, it is being applied in different forms to examine the behaviour of countries and enterprises and to guide policy making in developed and developing countries. Whilst the output of these exercises is of variable quality, a carefully defined concept of competitiveness remains useful for analysing comparative performance and formulating policies for industrialisation in the developing world.

Different approaches to competitiveness (e.g. the macroeconomic and business strategy perspectives) offer insights into these issues but the relatively recent technology and innovation perspective seems to provide the optimal framework for evaluating performance and designing policy remedies. This perspective emphasises that enterprises have to undertake conscious investments to convert imported technologies into productive use and they interact with different kinds of institutions within an NIS. In turn, sustained collective learning is associated with enhanced industrial competitiveness. The process of collective learning is itself affected by systemic weaknesses in NIS (e.g. market imperfections, systems failures and poor incentive and regulatory policies) and leads to inter-country differences in collective learning and competitiveness records. Remedying systemic weaknesses is the principal aim of an industrial competitiveness strategy.

The central elements in an industrial competitiveness strategy from a technology and innovation perspective are: (a) a national partnership involving complementary actions by the government and the private sector; (b) a liberalisation plus approach involving a mix of incentive and supply-side policy measures; and (c) where appropriate on economic grounds, policies to promote the competitiveness of particular industrial clusters.

The experience of successful developing countries suggests that the following incentive and supply-side measures are pertinent to a liberalisation plus approach:

- a stable, predictable macroeconomic environment characterised by low budget deficits, tight inflation control and competitive real exchange rates;
- an outward-oriented, market-friendly trade regime emphasising the dismantling of import controls and tariffs to send signals to industry to restructure, a strong export push (duty-free access to raw materials and export marketing support) and good international negotiations capabilities to gain market access;
- an effective domestic competition regime with free entry and exit at industry-level, a carefully managed programme of privatising SOEs and a strong regulatory authority to deal with anti-competitive practices;
- a pro-active foreign investment strategy which emphasises the targeting of a few realistic sectors and host countries, overseas promotion offices as public–private partnerships, competitive investment incentives and radically streaming investment approval processes;
- sustained investments in human capital at all levels (particularly tertiary-level scientific, information technology and engineering education) and increased enterprise training including (assistance for industry associations to launch training schemes, an information campaign to educate firms about the benefits of training and tax breaks for training);

- comprehensive technology support for quality management, productivity improvement, metrology and technical services for SMEs (including grants for SMEs to obtain ISO9000 certification, creating productivity centres and commercialisation of public technology institutions);
- access to ample industrial finance at competitive interest rates through prudent monetary policy management, competition in the banking sector, training for bank staff in assessing SME lending risks and specialist soft loans for SMEs;
- an efficient and cost-competitive infrastructure with respect to air and sea cargo, telecommunications, Internet access and electricity;
- an apex public–private sector body to formulate strategy and monitor its implementation.

Some of these measures such as macroeconomic management, outward-orientation and privatisation are a part of standard structural adjustment programmes (SAPs). Others – such as human development, technology support, targeted foreign investment promotion and comprehensive SME policies – go beyond SAPs but are still consistent with a market-friendly approach to industrial competitiveness.

Cluster policies emphasise detailed actions to improve the competitiveness of specific industrial clusters within a market-oriented developing country. Interventions at the micro-level are often directed towards acquiring technological capabilities, promoting upgrading and improving links between different parts of the cluster. Public policies might range from industry-specific tax measures to the provision of specialised institutional support facilities for a particular cluster. Joint actions (e.g. setting up a specialised training school) between a business association in the cluster and an aid donor or a government agency are also commonplace. The use of cluster policies should be guided by near future comparative advantage, be at a broad industry-level, be strictly time-bound and be measured by clearly defined performance criteria.

In a market economy, the main role for the private sector is to become productive and generate national wealth. Business associations assist industry by advocating the case for business and deregulation. With a rising share of private sector activity in GDP, however, the private sector needs to move beyond its traditional function of wealth creation and advocating the case for business. The private sector itself can make an important contribution to designing and implementing national competitiveness strategies in developing countries. This pro-active role can include: helping government to plug information gaps through participation in national policy making bodies and international trade negotiations; augmenting government capabilities via short-term secondment programmes of private sector managers and technicians; participating in infrastructure projects through joint finance and management skills; and helping weaker firms to help themselves via creating industry-specific training centres and other actions.

The effective management of an industrial competitiveness strategy is itself quite a demanding exercise but this aspect is often neglected in developing countries. The requisite management needs can be broken into three inter-related phases: (a) assessing competitiveness performance and the policy framework; (b) designing a new set of policies; and (c) implementing the new policies. Each of these headings involves a variety of sub-tasks – such as diagnostic studies, regular consultation with key stakeholders, developing a common vision and strategy, coordination of different parts and continuous monitoring – which have to be undertaken to ensure sound management of the strategy.

A well-designed and managed competitiveness strategy is necessary but not sufficient for industrial success in developing countries. Clearly, even the best-managed and coherent industrial competitiveness strategies require other factors to realise industrial success. These include: political stability, sound macroeconomic management, strong government capabilities to manage strategy, sustained government commitment to strategy implementation and good private sector capabilities and relations with government.

Notes

1 I would like to thank Sanjaya Lall, Stan Metcalfe and Frederic Richard for many discussions but I bear sole responsibility for the views expressed here.
2 For recent surveys of this literature see Boltho (1996), Kumar et al. (2000), Lall (2001).
3 There are other ways of classifying this literature. Corden (1994) identifies three different measures of national competitiveness in common usage: sectoral competitiveness, relative cost competitiveness (the real exchange rate) and productivity.
4 Corden (1994) pp. 284–286 summarises the dynamics of a real exchange appreciation problem within a specific factors trade model of a given economy with a tradeables sector and a non-tradeables sector. Using this model, he also analyses two other competitiveness cases – a sectoral competitiveness problem and a real wage rigidity problem.
5 The tradeable non-tradeable real exchange rate, often referred to as the trade theory definition, is derived from the Swan–Salter model. It has been rarely used in empirical work because of the lack of information on tradeable and non-tradeable prices. Following Harberger (1986), however, some analysts have begun to employ proxies for the two categories (notably, the wholesale price index for traded goods and the consumer price index for non-traded). The trade theory version of the real exchange rate and the PPP definition may not vary much in practice as the wholesale price index and consumer price index will be closely correlated.
6 See Bank of England (1982), Harberger (1986) and Boltho (1996) for a discussion on the merits of these deflators.
7 The IMF's *International Financial Statistics Yearbook 2001* provides real effective exchange rate indicators (based on relative unit labour costs and relative value added deflators) for developed countries while the World Bank's *World Development Indicators 2001* gives real effective exchange rate indicators (based only on relative consumer prices) for a range of developing, transition and developed countries. Other sources of such data for developing and transition countries are country reports of the IMF and World Bank and the annual reports of regional development banks such as the Asian Development Bank and the Inter-American Development Bank. Some central banks are also beginning to publish real exchange rate series.
8 Studies on the United Kingdom include Bank of England (1982); Eltis and Hingham (1995) while recent country studies on developing countries can be found in Helleiner (1994, 2001).

9 On the notion of unique resources see Barney (1991) and on leveraging see Hamel and Prahalad (1994).
10 See Porter (1990), p. 72. See also Porter (1998).
11 Porter (1990) summarises his view of the role of government as follows: 'Many treatments of national competitiveness assign government the pre-eminent role. Our study of ten nations does not support this view. National competitive advantage in an industry is a function of underlying determinants that are deeply rooted in many aspects of a nation. Government has an important role in influencing the diamond but its role is ultimately a partial one. It only succeeds when working with the determinants' (Porter, 1990, pp. 680–681).
12 Lall (2001) says 'Porter's business school approach has no place for market failures' (p. 18).
13 See, for instance, Dahlman et al. (1987), Katz (1987), Lall (1992), Bell and Pavit (1993), Hobday (1995), Ernst et al. (1998), Wignaraja (1998), Radosevic (1999), Romijn (1999) and Kim and Nelson (2000).
14 See, for instance, Lundvall (1992), Nelson (1993) and Edquist and McKelvey (2000).
15 Chapter 4 by Stan Metcalfe provides a detailed exposition of the evolutionary and neoclassical perspectives on technological change/innovation.
16 The concept of tacit knowledge is discussed in more detail in Radosevic (1999).
17 A firm's technology strategy will vary according to its own stage of technological development. Mytelka (1999) distinguishes between three strategies, namely catch-up strategy, keep-up strategy and get-ahead strategy, which correspond to the latecomer stage, quick follower stage and front-runner stage of a technological learning approach. She argues that the three different strategies, each with different objectives, require different types of knowledge for upgrading and different sources of such knowledge. This suggests that it is important for a learning firm to be able to assess its own technological capability objectively and adopt a phased approach to technology development.
18 Static versions of trade theories suggest that the export performance of a given developing country can be attributed to the comparative advantage that this economy would have over another in terms of access to certain factor inputs – such as capital, labour, skills, economies of scale and technology. Dynamic versions suggest that comparative advantages of developing countries will change under the impetus of the accumulation of capital, skills, economies of scale and technology. For a recent survey of the major theories of comparative advantage see Wignaraja (1998).
19 Humphrey and Schmitz (1996) suggest 'a cluster is defined as a sectoral and geographical concentration of enterprises', Humphrey and Schmitz (1996, p. 1863). A recent OECD study adds: 'clusters are characterised as networks of production of strongly interdependent firms (including specialised suppliers) knowledge producing agents (universities, research institutes, engineering firms), bridging institutions (brokers, consultants) and customers linked to each other in a value-adding production chain' (Boekholt and Thuriaux, 1999, p. 381).
20 Edquist (2001) states: 'Within a systems of innovation framework an identification of the causes behind the problems is the same as identifying deficiencies in the functioning of the system. It is a matter of identifying functions that are missing or inappropriate and which lead to the problem of comparative performance. Let us call these deficient functions systems failures' (p. 36). He points out three principal types of systems failures: (a) firms in the NIS might be inappropriate or missing; (b) institutions in the NIS might be inappropriate or missing; and (c) interactions or links between these elements in the NIS might be inappropriate or missing.
21 The technology literature suggests different policies, which can be grouped under these headings. See Nelson (1993), Esser et al. (1996), Lall and Teubal (1998), Wignaraja (1999), Boekholt and Thuriaux (1999), Edquist (2001).
22 See Humphrey and Schmitz (1996), Fischer and Reuber (Chapter 5 in this volume).

23 Drawing on OECD surveys of business associations in five transition economies including Russia and Poland, Levitsky (1994) comments that the degree to which organisations are able to offer such services at an acceptable quality level would differ according to circumstances and resources available. Typically in the early stages of their development, with limited resources and experience, fewer services would be offered and the main role would be advocacy.
24 See Dunning et al. (1998) on Northern Ireland; Singapore Ministry of Trade and Industry (1998) on Singapore; Lall and Wignaraja (1998) on Mauritius and UK DTI (1998, 1999) on the United Kingdom.
25 Teubal (2001) provides insights into the development of a common vision and strategy in Israel and other NIEs.
26 See Lall and Wignaraja (1998) for the outline of a national competitiveness council for Mauritius (probably the leading export-oriented economy in Africa) that has since been set up to monitor competitiveness and contribute to policy making.
27 The classic study on the impact of these and policy factors on industrialisation and growth is Chenery et al. (1986).

References

Amsden, A. (2001), *The Rise of the Rest: Challenges to the West from Late-Industrialising Economies*, Oxford: Oxford University Press.
Bank of England (1982), 'Measures of Competitiveness', *Bank of England Quarterly Bulletin*, 22(3), 369–375.
Barney, J.B. (1991), 'Firm Resources and Sustained Competitive Advantage', *Journal of Management*, 17(1), 99–120.
Bell, M. and Pavitt, K. (1993), 'Technological Accumulation and Industrial Growth: Contrasts Between Developed and Developing Countries', *Industrial and Corporate Change*, 2(2), 157–210.
Bhagwati, J.N. (1988), 'Export-Promoting Trade Strategy: Issues and Evidence', *World Bank Research Observer*, 3(1), 27–58.
Boekholt, P. and Thuriaux, B. (1999), 'Public Policies to Facilitate Clusters: Background, Rationale and Policy Practices in International Perspective', in OECD, *Boosting Innovation: The Cluster Approach*, Paris: OECD.
Boltho, A. (1996), 'The Assessment: International Competitiveness', *Oxford Review of Economic Policy*, 12(3), 1–16.
Cassen, R. and Wignaraja, G. (1997), 'Social Investment, Productivity and Poverty: A Survey', *UNRISD Discussion Paper* No. 88.
Chenery, H., Robinson, S. and Syrquin, M. (1986), *Industrialisation and Growth: A Comparative Study*, Oxford: Oxford University Press.
Corden, M. (1994), *Economic Policy, Exchange Rates and the International System*, Oxford: Oxford University Press.
Dahlman, C.J., Ross-Larson, B. and Westphal, L.E. (1987), 'Managing Technological Development: Lessons from Newly Industrialising Countries', *World Development*, 15(6), 759–775.
Dosi, G., Pavitt, K. and Soete, L. (1990), *Economics of Technical Change and International Trade*, Brighton: Wheatsheaf.
Dunning, J.H., Bannerman, E. and Lundan, S. (1998), *Competitiveness and Industrial Policy in Northern Ireland*, Belfast: Northern Ireland Economic Council.
Edquist, C. (2001), 'Systems of Innovation for Development', Background Paper Prepared for UNIDO World Industrial Report 2002, UNIDO: Vienna.

Edquist, C. and McKelvey, M. (eds) (2000), *Systems of Innovation: Growth, Competitiveness and Employment*, Cheltenham: Edward Elgar.

El-Agraa, A.M. (1997), 'UK Competitiveness Policy vs. Japanese Industrial Policy', *The Economic Journal*, 107, 1504–1517.

Eltis, W. and Hingham, D. (1995), 'Closing the UK Competitiveness Gap', *National Institute Economic Review*, No 154, 71–84.

Ernst, D., Ganiatsos, T. and Mytelka, L. (eds) (1998), *Technological Capabilities and Export Success in Asia*, London: Routledge.

Esser, K., Hildebrand, W., Messner, D. and Meyer-Stamer, J. (1996), *Systemic Competitiveness: New Governance Patterns for Industrial Development*, London: Frank Cass.

Fagerberg, J. (1988), 'International Competitiveness', *The Economic Journal*, 98(2), 355–374.

Fagerberg, J. (1996), 'Technology and Competitiveness', *Oxford Review of Economic Policy*, 12(3), 39–51.

Hamel, G. and Prahalad, C.K. (1994), *Competing for the Future*, Boston: Harvard Business School Press.

Harberger, A. (1986), 'Economic Adjustment and the Real Exchange Rate', in S. Edwards, and L. Ahmed, (eds), *Economic Adjustment and Exchange Rates in Developing Countries*, Chicago: University of Chicago Press.

Helleiner, G.K. (ed.) (1994), *Trade Policy and Industrialisation in Turbulent Times*, London: Routledge.

Helleiner, G.K. (ed.) (2001), *Non-Traditional Export Promotion in Africa*, Basingstoke (UK): Palgrave.

Hobday, M. (1995), *Innovation in East Asia: The Challenge to Japan*, Cheltenham: Edward Elgar.

Howes, C. and Singh, A. (1999), 'National Competitiveness, Dynamics of Adjustment, and Long-Term Economic Growth: Conceptual, Empirical and Policy Issues', *Accounting and Finance Discussion Papers* No. 00-AF43, Department of Applied Economics, University of Cambridge.

Humphrey, J. and Schmitz, H. (1996), 'The Triple C Approach to Local Industrial Policy', *World Development*, 24(12), 1859–1877.

IMF (2001), *International Financial Statistics Yearbook*, Washington, DC: International Monetary Fund.

Katz, J. (ed.) (1987), *Technology Generation in Latin American Manufacturing Industries*, New York: St Martins Press.

Kim, L. and Nelson, R. (eds) (2000), *Technology, Learning and Innovation: Experience of Newly Industrialising Economies*, Cambridge: Cambridge University Press.

Krugman, P. (1994), 'Competitiveness: A Dangerous Obsession', *Foreign Affairs*, 73(2), 28–44.

Krugman, P. (1996), 'Making Sense of the Competitiveness Debate', *Oxford Review of Economic Policy*, 12(3), 17–25.

Kumar, R., Chadee, D. and Beppu, F. (2000), 'Determinants of International Competitiveness of Asian Firms: A Framework for an Empirical Inquiry', Manila: Asian Development Bank (memo).

Lall, S. (1992), 'Technological Capabilities and Industrialisation', *World Development*, 20, 165–186.

Lall, S. (2001), *Competitiveness, Technology and Skills*, Aldershot (UK): Edward Elgar.

Lall, S., Barba-Navaretti, G., Teitel, S. and Wignaraja, G. (1994), *Technology and Enterprise Development: Ghana Under Structural Adjustment*, London: Macmillan Press.

Lall, S. and Teubal, M. (1998), '"Market Stimulating" Technology Policies in Developing Countries: A Framework with Examples from East Asia', *World Development*, 26(8), 1369–1386.

Lall, S. and Wignaraja, G. (1995), 'Building Export Capabilities in Textiles and Clothing: Case Studies of German and Italian Companies' Exports', in G.B. Navaretti, R. Faini and A. Silberston (eds), *Beyond the MultiFibre Agreement: Third World Competition and Restructuring Europe's Textile Industry*, Paris: OECD.

Lall, S. and Wignaraja, G. (1998), *Mauritius: Dynamising Export Competitiveness*, (Commonwealth Economic Paper No 33), London: Commonwealth Secretariat.

Levitsky, J. (1994), 'Business Associations in Countries in Transition to Market Economies', *Small Enterprise Development*, 5(3), 24–34.

Llewellyn, J. (1996), 'Tackling Europe's Competitiveness', *Oxford Review of Economic Policy*, 12(3), 87–96.

Lipsey, R.G. (1997), 'Globalisation and National Government Policies: An Economist's View', in J.H. Dunning (ed.), *Governments, Globalisation and International Business*, Oxford: Oxford University Press.

Lundvall, B.A, (ed.) (1992), *National Systems of Innovation: Towards a Theory of Innovation and Interactive Learning*, London: Pinter Publishers.

Moon, H.C., Rugman, A.M. and Verbeke, A. (1998), 'A Generalised Double Diamond Approach to the Globalised Competitiveness of Korea and Singapore', *International Business Review*, 7, 135–150.

Mytelka, L.K. (1999), *Competition, Innovation and Competitiveness in Developing Countries*, Paris: OECD.

Nelson, R. (1993), 'A Retrospective', in Nelson (ed.), *National Innovation Systems: A Comparative Analysis*, Oxford: Oxford University Press.

Nelson, R. (ed.) (1993), *National Innovation Systems: A Comparative Analysis*, Oxford: Oxford University Press.

Nelson, R. and Pack, H. (1999), 'The Asian Miracle and Modern Economic Growth Theory', *Economic Journal*, 109, 416–436.

OECD (1992), *Technology and the Economy: The Key Relationships*, Paris: OECD.

Ohmae, K. (1994), *The Borderless World*, London: Harper Collins Business.

Oughton, C. (1997), 'Competitiveness Policy in the 1990s', *The Economic Journal*, 107.

Porter, M.E. (1980), *Competitive Strategy*, New York: Free Press.

Porter, M.E. (1990), *The Competitive Advantage of Nations*, London: Macmillan Press.

Porter, M.E. (1998), 'The Microeconomic Foundations of Economic Development', in *The Global Competitiveness Report 1998*, Geneva: World Economic Forum.

Radosevic, S. (1999), *International Technology Transfer and Catch-up in Economic Development*, Cheltenham (UK): Edward Elgar.

Reinert, E.S. (1994), 'Competitiveness and its Predecessors: A 500 Year Cross National Perspective', Paper prepared for the Business History Conference, Williamburg, Virginia, USA, March 11–13.

Romijn, H. (1999), *Acquisition of Technological Capability in Small Firms in Developing Countries*, London: Macmillan Press.

Singapore Ministry of Trade and Industry (1998), *Committee on Singapore's Competitiveness*, Singapore: Ministry of Trade and Industry.

Stiglitz, J. (1996), 'Some Lessons from the East Asian Miracle', *World Bank Research Observer*, 11(2), 151–177.

Teubal, M. (2001), 'The Systems Perspective to Innovation and Technology Policy (ITP): Theory and Application to Developing and Newly Industrialised Economies',

Background Paper Prepared for UNIDO World Industrial Report 2002, UNIDO: Vienna.
UNIDO (2000), *Public–Private Partnerships for Economic Development and Competitiveness with Special Reference to the African Experience*, Vienna: UNIDO.
UNIDO (2002), *World Industrial Development Report 2002: Competing Through Innovation*, Vienna: UNIDO.
United Kingdom DTI (1998), *Our Competitive Future: Building the Knowledge Driven Economy*, UK Department of Trade and Industry, London: Her Majesty's Stationary Office.
United Kingdom DTI (1999), *Manufacturing in the Knowledge Driven Economy*, London: UK Government Department of Trade and Industry.
United States OTA (1990), *Making Things Better: Competing in Manufacturing*, Washington, DC: Office of Technology Assessment, US Senate.
Westphal, L.E. (2001), 'Technology Strategies for Economic Development in a Fast Changing Global Economy', Swarthmore (PA): Swarthmore College. (Processed)
Wignaraja, G. (1998), *Trade Liberalisation in Sri Lanka: Exports, Technology and Industrial Policy*, London: Macmillan Press.
Wignaraja, G. (1999), 'Tackling National Competitiveness in a Borderless World', *Commonwealth Business Council Policy Paper Series* No.1. London: Commonwealth Business Council.
Wignaraja, G. (2002), 'Firm Size, Technological Capabilities and Market-Oriented Policies in Mauritius', *Oxford Development Studies*, 30(1), 87–104.
Wignaraja, G. (forthcoming), 'Institutional Support and Policies for Technology Development in SMEs in Mauritius', in B. Oyenka (ed.), *Innovation and the African Enterprise*, Cheltenham (UK): Edward Elgar.
Wignaraja, G. and Ikiara (1999), 'Adjustment, Technological Capabilities and Enterprise Dynamics in Kenya', in S. Lall (ed.), *The Technological Response to Import Liberalisation in SubSaharan Africa*, London: Macmillan Press.
Williamson, J. (1994), 'In Search of a Manual for Technolopolis', in J. Williamson (ed.), *The Political Economy of Policy Reform*, Washington, DC: Institute for International Economics.
World Bank (1999), *World Development Report 1998/99: Knowledge for Development*, Oxford: Oxford University Press.
Yip, G. (1992), *Total Global Strategy*, New Jersey: Prentice Hall.

3 Benchmarking competitiveness
A first look at the MECI

Ganeshan Wignaraja and Ashley Taylor[1]

Introduction

There is a growing interest in policy circles in the developing world in comparing competitive performance across countries and obtaining guidelines for strategy.[2] Policy makers are typically concerned with how their economy has been performing in relation to: (a) countries at a similar level of economic development (or within the region) which they would like to outperform; and (b) countries at a higher level of economic development (e.g. Newly Industrialising Economies (NIEs) in East Asia) whose strategies they would like to emulate. Similarly, multinational companies constantly research the costs and benefits of production locations on a worldwide basis. This interest has fuelled several attempts to devise a competitiveness indicator at the national level, a composite measure ranking countries according to particular criteria. The number of rankings (published and unpublished) of national competitiveness prepared by governments, consultants and research organisations is growing and becoming increasingly influential in policy formulation.

Against this background, this chapter appraises some existing work and puts forward an alternative measure of competitiveness performance in the developing world. The section on 'The "Swiss" competitiveness indices' examines the two well-known indices contained in *The Global Competitiveness Report 2001* of the World Economic Forum (WEF) and *The World Competitiveness Yearbook 2001* of the International Institute for Management Development (IMD) and highlights weaknesses of such approaches. The section on 'Construction of the MECI' constructs a simple manufactured export competitiveness index (MECI) based on three subcomponents (current manufactured exports per capita, long-term growth in manufactured exports and the share of technology-intensive exports) and presents the results by country, region and income group. The country coverage of the MECI (eighty developing and transition economies) provides a more comprehensive representation of the developing world than the WEF and IMD exercises. The section on 'Factors affecting manufacturing export competitiveness' examines potential determinants of manufactured export competitiveness and concludes that sound macroeconomic conditions, trade liberalisation and an emphasis on supply-side factors (such as foreign direct investment (FDI), technological effort, human capital and communications infrastructure) are closely associated with better manufactured competitiveness performance in the developing world.

The 'Swiss' competitiveness indices

The global competitiveness benchmarking industry is dominated by two Swiss institutions.[3] The Geneva WEF and Lausanne International IMD have attracted media attention and generated debate for their annual rankings of the competitiveness of leading developed and developing countries. The WEF has been publishing its *Global Competitiveness Report* since 1979 while the IMD *World Competitiveness Yearbook* has been available since 1990. At the heart of the two reports is a country-level competitiveness indicator that shows the relative positioning of a sample of countries (see IMD, 2001; WEF, 2001). The WEF terms its indicator the Competitiveness Index while the IMD version is called the Competitiveness Scoreboard.

Core elements of the WEF and IMD indices

Both reports suggest that their work involves a thorough assessment of the existing level of competitiveness in the sample countries and provides a realistic basis for designing future economic policies to promote competitiveness. The focus of these exercises is not to compute a measure of competitiveness performance of the productive sector *per se*; instead they adopt a more indirect approach of looking at a large assortment of factors affecting national competitiveness in a broad sense.[4] Using their respective measures, these reports list countries in descending order from the most competitive to the least competitive for the current year. The change in country rankings over the previous year is also shown which underlines the fact that these reports are concerned with very short-term improvements in the ranking of countries.

There is a healthy rivalry between the two Swiss organisations. Each claims to be at the cutting edge of competitiveness thinking – Michael Porter and Jeffrey Sachs of Harvard University are advisors to the WEF report whilst Stephane Garelli of the International Institute for Management Development advises the IMD report – and to have a more authoritative competitiveness indicator than the other. While the WEF exercise may be viewed as more rigorous and up to date than the IMD effort, the methodology underlying both competitiveness indices is broadly similar which suggests considerable cross-fertilisation of ideas between them over time. This is somewhat inevitable given that the two reports are in the public domain, that both organisations are headquartered in Switzerland and that there have been occasional movement of research staff between the organisations.

The following similarities are visible in the methodology adopted by the WEF Competitiveness Index and IMD Competitiveness Scoreboard.

First, a micro-level business strategy perspective to national competitiveness underpins the two indices. As opposed to emphasising some proxy for national performance (such as productivity, exports and the real effective exchange rate), these measures give prominence to the rôle of company operations and strategy and the factors affecting the performance of the productive sector. Many of these ideas originate in Michael Porter's book *The Competitive Advantage of Nations* (1990) which links corporate strategy analysis to the competitive advantage of nations.[5] Strongly influenced by Porter (one of the WEF report's main authors) the WEF index focuses on the nature of the economic environment (e.g. markets, institutions and economic policies) underlying the performance of the enterprise sector. The WEF report states that it seeks 'to measure the set of institutions, market structures and economic policies supportive of high current levels of prosperity' (WEF, 2001, p. 16). While it does not explicitly acknowledge the impact of Porter's work, the IMD report also says that its Competitiveness Scoreboard 'ranks nations' environments and analyses their ability to provide an environment in which enterprises can compete' (IMD, 2001, p. 7).

Second, both competitiveness indices are computed from a mixture of published statistics and surveys of businessmen. The WEF report explains the rationale for incorporating an executive opinion survey as follows: 'the Survey is indispensable to the Report, since no reliable hard data sources exist for many of the most important aspects of an economy such as the efficiency of government institutions, the sophistication of local supplier networks or the nature of competitive practices. Even where hard data are available, the data often do not cover all the countries in our sample' (WEF, 2001, p. 17). Both indices use many of the same hard data and survey data indicators to highlight the nature of the economic environment facing the productive sector in a country and there is a particular emphasis on measures relating to trade openness, government and fiscal policy, labour market flexibility and physical infrastructure. Furthermore, the hard data largely comes from the same well-known national and international sources (e.g. UN agencies) and the executive opinion surveys are based on a similar sample population (business leaders) and size (4,600 respondents for the WEF and 3,678 for the IMD).

Third, until recently, there were only subtle differences in the method of computation of the two indices that relate to the standardisation of the data and the weighting of components. These differences seem to have become somewhat more pronounced with the unveiling of two separate indices (the Growth Competitiveness Index (GCI) and the Current Competitiveness Index (CCI)) in the 2000 edition of the WEF report.[6] Meanwhile, the latest IMD report (2001) still contains an overall competitiveness index, which is discussed further below. Given the history of cross-fertilisation of ideas between the two Swiss organisations, however, it would not be suprising to see the IMD develop multiple indices along the lines of the WEF's GCI and CCI.

Fourth, the country coverage of the WEF and IMD indices is typically oriented towards high-income economies amongst developed, transition and developing countries. WEF (2001) lists seventy-five countries (up from fifty-nine in the 2000 report) and IMD (2001) shows forty-nine countries but few low-income African or South Asian developing countries are present in either report.[7] There are also hardly any of the smaller African, Caribbean and the Pacific economies represented. The WEF and IMD reports thus present an incomplete picture of the diversity of competitiveness performance in developing countries. The omission of many of the world's poorest countries is particularly glaring. These factors suggest that the WEF and IMD reports can offer little guidance on the formulation of competitiveness policies in low-income or small developing economies. The neglect of large swaths of the developing world may be partly due to the readership of these reports being concentrated in high-income countries and the practical difficulties of collecting reliable data from low-income developing countries as well as small economies. Indeed the IMD notes that: 'All countries were chosen because of their impact on the global economy and the availability of comparable international statistics'.[8]

Given these similarities, it should not come as a great surprise then that the country rankings are quite closely correlated in the 2001 editions of both reports.[9] With some notable exceptions, developed countries tend to be concentrated towards the top of the list and developing countries towards the bottom. The United States is at the top of the IMD list while Finland has the highest WEF ranking. Six countries are common to the top ten lists of both reports with the United States, Singapore and Finland in the top four in both lists (the other notable performers include Canada, the Netherlands and Luxembourg).

Reviewing the IMD World Competitiveness Scoreboard

Closer scrutiny of IMD *World Competitiveness Yearbook 2001* is quite revealing about the methodology underlying both the WEF and the IMD competitiveness indices.[10] In order to analyse and rank the competitiveness of forty-nine leading developed and developing countries, IMD (2001) employs 224 criteria per country (another sixty-two criteria are provided as background information but are not used in the ranking calculations). Of these 224 criteria, 118 are statistical indicators from published sources and 106 capture perceptions of businessmen from an executive opinion survey (covering 3,678 top and middle-level managers). The statistical and executive survey data are then standardised, ordered and grouped under four broad competitiveness input factors: (a) economic performance which represents a macroeconomic evaluation of the economy; (b) government efficiency which captures the extent to which government policies are conducive to competitiveness; (c) business efficiency which depicts the extent to which enterprises are performing in an innovative and responsible manner; and (d) infrastructure which captures the extent to which basic, technological, scientific and human resources meet the needs of business. These four competitiveness input factors contain five further sub-factors.[11] The twenty sub-factor scores each have an equal 5 per cent weighting in the final index.

A huge amount of collective effort has gone into the preparation of IMD *World Competitiveness Yearbook 2001*. The main contribution of the largely non-technical report is the preparation of a large and relatively up to date database on different aspects of the economic development of the forty-nine sample countries. The report contains over five hundred pages of information including several pages of statistics per country. Along with the WEF (2001), it is one of the few annual sources of qualitative information on the nature of the economic environment facing businesses in these countries. It also presents a collection of internationally available information from published sources. IMD (2001) could be thus viewed as a useful reference tool for certain types of published information and enterprise perception data on the sample countries. The exercise would have been meaningful if IMD had limited itself to this simple purpose. This is not the case in practice, however. IMD (2001) attempts to ambitiously manipulate this extensive database to reduce 224 criteria into a summary statistic and make comparisons across countries. There are several weaknesses with the methodology underlying the IMD World Competitiveness Scoreboard and these can be listed as follows:

1 *Ambiguous theoretical basis.* The exercise is premised on the fact that the determinants of the index are a proxy for national competitiveness performance, however, there is little attempt to justify this in theoretical terms. Most of IMD (2001) is devoted to the presentation of information on the four competitiveness input factors at country level (as well as a composite index) and it is reasonable to expect this to have been accompanied by a detailed discussion on how each of these is related to national competitiveness performance. But a detailed theoretical discussion on the link between input factors and competitiveness is absent; instead IMD (2001) provides a six page discussion of 'Competitiveness of Nations: The fundamentals' (pp. 43–49). The 'theory' behind the index, the 'four fundamental forces of competitiveness' – attractiveness vs aggressiveness, proximity vs globality, assets vs processes and individual risk-taking vs social cohesiveness – appears more a schema than a theory.

2 *Problems of index construction.* The theoretical basis for the appropriate weights of sub-components is limited given the broad basis of competitiveness that the composite indicator attempts to capture. Nevertheless, it is useful for practical purposes to have as clear a weighting structure as possible. The IMD weighting scheme has recently been modified to fix the weightings on each sub-factor to 5 per cent of the total index (whatever the number of criteria in a sub-factor). This modification was made to make the index more transparent and user-friendly. Although the IMD indicates the twenty strongest criteria for each economy in each factor overall the transparency of the final index ranking remains an issue. Another difficulty in interpreting the IMD index relates to the distinction between indicators of competitiveness and its determinants. Some of the components of the index appear to conflate these two elements. For example, growth and investment appear as criteria under macroeconomic performance. However, the former is often taken as an indicator of national performance with the latter taken as a determinant. This problem is perhaps reflected in the fact that the IMD report makes no attempt to empirically correlate the index (or the input factors, individually) with some measures of national performance (e.g. export performance or GDP per capita).

3 Ad hoc *data and proliferation of components.* There is an *ad hoc* use of data on the perceptions of businessmen about the economic environment as well as other statistical information to construct the IMD index. Qualitative, executive perception data about national economic conditions are often regarded as being less reliable than quantitative, statistical information (see Llewellyn, 1996). In principle, there may be some merit in asking multinational corporations (which operate similar plants in the same industry in different countries) to compare plant productivity levels and the national economic environments in which they operate. Leading international companies regularly conduct technical studies to benchmark plant-level performance and incentive regimes across countries and use this information to determine the profitability of their overseas investments.[12] IMD (2001), however, does not do this. Instead, the IMD report follows the misleading procedure of asking local and foreign firms to rank their national environment in isolation and then compares these views across countries via a composite competitiveness index based on survey data and statistical information. Quite apart from this problem, as mentioned above, the IMD index incorporates 118 statistical indicators per country with a further sixty-two for illustrative purposes. It does not explain why such a large number of indicators are necessary for the exercise, what hypotheses are being explored by specific statistical indicators or the informational content of additional indicators (given their combination into a single final index). There is also no discussion on the reliability of its database (although original data sources are clearly indicated).

4 *Short-run focus of data.* The country rankings are computed annually indicating an implicit assumption that many or nearly all the determinants vary in the short run. Some of the criteria, particularly macroeconomic indicators, do change on a yearly basis (e.g. GDP growth, inflation, exchange rates and interest rates). But many of the criteria are more structural in nature (e.g. R&D expenditure and other technology criteria; educational enrolments and other labour force criteria; and information, communications telecommunications infrastructure and other infrastructure criteria) and vary in the medium to long run. In this vein, emphasis on annual rankings might be misplaced and it would be more useful to conduct the exercise every five or ten years.

In sum, the IMD's *World Competitiveness Report* claims to be at the forefront of national competitiveness measurement and its work has attracted considerable interest in government and business circles. But close scrutiny of its World Competitiveness Scoreboard suggests little to recommend it as a benchmarking tool of national competitiveness in the specific context of developing countries. The theoretical principles that underlie it and its empirical relationship to a national performance measure are unclear. There is selective use of hard and corporate perception data to make invalid comparisons across developed and developing countries. Its focus on annual rankings is misplaced. Its limited country coverage neglects large parts of the developing world. Interestingly, many of the weaknesses of the IMD report also appear to be present in the WEF report. Following a comprehensive review of the 2000 edition of the WEF report,

Lall (2001) concludes: 'While the *Global Competitiveness Report* is well written and contains useful material, its competitiveness indices do not merit the attention they attract and the policy concern and debate they generate' (Lall, 2001, p. 1519).[13] At best the two leading global competitiveness organisations thus only make a partial contribution to understanding national competitiveness performance and its determinants in the developing world.

Construction of the MECI

Methodological issues

While the IMD and WEF indices have notable methodological problems, benchmarking exercises remain of considerable value in manufactured export competitiveness policy analysis in the developing world. Among other uses, they permit governments to compare national performance with countries they wish to emulate. In addition, they are an important input into the production location decision-making process of multinational corporations. Furthermore, they can assist aid donors to evaluate the impact of industrial development assistance. Considerable care, however, needs to go into designing indices and interpreting the results from such exercises. Lall (2001) offers some insights into this:

> To be analytically acceptable, however, all such efforts should be more limited in coverage, focusing on particular sectors rather than economies as a whole and using a smaller number of critical variables rather than putting in everything the economics, management, strategy and other disciplines suggest. They should also be more modest in claiming to quantify competitiveness: the phenomenon is too multifaceted and complex to permit easy measurement.
>
> (Lall, 2001, p. 1520)

Taking this suggestion on board, we develop a simple MECI based on the technology and innovation perspective approach to national competitiveness outlined in Chapters 1 and 2. This approach emphasises the rapid pace of technological progress internationally, the continuous process of firm-level learning to absorb imported technologies and the role of public policies in facilitating shifts in industry-level comparative advantage over time in developing countries. The strong microeconomic roots of the technology and innovation perspective can be readily extended to the national level. Following an influential OECD report, competitiveness at the macro level can be defined as 'the degree to which, under open market conditions, a country can produce goods and services that meet the test of foreign competition while simultaneously maintaining and expanding domestic real income' (OECD, 1992, p. 237).

At first glance, this seems like quite a difficult notion to translate into empirical terms as the OECD definition emphasises both the capacity to deal with foreign competition and the expansion of real income. Yet, trade performance offers a convenient yardstick to try to capture country-level competitiveness performance in technological terms (Fagerberg, 1988, 1996). Our competitiveness index focuses on manufactured exports given that the technology and innovation perspective to competitiveness is primarily concerned with the process of industrial growth and structural change in open developing economies. Accordingly, a more competitive developing economy is characterised by rapid manufactured export growth combined with sustained technological upgrading and diversification (Wignaraja, 1999; Lall, 2001). In conjunction with the absolute level of exports, which can be viewed as a proxy for the extent to which a developing country can meet the test of foreign competition, this provides a multi-faceted yet relatively simple notion of export competitiveness. In using this simple approach, we hope to address some of the criticisms made of the WEF and IMD indices.

Given the focused notion of technological competitiveness adopted above, and the desire for transparency of the final index, the MECI constructed in this chapter is composed of just the following three components:

a   Manufactured export value per capita in 1999 (US$);
b   Average manufactured export growth per annum 1990–1999;
c   Technology-intensive manufactures exports as a percentage of total merchandise exports in 1998.[14]

Thus the MECI incorporates the current position of a developing country in export markets (scaled by population), the long-term export growth that led to this position and the extent to which the developing country's exports are technology-intensive. The MECI takes values between 0 and 1 with higher values indicating greater levels of competitiveness at macro level. For instance, Taiwan Province of China (henceforth Taiwan), with an MECI of 0.79, is perceived to be more competitive than Nigeria (0.13) in Table 3.2.

Following a similar approach to the Human Development Index (HDI) an MECI may be constructed through a weighted sum of three indices representing manufactured export value per capita, manufactured export growth and technology-intensity of exports indices (see Box 3.1). In doing so the following two questions must be addressed:

- At what level should the fixed maximum and minimum values within each of the three indices be set?
- What weightings should be used to combine the three indices?

In the absence of a theoretical rationale suggesting alternative values, the sample maxima and minima have been used as the fixed values in each of the three indices forming the MECI. We have also discounted the extreme values of manufactured export per capita values by taking logarithms in the calculation of this sub-index.[16] The weightings used to combine the three indices are 0.3 on both

Box 3.1 Constructing a simple composite performance index: the HDI

> The use of a composite index to measure export competitiveness is analogous to the index used by the United Nations Development Programme to represent different features of human development. In constructing the Human Development Index (HDI), the UN calculates a weighted sum of three indices representing life expectancy, educational attainment and adjusted GDP per capita respectively.[15] The general formula used to calculate these indices is as follows:
>
> $$\text{Index} = \frac{\text{Actual value} - \text{minimum value}}{\text{Maximum value} - \text{minimum value}}$$
>
> The educational attainment index is complicated somewhat by the weighted combination of two separate sub-indices whilst the GDP per capita index uses the logarithms of the actual values in the above formula in order to discount higher levels of income. The minimum and maximum values are fixed at given levels. In the case of educational enrolment and literacy rates these are 0 and 100 per cent, respectively. For GDP per capita and longevity the maximum and minimum fixed values are above and below the sample maximum and minimum, respectively. The three indices are combined with equal weights to form the HDI.

the export per capita value and the export growth indices and 0.4 on the technology-intensity of exports index. This approach has been adopted given the particular interest in technological upgrading within the notion of competitiveness and the policies that have been adopted to achieve such upgrading. As noted below the ranking of the index is robust to the use of an equal weighting on the three components. The simple and transparent weighting of just three sub-components facilitates investigation of the importance across economies of the different sub-components in determining respective MECI performance.

Table 3.1 provides an illustrative example for the construction of the MECI for Thailand.

Results

Using the approach illustrated above the MECI was calculated for a sample of eighty countries. The countries were classified by region and income level according to the World Bank's *World Development Indicators 2001* (see Appendices 'Details of regional grouping' and 'Details of income grouping'). A number of countries classified as high income by the World Bank have been included in the sample for further interest, for example, Cyprus, Greece, Hong Kong, Israel, Korea, Kuwait, Portugal and Singapore.

Table 3.1 Construction of MECI for Thailand

	Thailand	Sample maximum	Sample minimum
Manufactured exports per capita, 1999 (US$)	719	25,039 (Singapore)	0.5 (Nigeria)
Manufactured export growth rate 1980–1999, % per annum (geometric average)	18.8	23.4 (Mexico)	−18 (Yemen, Rep.)
Technology-intensive exports as % of total merchandise exports	42	70 (Singapore)	0 (15 countries)

Notes
Export value index = [log(719) − log(0.5)]/[log(25,039) − log(0.5)] = 0.67
Export growth index = [18.8 − (−18)]/[23.4 − (−18)] = 0.42
Technology-intensity of exports index = [42−0]/[70−0] = 0.60
MECI for Thailand = [0.3 * 0.67] + [0.3 * 0.42] + [0.4 * 0.60] = 0.71

Table 3.2 provides the MECI ranks of the sample countries, the component indices and underlying data. The mean MECI value is 0.40 with the median at 0.37. The value and ranking of the MECI appears robust to the use of a weighting of 0.4 on the technology-intensive export index as opposed to equal weights. The rank correlation for the MECI calculated on these two different weightings is 0.998.

The country-level results reveal an interesting picture of competitiveness performance in the developing world over the last two decades. Countries in the East Asia and Pacific region account for seven out of the top ten countries and have particularly strong performance on the technology-intensive exports subindex. Singapore has the highest MECI level, reflecting the fact that it has the highest proportion of technology-intensive exports and the highest manufactured exports per capita in the whole sample. Malaysia, Taiwan, Philippines and Korea (with significant shares of high technology exports and good export growth rates) closely follow Singapore.

Interestingly, the East Asian economies performed better than European and Central Asian economies owing to their better high technology shares of exports and, possibly, somewhat higher export growth. Hungary is just in the top ten performers and along with Portugal and Turkey (ranked thirteenth and seventeenth) leads the European and Central Asian grouping.

Mexico is the highest performer from the Americas, ranked six, and had the sample's highest manufactured export growth rate over the period. Costa Rica, Trinidad and Tobago, and Chile come next in this region whose members are largely concentrated in the top half of the sample of eighty countries. At the tail end of the American region come Nicaragua and Haiti (ranking seventieth and seventy-first, respectively).

Strong performance in different sub-factors can offset poor performance in others, for example, China's strong export growth and relatively high proportion of technology-intensive exports more than outweighs its relatively

Table 3.2 MECI ranking by country

Country	MECI index	Manufactured exports per capita (US$), 1999[a,c]		Manufactured export growth, 1980–1999, %[c,b]		Technology-intensive exports (% of total merchandise exports), 1998[b]	
		Value	Rank	Value	Rank	Value	Rank
1 Singapore	0.93	25,039	1	13.4	13	70	1
2 Malaysia	0.82	2,988	5	19.2	3	55	4
3 Taiwan	0.79	5,477	3	9.4	31	58	3
4 Philippines	0.78	204	31	14.2	11	67	2
5 Korea, Rep.	0.76	2,825	6	11.9	17	54	5
6 Mexico	0.74	1,206	11	23.4	1	40	9
7 Israel	0.73	3,936	4	9.2	33	49	6
8 Thailand	0.71	719	14	18.8	4	42	8
9 Hong Kong, China	0.70	24,651	2	11.9	16	32	10
10 Hungary	0.67	2,121	7	7.3	41	45	7
11 Costa Rica	0.60	1,246	10	15.6	8	24	12
12 China	0.55	137	36	15.1	10	27	11
13 Portugal	0.54	2,071	8	10.2	26	19	19
14 Tunisia	0.53	498	20	9.9	27	24	12
15 Trinidad and Tobago	0.52	645	16	7.7	37	23	14
16 Indonesia	0.51	128	37	23.2	2	10	30
17 Turkey	0.50	317	25	18.7	5	10	30
18 Morocco	0.47	127	38	11.0	22	19	19
19 Chile	0.47	173	34	10.5	25	18	21
20 Poland	0.47	545	17	3.8	56	20	16
21 Bolivia	0.46	52	48	15.5	9	16	24
22 South Africa	0.46	349	24	6.2	47	18	21
23 Brazil	0.45	155	35	6.8	45	20	16
24 Mauritius	0.45	984	12	12.8	15	3	43
25 Oman	0.45	520	19	13.6	12	5	40
26 Cyprus	0.45	684	15	3.1	62	17	23
27 Saudi Arabia	0.44	262	28	11.7	18	10	30
28 Sri Lanka	0.44	183	33	16.3	7	5	40
29 Greece	0.43	525	18	4.4	53	13	26
30 Bahrain	0.42	953	13	11.6	19	0	65
31 Argentina	0.41	202	32	7.5	38	10	30
32 Bulgaria	0.40	363	23	−4.2	77	20	16
33 Venezuela, RB	0.39	98	39	10.8	23	7	37
34 Romania	0.39	297	27	−0.1	70	15	25
35 Zimbabwe	0.39	48	51	0.6	68	23	14
36 Colombia	0.39	85	42	8.3	35	10	30
37 India	0.38	26	56	9.4	30	13	26
38 Dominica	0.38	393	21	9.2	34	0	65
39 Kuwait	0.38	1,289	9	1.0	66	4	42
40 Uruguay	0.37	258	29	4.1	55	7	37
41 Jordan	0.37	217	30	9.5	29	0	65
42 Nepal	0.36	19	61	17.2	6	1	58
43 El Salvador	0.36	95	40	2.8	63	11	29
44 Bangladesh	0.35	37	54	13.1	14	2	49
45 Jamaica	0.35	377	22	2.8	64	3	43
46 Guatemala	0.35	74	43	4.3	54	9	35
47 Senegal	0.35	62	45	11.5	20	0	65

(Continued)

Table 3.2 (Continued)

Country	MECI index	Manufactured exports per capita (US$), 1999[a,c]		Manufactured export growth, 1980–1999, %[c,b]		Technology-intensive exports (% of total merchandise exports), 1998[b]	
		Value	Rank	Value	Rank	Value	Rank
48 Ghana	0.34	20	59	6.0	49	12	28
49 Pakistan	0.34	55	47	9.8	28	2	49
50 St Kitts and Nevis	0.33	300	26	3.8	57	0	65
51 Panama	0.33	48	50	7.9	36	3	43
52 Honduras	0.33	63	44	7.3	40	2	49
53 Egypt, Arab Rep.	0.33	21	57	7.5	39	7	37
54 Ecuador	0.32	32	55	9.2	32	2	49
55 Grenada	0.31	45	52	7.2	42	0	65
56 Algeria	0.30	11	67	11.2	21	1	58
57 Peru	0.30	51	49	3.6	59	3	43
58 Belize	0.29	86	41	0.4	69	2	49
59 Paraguay	0.28	21	58	6.1	48	1	58
60 Kenya	0.28	14	64	5.5	51	3	43
61 Guyana	0.27	37	53	0.9	67	3	43
62 Central African Republic	0.26	19	60	4.6	52	0	65
63 Syrian Arab Republic	0.26	16	63	3.3	61	2	49
64 Cameroon	0.26	10	68	6.5	46	0	65
65 Côte d'Ivoire	0.25	18	62	3.6	60	0	65
66 Madagascar	0.25	6	71	7.2	43	0	65
67 Tonga	0.24	6	72	5.9	50	0	65
68 Gabon	0.24	57	46	−3.2	76	0	65
69 Uganda	0.23	1	77	10.6	24	2	49
70 Nicaragua	0.21	9	69	−1.5	72	2	49
71 Haiti	0.21	13	65	−2.0	73	1	58
72 Malawi	0.21	3	73	3.7	58	0	65
73 Mozambique	0.20	2	75	−2.9	75	9	35
74 Tanzania	0.19	3	74	1.1	65	1	58
75 Congo, Rep.	0.19	12	66	−4.2	78	0	65
76 Sudan	0.18	1	79	7.1	44	0	65
77 Zambia	0.17	8	70	−6.7	79	2	49
78 Congo, DR	0.15	1	76	−2.1	74	1	58
79 Nigeria	0.13	1	80	−1.2	71	1	58
80 Yemen, Rep.	0.00	1	78	−18.0	80	0	65

Sources: Data on manufactured exports and population are from World Bank, *World Development Indicators*, various, Asian Development Bank, *Key Indicators of Developing Asian and Pacific Countries*, various and Wignaraja (1999) while data on technology-intensive exports as percentage of total merchandise exports are from the ITC web site (www.itc.org) and Wignaraja (1999). Data for Taiwan and 1980 data for China are from ADB *Key Indicators of Developing Asian and Pacific Countries*, various.

Notes

a Manufactured exports defined by World Bank *World Development Indicators* as SITC 5 (chemicals), 6 (basic manufactured), 7 (machinery and transport), 8 (miscellaneous manufactured goods) minus 68 (nonferrous metals).

b Technology-intensive manufactured exports (as a percentage of total merchandise exports for 1998) are defined by UNCTAD/WTO International Trade Centre following UNCTAD's SITC three digit classification by factor intensity.

c Data on manufactured exports per capita are for 1999 while manufactured exports growth in current US$ are for 1990–1999 (with exceptions listed in Appendix on 'Details of Table 3.2: MECI rankings').

low manufactured exports per capita. Thus, China ranks twelfth in the overall list. On the other hand, Kuwait, for example, although having the ninth highest manufactured exports per capita performs poorly on export growth and technology-intensive exports and so has an overall ranking of thirty-nine.

South Asian economies are mainly in the middle of the rankings. Typically, these economies are characterised by reasonable manufactured export growth rates but with relatively small shares of high technology exports and low per capita manufactured export values. Sri Lanka, at twenty-eighth in the overall list, is the leading South Asian economy owing to its manufactured export growth rate and value of per capita manufactured exports. India (thirty-seventh) comes next and has the highest share of high technology exports in the South Asian region.

Greater variation is found in the case of Sub-Saharan African economies. South Africa (ranking twenty-second) and Mauritius (ranking twenty-fourth) are the best performers in this region. By regional standards, South Africa has the largest export base and a reasonable share of high technology exports while Mauritius has strong manufactured export growth rates with a limited high technology content of exports. Notwithstanding these exceptions, Sub-Saharan African countries dominate the lower rankings, for example, occupying eight of the bottom ten positions.

While Mauritius and Trinidad and Tobago are among the top twenty-five performers, small developing economies (largely in the Caribbean and the Pacific) such as Jamaica, St Kitts and Nevis, Grenada, Belize and Tonga are typically scattered throughout the bottom half of the whole sample. This indicates that small states seem to have done less well on MECI performance than larger economies.

By way of summary, regional MECI performance is provided in Table 3.3 according to World Bank categories. Given the dominance of East Asian economies in the top ten ranking by country, it is unsurprising that the MECI level for East Asia and the Pacific region is considerably larger than for other regions (primarily due to the higher proportion of technology-intensive exports). Whilst the Americas and South Asia also have average export growth rates over 10 per cent and the level of manufactured exports per capita is greater in Europe and Central Asia, it is the much higher proportion of technology-intensive exports that distinguishes the East Asia and Pacific grouping. Differences within regions are also noticeable. Regional outliers, in terms of export value and technology-intensity, include Mexico, South Africa and Israel in the Americas, Sub-Saharan Africa and North Africa and Middle East groupings respectively. The impact of outliers can be seen clearly if they are excluded from the regional samples. For example, if South Africa is excluded, exports per head for the rest of Sub-Saharan Africa fall from US$ 40 to US$ 10 and the regional MECI falls from 0.35 to 0.24.

The country-level MECI data can also be arranged by national income per head (see Table 3.4) following World Bank categories. The high-income group (per capita incomes of above US$ 9,266) consists of just eight economies (Hong Kong, Singapore, Taiwan, Cyprus, Israel, Kuwait, Greece and Portugal).

Table 3.3 Performance by region

Regional groupings[a]	MECI[b]	Manufactured exports per capita (US$), 1999	Manufactured export growth, 1980–1999, (%)	Technology-intensive exports (% total), 1998
Americas	0.55	352	13.1	24
East Asia and Pacific	0.66	496	12.9	42
Europe and Central Asia	0.49	600	5.5	21
Middle East and North Africa	0.46	254	8.4	17
South Asia	0.37	34	12.0	9
Sub-Saharan Africa	0.35	41	5.7	10

Sources: See Table 3.2 and authors' calculations.

Notes
a Regional groupings according to World Bank *World Development Indicators 2001* (see Appendix on 'Details of regional groupings' for details).
b Regional values calculated from weighted components of sub-indices for members of each region. Where original data for manufactured exports for 1980 and 1999 is not available, data for these years has been extrapolated using average growth rates. MECI value calculated using the country sample maximum and minimum levels in each sub-index.

Table 3.4 Performance by income grouping

Income groupings[a]	MECI[b]	Manufactured exports per capita (US$), 1999	Manufactured export growth, 1980–1999, (%)	Technology-intensive exports (% total), 1998
High income	0.74	7,073	18.0	46
Middle income – upper	0.60	793	11.6	32
Middle income – lower	0.53	163	11.9	27
Low income	0.38	37	11.1	8

Sources: See Table 3.2 and authors' calculations.

Notes
a Income groupings according to World Bank *World Development Indicators 2001* (see Appendix on 'Details of income groupings' for details).
b Group values calculated from weighted components of sub-indices for members of each income grouping. Where original data for manufactured exports for 1980 and 1999 is not available, data for these years has been extrapolated using average growth rates. MECI value calculated using the country sample maximum and minimum levels in each sub-index.

The other seventy-two economies fall into the three remaining categories: upper middle income, lower middle income and low-income economies. There is a clear divergence in MECI performance between higher and lower income economies. All three components – manufactured exports per head, export growth rates and technology-intensive proportion of exports – are considerably higher in the high-income economies than the other three groups. It is interesting to note that the MECI value for the high-income grouping (0.74) is nearly double that of the low-income grouping (0.38). However, there is a smaller gap in MECI values between the high-income group and the two middle-income groups.

Comparison with WEF and IMD results

Whilst the overall sample size of the MECI is only slightly larger than for the WEF sample there are considerably more lower income countries in our index (see Table 3.5). This feature (along with the wider regional sample of the MECI index, particularly in relation to Sub-Saharan Asia) is a key advantage of the MECI index.[17] Whilst the competitiveness of high-income economies is clearly of interest we aim to provide an index that will be of value to policy makers and the private sector throughout the developing world.

There are forty-four common members in the MECI and WEF indices and twenty-two common members in the MECI and the IMD index. The rank correlation between the WEF current competitiveness index and the IMD index (for common members of the two indices) is 0.88. There are some similarities between the MECI rankings and the WEF current competitiveness and IMD rankings (adjusted for common membership with the MECI sample). For example, Singapore is at the top of all three adjusted rankings and there are a number of common members of the top ten. Table 3.6 indicates the top ten members of the MECI and the top ten countries included in the MECI sample within the WEF current competitiveness and IMD indices samples (the actual, unadjusted, rankings are indicated in brackets). However, the similarities between the

Table 3.5 Sample composition of MECI, WEF (2001) and IMD (2001) indices

	MECI	WEF	IMD
Total	80	75	49
By region			
Americas	25	22	8
Europe and Central Asia	8	30	26
East Asia and Pacific	10	14	12
South Asia	5	3	1
Middle East and North Africa	12	2	1
Sub-Saharan Africa	20	4	1
By income			
High income	8	28	28
Middle income – upper	21	17	13
Middle income – lower	26	22	6
Low income	25	8	2

Sources: Authors' calculations, WEF (2001) and IMD (2001).

Note
Based on classifications in World Bank *World Development Indicators 2001*.

Table 3.6 Rankings of MECI, WEF current competitiveness index (2001) and IMD index (2001)

MECI index	WEF common sample	IMD common sample
1 Singapore	Singapore (10)	Singapore (2)
2 Malaysia	Israel (17)	Hong Kong, China (6)
3 Taiwan	Hong Kong, China (18)	Israel (16)
4 Philippines	Taiwan (21)	Taiwan (18)
5 Korea, Rep.	South Africa (25)	Chile (24)
6 Mexico	Hungary (26)	Hungary (27)
7 Israel	Korea, Rep. (28)	Korea, Rep. (28)
8 Thailand	Chile (29)	Malaysia (29)
9 Hong Kong, China	Brazil (30)	Greece (30)
10 Hungary	Portugal (31)	Brazil (31)

Sources: Authors' calculations, WEF (2001) and IMD (2001).

Note
WEF and IMD rankings are adjusted to reflect common membership with MECI. Unadjusted rankings are presented in brackets.

rankings of the WEF current competitiveness index and the IMD index are much higher than their respective rank correlations with the MECI index. The rank correlations between the MECI index and the WEF current competitiveness index and IMD index are 0.65 and 0.54, respectively (ranks adjusted to reflect common membership). This lower correlation is perhaps unsurprising given the much broader approach and composition of both of the 'Swiss' indices compared to the MECI index.

Next we examine the determinants of manufactured export competitiveness in the developing world using statistical analysis.

Factors affecting manufacturing export competitiveness

Previous empirical studies

The review of perspectives on manufacturing export competitiveness in developing countries in Chapter 2 suggested that determinants fall into two broad categories: (a) incentive factors and policies made up of issues like macroeconomic conditions (e.g. price stability, growth and savings and investment), import liberalisation and regulations for domestic competition; and (b) supply-side factors and policies (e.g. industrial skills and labour productivity, research and development and other forms of technological effort, foreign investment and other sources of imported technology and physical infrastructure). The former set frames the competitive environment for business and sends specific price signals for firm-level technological activity while the latter provide inputs and support for this process. Industrial success in developing countries occurs as a result of the interaction of these two sets of determinants.

While there is a considerable empirical literature on trade policy and economic growth across countries,[18] little attempt has been made to statistically analyse the broader issue (and arguably more fundamental one in the context of pervasive globalisation) of export competitiveness policies and performance. The handful of available empirical studies on the determinants of export competitiveness seem to differ considerably in the variables being investigated, the statistical methods employed and the coverage of countries. But they collectively offer useful insights into the influences on export competitiveness across countries. The main findings can be summarised as follows.

One of the earliest studies by Fagerberg (1988) developed a model of international competitiveness, which relates growth in market shares to three sets of factors: the ability to compete in technology (a proxy combining R&D expenditures as a percentage of GDP and external patents per capita), the ability to compete on price and the ability to compete in delivery capacity. Using a two-stage least squares regression analysis, he tested this model on pooled cross-sectional and time-series data for fifteen OECD countries. He concluded that: 'the results of this paper suggest that the main factors influencing differences in international competitiveness and growth across countries are technological competitiveness and the ability to compete on delivery' (Fagerberg, 1988, p. 371).

Ul Haque (1995) used cross-section regression analysis to look (a) at the relationship between export performance (share of a country's manufactured exports in total developing country manufactured exports) and productivity growth (industrial value added per worker) and (b) then to examine the determinants of productivity differences in a sample of fifteen developing countries. He found a strong association between the two variables in the first case. In the second case, he reported that secondary school enrolment and, to a lesser extent, the ratio of investment to GDP had an influence on inter-country differences in productivity. He concluded that education and investment are notable influences on national productivity and export competitiveness in developing countries.

James and Romijn (1997) used cross-section regression analysis to test an index related to the complexity involved in the manufacture of engineering goods against several country-level characteristics (market size, number of scientists and engineers, FDI stock and trade policy orientation) in a sample of forty-nine developing countries. They found that market size (population size and GDP per capita), the stock of scientists and engineers and trade policy orientation (a dummy variable to represent inward or outward-orientation) turn up as important determinants of cross-country differences in this index. They conclude that while market size matters for the creation of domestic competitiveness, 'countries with a small market can to some extent use supply-side variables to compensate in part for the difficulties they confront on the demand side' (James and Romijn, 1997, p. 201). In this vein, they emphasise expanding scientific and engineering education as well as creating institutions to promote the acquisition of domestic technological capabilities at firm-level.

Noorbakhsh and Paloni (1998) employed cross-section regression analysis to test the ratio of manufactured exports to total exports against macroeconomic conditions (fiscal deficit to GDP, external debt to GDP, investment to GDP), trade openness and terms of trade in a sample of eighty-six developing countries. They found that the ratio of manufactured exports to total exports is positively associated with trade openness and prudent macroeconomic conditions (especially high levels of investment and low fiscal deficits and external debt). They conclude with a note of caution: 'in order to reach more solid conclusions, cross-section studies should be complemented by country studies and by analyses at a more disaggregate level to highlight the relevance of specific institutional structures and other characteristics' (Noorbakhsh and Paloni, 1998, p. 569).

Wignaraja (1999) conducted a t-test on the means of the individual variables to examine factors affecting manufactured export competitiveness in a sample of thirty-nine strong (Asian economies) and weak performers (African economies) in the developing world. The results (at the 5 per cent statistical significance level) suggested that the Asian countries had better macroeconomic conditions (ratios of savings and investment in GDP) than African economies as well as higher levels of FDI (cumulative FDI inflow), technological effort (total factor productivity growth) education (secondary school enrolment and tertiary technical enrolment) and physical infrastructure (telephone lines per 1,000 population). He concluded that good macroeconomic conditions, attracting foreign investment and investments in supply-side factors (notably technical skills and technological effort) had a strong influence on manufactured export competitiveness in the developing world.

Statistical test, variables and results

Drawing on previous work, a simple statistical test of the influences on cross-national competitiveness in developing countries is conducted below. One of the challenges facing empirical research in this area is the dearth of information on relevant national variables even on a cross-section basis. Data on inflation rates, GDP growth and other proxies for macroeconomic conditions are readily available from international sources for a large sample of developing countries but suitable proxies for the trade and domestic competition regime (e.g. effective rates of protection, export bias indices and three firm concentration ratios) are hard to find. Similarly, some supply-side indicators, particularly those for general education (e.g. primary and secondary school enrolment) and infrastructure (telephones per 1,000 population), are available while technological indicators (e.g. R&D expenditures or external patents filed) and technical manpower (e.g. tertiary-level enrolment of scientists and engineers and engineers and technicians per 1,000 population) are more problematic for many poor countries. Where point estimates are available for the technological and technical manpower indicators, they are sometimes quite dated which can diminish their value in statistical analysis. The varying quality and coverage of proxies for incentive and supply-side factors for developing countries has affected the type of statistical analysis that can be conducted on national competitiveness.

Empirical studies have responded to this issue in one of two ways. First, multiple regression analysis has been used to examine a limited selection of largely macroeconomic and trade policy determinants, which are readily available, and neglected supply-side factors (with the exception of primary and secondary schooling) or vice versa. Noorbakhsh and Paloni (1998) and James and Romijn (1997) fall into this camp. Second, a series of simple linear regressions have been employed to relate a given independent variable to a single dependent variable. Ul Haque (1995) is an example of this approach.[19] Neither method is entirely satisfactory. The first approach typically downplays the complementary roles of incentive and supply-side factors in determining competitiveness while the second gives prominence to single determinants. Until a comprehensive and up-to-date database of incentive and supply-side factors becomes available for a large sample of developing countries, it will be a challenge to use multiple regression analysis to fully examine the determinants of manufactured export competitiveness.

Hence, this research relies on simpler statistical techniques to shed preliminary light on the factors associated with manufacturing export competitiveness in the developing world. One such method of comparative statistical analysis of the determinants of national competitiveness is a two-sample t-test of the variable means.[20] The purpose of the t-test is to determine whether the two sample means are equal and when the two groups under study are distinct in statistical terms. The two samples examined are distinguished by their MECI values: high performing countries with MECI greater than the mean and low performing countries with MECI below the mean. Then the mean values of a particular country characteristic (e.g. R&D expenditures as a percentage of GDP) are computed separately for the high performing sample and the low performing sample and examined for statistical significance. Thereafter, this exercise is repeated for each of the remaining country characteristics.

A similar t-test was also conducted on the low performing country sub-sample alone (i.e. developing countries with MECI values below the total sample mean) to see whether the factors from the first test provide insights into the competitiveness behaviour within the second group. This test divides the low performing country sample into two groups distinguished by their sample mean and follows the testing procedure mentioned above.[21]

The following variables were used to explore the major incentive (e.g. macroeconomic environment and trade liberalisation) and supply-side factors (e.g. human capital, FDI, technological effort and physical infrastructure) associated with manufactured competitiveness success in the developing world in the two types of t-tests described above.[22]

- A stable, predictable *macroeconomic environment* – characterised by low inflation, steady GDP growth and high levels of savings and investment – is widely accepted as a fundamental condition for business activity, industrial technological activity and competitiveness. A macroeconomic climate based on low inflation and good growth prospects sends positive signals about the profitability of resource allocation to the private sector and can induce a strong supply response. Macroeconomic stability is proxied by four variables: average annual GDP growth in 1990–1998, average annual inflation in 1990–1998 as well as savings and investment to GDP ratios in 1990.

- *Trade liberalisation* exposes the business sector to competition from imports, provides access to new technologies and skills from abroad, facilitates the realisation of economies of scale in production and stimulates industrial technological activities and competitiveness. Trade liberalisation is represented by two variables. The first is based on Sachs and Warner's (1995) classification of an economy as open. An index was calculated representing the proportion of the period 1980–1994 for which a country's economy was classified on this measure to be 'open'. The second trade liberalisation measure is the import to GDP ratio in 1990.

- Attracting export-oriented *FDI* is a short cut method to enter the production of manufactures for export and to upgrade export competitiveness over time. Among other reasons, FDI facilitates technology transfer from abroad, inflows of managerial and technical skills, access to marketing connections and linkages with international capital markets. It can also promote local technological development via subcontracting and supplier relationships between foreign and locally owned firms. FDI is captured by recent figures for inward FDI stock (1998).

- Sustained *technological effort* is vital to put imported technologies to productive use within industry. Imported technologies contain many tacit elements, which require conscious investments in creating skills, information and research to acquire industrial technological capabilities. In turn, the process of acquiring industrial technological capabilities is closely associated with enhanced competitiveness at firm and national levels. Active technological effort is proxied by: (a) the most recent estimates of R&D expenditure as a percentage of GNP and (b) the most recent estimates for the absolute number of R&D researchers. Manufacturing value added (MVA) as a percentage of GDP in 1990 has also been added as a crude proxy for learning by doing and experience-based learning in the manufacturing sector.

- A strong base of *productive human capital* at all levels (primary, secondary, tertiary, technical and vocational education) is increasingly recognised as the bedrock for industrial technological activity and competitiveness. Primary and secondary educated workers help to develop productive, numerate workers who are critical inputs to labour-intensive industrialisation. As industrial upgrading occurs, vocational and tertiary-level technical skills become important entry requirements into more complex industries. Secondary and tertiary gross enrollment ratios are taken as proxies for human capital while value added per worker in manufacturing (for 1990) has been used as a proxy for labour productivity.

- An efficient and cost-competitive *physical infrastructure* is closely associated with industrial technological activity and competitiveness. In a world of accelerating technological progress and explosive growth in FDI, there is a premium on having a well-developed information and communications technologies (ICT) infrastructure. Hence, infrastructure is represented by main phone lines per 1,000 people for 1990 and the number of personal computers (PCs) per 1,000 people for 1998.

Table 3.7 shows the results of the *t*-tests on the means of the variables for the high performing sample countries (MECI values above the mean) and the low performing sample countries (MECI values below the mean).[23] Data availability determined the sample size for a given *t*-test. The largest sample for these *t*-tests

Table 3.7 *t*-Tests to compare the means of high performing countries and low performing countries

	High performers (MECI above mean)		Low performers (MECI below mean)		t-Stat
	Mean	Observations	Mean	Observations	
Macroeconomic environment					
GDP annual % growth 1990–1998	4.7	31	3.6	49	1.81**
CPI annual % growth 1990–1998	21.8	31	69.1	48	−1.06
Savings as % GDP 1990	25.0	31	15.0	49	4.31***
Investment as % GDP 1990	25.0	31	20.5	49	2.47***
Trade liberalisation					
Sachs and Warner Index	0.65	28	0.22	41	5.62***
Import as % GDP 1990	25	29	13.1	46	3.96***
FDI					
Logarithm of inward FDI stock 1998 US$m	4.14	31	3.02	49	7.31***
Technological effort					
R&D expenditure as % GNP	73	25	35	24	2.20**
Logarithm of absolute numbers of R&D researchers	3.50	25	2.81	25	2.86***
MVA as % GDP 1990	21.1	30	14.3	47	4.32***
Human capital and productivity					
Secondary education gross enrolment ratio	74.6	31	41.4	41	6.66***
Tertiary education gross enrolment ratio	24.5	31	19	41	4.29***
Productivity					
Logarithm of value added per worker in manufacturing (US$ current) 1990	9.48	27	9.12	37	1.75**
Physical infrastructure					
PCs (per 1,000 people) 1998	43.4	27	2.7	43	2.94***
Logarithm of 1990 main phone lines per 1,000 people	1.87	30	1.20	49	4.76***

Sources: Data from *World Development Indicators*, World Bank, unless indicated. Sachs and Warner Index calculated from data in Sachs and Warner (1995). FDI data from UNCTAD (1999), *World Investment Report*. Educational attainment data from UNESCO (1999) and UNESCO website. MVA and value added per worker data from UNIDO, *Yearbook of Industrial Statistics and Industry and Global Development Report* (various issues). Data for Taiwan from Asian Development Bank *Key Indicators of Developing Asian and Pacific Countries* and national sources.

Notes
* Statistically significant at the 10 per cent level.
** Statistically significant at the 5 per cent level.
*** Statistically significant at the 1 per cent level.
MVA is manufactured value added. Low performers have MECI below sample mean of 0.397.

had thirty-one high performing countries and forty-nine low performing countries while the smallest sample had twenty-five high performing countries and twenty-four low performing countries. These sample sizes are reasonable by the standards of cross-national statistical analysis. The main findings are as follows.

- *Macroeconomic environment.* The higher performing sample countries have significantly higher savings ratios (at the 1 per cent confidence level) with a mean of 25 per cent compared to 15 per cent for the low performing country sample. The investment ratio is also significant at the 1 per cent confidence level. These two variables can be seen to represent capital accumulation, perhaps one of the drivers behind the higher MECI performance. The means of GDP growth of the two samples are statistically different at the 5 per cent level with higher performing sample experiencing on average a higher GDP growth (4.7 per cent compared with 3.6 per cent for the low performing sample). Whilst the high performing sample countries do have a lower mean of inflation, the difference is not statistically significant at the 10 per cent level.
- *Trade liberalisation.* The difference in the mean of the Sachs and Warner Index between the two samples is statistically significant at the 1 per cent level suggesting greater openness is related to higher MECI values. There also appears to be a statistically significant difference for the import to GDP ratio.
- *FDI.* The high performing countries have significantly greater mean (log) FDI stocks than the low performing sample countries suggesting the role of foreign investment in promoting export competitiveness.
- *Technological effort.* The sample means for the R&D variables are both significantly greater for the higher performing sample. Manufacturing experience, as represented by MVA as a percentage of GDP also appears to be related to our measure of export competitiveness.
- *Human capital and productivity.* The importance of human capital in export competitiveness may be suggested by the significantly higher average education attainment at both the secondary and tertiary levels for the higher performing country sample. For instance, the mean values for secondary education are 74.6 and 41.4 per cent respectively, for the two samples. In addition, the difference in mean value added per worker between the two samples does appear to be statistically significant, although only at the 5 per cent level.
- *Infrastructure.* Both the means of the telephone mainlines and PCs variables are significantly greater for the high performing country sample suggesting that modern infrastructure is associated with greater competitiveness. A striking result is particularly visible for the PC variable where the higher performing sample has 43.4 computers per 1,000 people compared with only 2.7 for the low performing country sample.

Table 3.8 t-Tests to compare the means within the low performing sample

	Relatively better low MECI performers (MECI above mean of sample of lower performers)		Other (MECI below sample of mean of lower performers)		t-Stat
	Mean	Observations	Mean	Observations	
Macroeconomic environment					
GDP annual % growth 1990–1998	4.1	26	3.1	23	1.20
CPI annual % growth 1990–1998	25.6	26	12.5	22	−1.01
Savings as % GDP 1990	15.9	26	14.1	23	0.63
Investment as % GDP 1990	22.9	26	17.8	23	2.12**
Trade liberalisation					
Sachs and Warner Index	0.28	21	0.15	20	1.60*
Import as % GDP 1990	14.2	24	11.9	22	1.28
FDI					
Logarithm of inward FDI stock 1998 US$ m	3.24	26	2.77	23	2.61***
Technological effort					
MVA % GDP 1990	16.2	24	12.2	23	2.23**
Human capital and productivity					
Secondary education gross enrolment ratio	53.0	22	29.3	19	3.67***
Tertiary education gross enrolment ratio	16.3	22	4.7	19	4.18***
Physical infrastructure					
PCs (per 1,000 people) 1998	4.74	23	0.30	20	2.32**
Logarithm of 1990 main phone lines per 1,000 people	1.53	26	0.84	23	4.21***

Sources: As Table 3.7.

Notes
* Statistically significant at the 10 per cent level.
** Statistically significant at the 5 per cent level.
*** Statistically significant at the 1 per cent level.
MVA is manufactured value added. Relatively better low MECI performers had an MECI above 0.289 and below 0.397 with others having an MECI below 0.289.

Table 3.8 shows the results of the t-tests on the means of the variables for the stronger low performing sample countries and the weaker low performing sample countries.[24] The largest samples for these t-tests had twenty-six stronger low performing countries and twenty-three weaker low performing countries. A lack of data meant that the t-tests for the R&D variables and labour productivity could not be conducted. With this qualification, the main findings are as follows.

- *Macroeconomic environment.* The stronger low performing countries have significantly higher investment-to-GDP (at the 5 per cent level) but the other three macroeconomic variables (inflation, growth and savings ratios) do not show any statistically significant difference.
- *Trade liberalisation.* The stronger low performing countries have a significantly higher trade openness, measured by the Sachs–Warner Index, though only at the 10 per cent level while the ratio of imports to GDP does not show any statistically significant difference.
- *FDI.* The stronger low performing countries had significantly higher inward FDI stocks (1 per cent confidence level).
- *Technological effort.* Manufacturing experience (MVA as a percentage of GDP) shows a statistically significant difference between the two sub-samples of the low performing group.
- *Human capital.* The stronger low performing countries had significantly higher secondary and tertiary enrolment ratios (at the 1 per cent level).
- *Infrastructure.* Both the means of the telephone mainlines and PCs variables are significantly greater for the high performing country sample.

Comparing the two sets of t-tests, it is interesting to note that whilst all the macroeconomic variables (with the exception of inflation) have statistically different means between the low and high performing countries it is only investment that has a statistically different mean between the relatively better low performers and other low performers. In contrast, the supply-side variables have statistically different means in both sets of t-tests.

Conclusions

Benchmarking competitiveness performance across countries has become an increasingly popular pursuit among applied economists, business strategists and policy makers. While early work focused on differential performance among developed countries, there is greater attention being paid to benchmarking competitiveness performance within the developing world. The WEF and the IMD currently dominate a growing benchmarking industry with their annual rankings of leading developed and developing countries. The WEF and IMD reports contain a wealth of published and survey data on several developing countries. However, close scrutiny of the IMD report in particular revealed several weaknesses in the methodology underlying their competitiveness indices including ambiguous theoretical principles and empirical relevance; *ad hoc* use of survey and statistical data; and limited coverage of poor countries in the developing world.

In an attempt to deal with these gaps, this chapter presents a simpler but clearer and more intuitive MECI based on three sub-components (manufactured exports per capital, long-term growth in manufactured exports and the share of technology-intensive exports) and reports the findings by country, region and income group. East Asian economies emerge as the leading performers in the sample. Economies from the Americas and Europe and Central Asia come next. Middle East, North African and South Asian economies are around the middle of the sample while Sub-Saharan African economies are towards the bottom. The results by income group also point to a varied pattern of competitiveness performance.

A lack of data on key variables hampers the use of econometric techniques to analyse the cross-national determinants of competitiveness performance in the developing world. Nevertheless, simpler statistical analysis involving two-sample *t*-tests of the means of variables suggest some interesting results. Competitiveness success in the developing world is associated with sound macroeconomic conditions, trade openness, inflows of foreign investment, sustained technological effort, investments in human capital and productivity and an efficient and cost-competitive infrastructure. Thus, focusing on incentive factors and policies alone is insufficient to ensure competitiveness success in the developing world. Instead, the available evidence points to the need for a combination of incentive and supply-side factors and policies to deliver a positive outcome.

Appendix: Data characteristics

Details of Table 3.2: MECI ranking

Manufactured exports per capita for 1999 with exceptions of: 1998 – Bangladesh, Chile, India, Jordan, Madagascar, Nepal, Saudi Arabia, Sudan, Yemen, Rep.; 1997 – Bulgaria, Congo, DR, Grenada, Guyana, Haiti, Jamaica, Morocco, St Kitts and Nevis; 1996 – Bahrain, Cameroon, Central African Republic, Côte d'Ivoire, Gabon, Mozambique; 1995 – Congo, Rep., Malawi, Tonga, Zambia.

Manufactured exports growth 1990–1999 with exception of: 1980–1998 – Bangaldesh, Chile, India, Jordan, Madagascar, Nepal, Saudi Arabia, Sudan, Yemen, Rep.; 1980–1997 – Bulgaria, Congo, DR, Grenada, Guyana, Haiti, Jamaica, Morocco; 1980–1996 – Bahrain, Cameroon, Central African Republic, Côte d'Ivoire, Mozambique; 1980–1995 – Congo, Rep., Malawi, Tonga, Zambia; 1981–1996 – Gabon; 1981–1997 – St Kitts and Nevis; 1981–1999 – Taiwan and China. Average growth rate calculated as geometric mean between start and end values.

Details of regional groupings

Regional groupings used are from World Bank *World Development Indicators 2001*. Taiwan is not classified by WDI but is placed in the East Asia and Pacific regional grouping.

Americas – Argentina, Belize, Bolivia, Brazil, Chile, Colombia, Costa Rica, Dominica, Ecuador, El Salvador, Grenada, Guatemala, Guyana, Haiti, Honduras, Jamaica, Mexico, Nicaragua, Panama, Paraguay, Peru, St Kitts and Nevis, Trinidad and Tobago, Uruguay, Venezuela, RB.

East Asia and Pacific – China, Hong Kong, China, Indonesia, Korea, Rep., Malaysia, Philippines, Singapore, Taiwan, Thailand, Tonga.

Europe and Central Asia – Bulgaria, Cyprus, Greece, Hungary, Poland, Portugal, Romania, Turkey.

Middle East and North Africa – Algeria, Bahrain, Egypt, Arab Rep., Israel, Jordan, Kuwait, Morocco, Oman, Saudi Arabia, Syrian Arab Republic, Tunisia, Yemen, Rep.

South Asia – Bangladesh, India, Nepal, Pakistan, Sri Lanka.

Sub-Saharan Africa – Cameroon, Central African Republic, Congo, DR, Congo, Rep., Côte d'Ivoire, Gabon, Ghana, Kenya, Madagascar, Malawi, Mauritius, Mozambique, Nigeria, Senegal, South Africa, Sudan, Tanzania, Uganda, Zambia, Zimbabwe.

Details of income groupings

Income groupings used are from World Bank *World Development Indicators 2001*. Groupings based on 1999 Gross National Income per capita. The groups are: Low income has GNI per capita of US$ 755 or less; lower middle income between US$ 756 and US$ 2,995; upper middle income between US$ 2,996 and US$ 9,265; and high income greater than US$ 9,266 or more. Taiwan is not classified by WDI but fits the description of a high income economy.

High income – Hong Kong, China, Singapore, Taiwan, Cyprus, Israel, Kuwait, Greece, Portugal.

Middle income – upper – Argentina, Brazil, Chile, Dominica, Grenada, Mexico, Panama, St Kitts and Nevis, Trinidad and Tobago, Uruguay, Venezuela, RB, Korea, Rep., Malaysia, Hungary, Poland, Bahrain, Oman, Saudi Arabia, Gabon, Mauritius, South Africa.

Middle income – lower – Belize, Bolivia, Colombia, Costa Rica, Ecuador, El Salvador, Guatemala, Guyana, Honduras, Jamaica, Paraguay, Peru, China, Philippines, Thailand, Tonga, Bulgaria, Romania, Turkey, Algeria, Egypt, Arab Rep., Jordan, Morocco, Syrian Arab Republic, Tunisia, Sri Lanka.

Low income – Haiti, Nicaragua, Indonesia, Yemen, Rep., Bangladesh, India, Nepal, Pakistan, Cameroon, Central African Republic, Congo, DR, Congo, Rep., Côte d'Ivoire, Ghana, Kenya, Madagascar, Malawi, Mozambique, Nigeria, Senegal, Sudan, Tanzania, Uganda, Zambia, Zimbabwe.

Characteristics of samples for t-tests

Table 3.9 High performing and low performing samples

	Mean for high performers	Mean for low performers	t-Stat
MECI Index	0.57	0.29	9.65***
Manufactured exports per capita (US$), 1999	2,578	97	2.27**
Manufactured export growth, 1980–1999, %	12.1	4.1	6.29***
Technology-intensive exports (% total merchandise exports), 1998	27	4	6.34***
Memo			
Sample size	31	49	

Notes
Low performers have MECI below sample mean of 0.397.
* Statistically significant at the 10 per cent level.
** Statistically significant at the 5 per cent level.
*** Statistically significant at the 1 per cent level.

Table 3.10 Low performing sub-samples

	Relatively better low MECI performers (MECI above mean of sample of lower performers)	Other (MECI below mean of sample of lower performers)	t-Stat
MECI Index	0.35	0.22	9.57***
Manufactured exports per capita (US$), 1999	169	15	2.99***
Manufactured export growth, 1980–1999, %	6.7	1.1	3.59***
Technology-intensive exports (% total merchandise exports), 1998	6	1	3.58***
Memo			
Sample size	26	23	

Notes
* Statistically significant at the 10 per cent level.
** Statistically significant at the 5 per cent level.
*** Statistically significant at the 1 per cent level.
Relatively better low MECI performers had an MECI above 0.289 and below 0.397 with others having an MECI below 0.289.

Notes

1 We would like to thank Michael Chui for helpful comments on the chapter and Friedrich von Kirchbach for access to International Trade Centre export data and many discussions. The views expressed here are ours and should not be attributed to the organisations to which we belong.
2 Developed countries have used benchmarking methods for many years. To quote a recent UK Government report: 'We find that in many sectors, such as pharmaceuticals, there are UK firms whose achievements match the world's best.... The challenge is to find ways to enable other firms to reach the standards achieved by the best. This benchmarking document provides a stimulus for business to review its own performance and a basis for the development of new policies to help business help itself. These policies will be developed in close partnership with business and others' (UK DTI, 1998, p. 4).
3 Other notable players include London based magazines such as *Corporate Location* and *The Economist* that provide cost information on international production locations and the IMF's *International Financial Statistics Yearbook* that provides real effective exchange rates calculated using relative unit labour costs data. More recently, the UNCTAD/WTO International Trade Centre and UNIDO's 2002 *World Industrial Development Report* have put forward indices based on trade performance.
4 Similar ideas can be found in Alavi (1990) who focuses on the determinants of national competitiveness and lists six categories of factors that affect national performance: macroeconomic dynamism, financial dynamism, infrastructural elements, human resources and firm-level elements. He also proposes a mix of hard and survey data indicators to capture these factors at national level but does not attempt to develop a cross-country composite competitiveness index. For other attempts along these lines see Dominique and Oral (1986) and Pietrobelli (1994).
5 As discussed in Chapter 2, Porter suggests that competitiveness advantages of nations arise from firm-level efforts to innovate in a broad sense (i.e. develop new products, improve production processes and introduce new brands). In turn, he suggests that innovations can take place in any industry as a result of four elements of the diamond framework: factor conditions, demand conditions, related and supporting industries and the context for firm strategy and rivalry.
6 The CCI evaluates the factors defining the current level of productivity, measured by the level of GDP per person. CCI looks at microeconomic influences including the sophistication of company operations and strategy as well as the quality of the national business environment. The GCI – the traditional WEF Competitiveness Index which was presented prior to 2000 – aims to provide a ranking of the factors affecting medium term (five year) growth, measured by the change in GDP per person. GCI is comprised of three sub-indexes: the level of technology in an economy, the quality of public institutions and macroeconomic conditions related to growth. A mix of hard and survey data (4,600 businessmen) is used to compute the CCI and the GCI. See WEF (2001).
7 The 2001 WEF report is marginally better than the 2001 IMD report in this regard. It contains three low income South Asian economies (India, Sri Lanka and Bangladesh) as well as three middle-income and two low-income African economies (Egypt, South Africa and Mauritius and Nigeria and Zimbabwe, respectively).
8 See *http://www02.imd.ch/wcy/methodology/methodology.cfm*.
9 Our estimates suggest that for the forty-four common members of the two indices, the ranks (adjusted for non-inclusion) have a correlation of 0.88. This calculation was based on the CCI of the WEF and the IMD single index.
10 The WEF and IMD reports have historically been quite secretive about the methodology used to compute country rankings and this has shielded them from academic

scrutiny. Oral and Chabchoub (1997) argue that past WEF reports did not provide many details of how the country ranking were derived and go on to replicate these using an estimation model based on mathematical programming. The 2001 editions of both reports are more transparent in this regard.

11 These can be listed as follows. *Economic performance* contains domestic economy, international trade, international investment, employment and prices. *Government efficiency* has public finance, fiscal policy, institutional framework, business framework and education. *Business efficiency* contains productivity, labour market, financial markets, management practices and impact of globalisation. *Infrastructure* has basic infrastructure, technological infrastructure, scientific infrastructure, health and environment and value system. See IMD (2001), p. 50.

12 As such plant-level studies contain commercially sensitive information which could be of use to competitors, they are rarely published. A published study by Andersen Consulting (in collaboration with the University of Cambridge and the Cardiff Business School) of manufacturing performance in the Japanese and UK automotive component industries sheds some light on the approach used by these studies (See Andersen Consulting, 1993).

13 One of the major problems Lall (2001) points out is the broad definition of competitiveness in terms of GDP per person in the 2000 WEF report, which diverts it 'from its legitimate focus on direct competition between countries, taking it into areas where competitiveness analysis is both unwarranted and has little analytical advantage' (WEF, 2000, p. 1519). He also highlights additional problems of 'model specification, the choice of variables, the identification of causal relations and the use of data' (WEF, 2000, p. 1520).

14 Most recent data available used if data for these periods are not available.

15 See Technical Note p. 269 to UNDP (2000) for further details.

16 The use of a logarithmic scale makes relatively little difference to the ranking of the index – the rank correlation between the MECI based on a logarithmic and a non-logarithmic approach is 0.949. However, using a logarithmic scale does reduce the skewness of the index values (to 0.85 compared to a skewness of 1.87 when a non-logarithmic approach is used).

17 The MECI coverage in Europe and the Middle East is reduced due to data availability for transition economies (particularly for export growth from 1980 since when there has been a considerable change in states in eastern Europe).

18 See Rodriguez and Rodrik (2000) for a critical survey of recent cross-country econometric studies on trade and growth. Helleiner (1994) contains detailed country studies of the influence of trade policies on growth and industrialisation in developing countries.

19 WEF (2001) also employed simple regression analysis to test its composite competitiveness indicator.

20 A one-tail test is used in Tables 3.7 and 3.8. This is because, *a priori*, it would be expected that each of the above variables may be unidirectional correlated with MECI values. For example, price stability may be conducive to economic performance and export growth, that is, the mean of CPI growth is expected to be lower for the higher MECI sample. On the other hand, a higher level of investment in capital may stimulate export performance and so the mean of the investment ratio to GDP may be expected to be greater for the higher MECI sample.

21 One qualification about the testing procedure should be noted. The simple t-test shows significantly different means between the two samples for individual variables. It does not indicate causal relationships between variables and is less powerful than econometric analysis. However, it can provide insights into those underlying factors correlated with competitive success in comparisons of strong and weak national performance. Summary statistics for the different sub-samples by MECI value are provided in Appendix on 'Characteristics of samples for t-tests'.

22 In general, post-1990 period averages and 1990 point estimates have been used to represent the various potential determinants. This base was chosen because of data availability and the assumption that some determinants will affect competitiveness performance with a lag (e.g. investment). Where 1990 data have not been widely available, the most recent data have been used.
23 The means of the MECI values (and the values of each of the sub-components) of the high performing and low performing sub-samples were statistically different at the 1 per cent significance level (see Appendix on 'Characteristics of samples for t-tests').
24 Again, the means of the MECI values (and the values of each of the sub-components) of the stronger and weaker low performing sub-samples were statistically different at the 1 per cent significance level (see Appendix on 'Characteristics of samples for t-tests').

References

Alavi, H. (1990), 'International Competitiveness: Determinants and Indicators', *Industry and Energy Department Working Paper Series No. 29*, World Bank.

Andersen Consulting (1993), *Worldwide Manufacturing Competitiveness Study: The Second Lean Enterprise Report*, London: Andersen Consulting.

Asian Development Bank (various), *Key Indicators of Developing Asian and Pacific Countries*, Manila: Asian Development Bank.

Council for Economic Planning and Development, Republic of China (1999), *Taiwan Statistical Data Book*, Taipei: Council for Economic Planning and Development, Republic of China.

Dominique, C.R. and Oral, M. (1986), 'Exporting to Northern Markets: The Making of an Industrial Competitiveness Index', *Industry and Development*, 18, 1–17.

Fagerberg, J. (1988), 'International Competitiveness', *Economic Journal*, 98 (June), 355–374.

Fagerberg, J. (1996), 'Technology and Competitiveness', *Oxford Review of Economic Policy*, 12(3), 39–51.

Helleiner, G.K. (ed.) (1994), *Trade Policy and Industrialisation in Turbulent Times*, London: Routledge.

IMD (2001), *The World Competitiveness Yearbook 2001*, Lausanne: International Institute for Management Development.

International Trade Centre UNCTAD/WTO website, *www.itc.org* (data availability has subsequently been restricted).

James, J. and Romijn, H. (1997), 'Determinants of Technological Capability: A Cross-Country Analysis', *Oxford Development Studies*, 25(2), 189–207.

Lall, S. (2001), 'Competitiveness Indices and Developing Countries: An Economic Evaluation of the Global Competitiveness Report', *World Development*, 29(9), 1501–1525.

Llewellyn, J. (1996), 'Tackling Europe's Competitiveness', *Oxford Review of Economic Policy*, 12(3), 87–96.

Noorbakhsh, F. and Paloni, A. (1998), 'Structural Adjustment Programmes and Export Supply Response', *Journal of International Development*, 10, 555–573.

OECD (1992), *Technology and the Economy: The Key Relationships*, Paris: OECD.

Oral, M. and Chabchoub, H. (1997), 'An Estimation Model for Replicating the Rankings of the World Competitiveness Report', *International Journal of Forecasting*, 13(4), 527–537.

Pietrobelli, C. (1994), 'Technological Capabilities at the National Level: An International Comparison', *Development Policy Review*, 12, 115–148.
Porter, M.E. (1990), *The Competitive Advantage of Nations*, London: Macmillan Press.
Rodriguez, F. and Rodrik, D. (2000), *Trade Policy and Economic Growth: A Skeptic's Guide to the Cross-National Evidence*, Cambridge, MA: John F. Kennedy School of Government, Harvard University.
Sachs, J. and Warner, A. (1995), 'Economic Reform and the Process of Global Integration', *Harvard Institute of Economic Research Discussion Paper* 1733, August.
UNESCO (1999), *Statistical Yearbook*, Paris: UNESCO.
UK DTI (1998), *Competitiveness: Our Partnership with Business*, UK, London: Department of Trade and Industry.
Ul Haque, I. (1995), 'Introduction', in Ul Haque (ed.), *Trade, Technology and International Competitiveness*, Washington, DC: World Bank, Economic Development Institute.
UNCTAD (1999), *World Investment Report 1999*, Geneva: UNCTAD.
UNIDO (2002), *World Industrial Development Report 2002: Competing Through Innovation*, Vienna: UNIDO.
UNIDO (various years), *Industry and Development Global Report*, Vienna: UNIDO.
UNIDO (various years), *International Yearbook of Industrial Statistics*, Vienna: UNIDO.
United Nations Development Programme (UNDP) (2000), *Human Development Report 2000*, New York and Oxford: Oxford University Press and website *www.undp.org*.
WEF (2001), *Global Competitiveness Report 2001–2002*, New York: Oxford University Press for the World Economic Forum, *www.weforum.org*.
Wignaraja, G. (1999), 'Tackling National Competitiveness in a Borderless World', *Commonwealth Business Council Policy Paper Series No. 1* (2nd edition), London: Commonwealth Business Council.
World Bank (various years), *World Development Indicators (CD-ROM)*, Washington, DC: World Bank.

Part II
Supply-side issues and policies for competitiveness

4 Science, technology and innovation policy

Stan Metcalfe[1]

Introduction

The central theme of this chapter is the nature and role of science, technology and innovation (STI) policies in developing countries. Perhaps paradoxically, a good deal of attention is given to the formulation and implementation of innovation policies in developed, Western economies, for reasons which will become clear as we proceed. Most notably because the focus and content of STI policy has changed fundamentally in the past two decades. Three themes dominate our discussion:

- the factors influencing innovation;
- the distributed form of modern innovation processes within a division of labour between multiple kinds of knowledge and multiple organisational sources of knowledge; and
- the elements of workable innovation policies.

The importance of this topic for the achievement of international competitiveness and industrial development should not be underestimated. Competition and development are knowledge-driven processes and the conditions and contexts in which knowledge is accumulated and applied in the modern world are changing rapidly.

We must be clear from the outset, that, as a general rule, the STI policy of developing economies should not be directed at reaching the world STI frontier. Rather, the central concern should be with absorption and adaptation of established practice to suite local resource endowments and market prospects. As we will see these are non-trivial tasks. Even imitation and adaptation far from the technological frontier can require major investments in organisations and capabilities.

In addressing this topic we must face a number of difficulties. The first is the vast range of economic performance in developing economies. In terms of GDP per head, or the scale and composition of economic activity, or the relative contributions of the public and private sectors, or the levels of education in general and in relation to science and technology in particular, or in relation to the institutional infrastructure and business culture there are enormous differences between developing countries. Just as there are between the so-called advanced economies. At one end, we see the great success of the newly industrialised economies of South East Asia; at the other end, we observe the continual problems of many of the predominantly agrarian and mineral exporting African economies. In between is a vast range of performance. A moment's reflection is enough to establish that the nature of STI policy will differ widely across this range of developing economies. South Korea will differ from Colombia and what is appropriate for Colombia will not be appropriate to an economy such as that of Mauritius. Similarly, the appropriate science and technology policy for South Korea in the 1990s is quite different from that which was appropriate in the 1960s. Developing economies are adapting systems working within an evolving world situation and the evolution of new policy frameworks is an important part of that development process. A 'one policy fits all' for all time approach clearly will not suffice. There is consequently no best policy independent from time, place and the legacy of the past. Context is fundamental to all appropriate policy endeavour.

Second, the world economic system continues to develop at a rapid pace. For the past three decades in particular, we have seen world trade grow more quickly than real world GDP, and world direct foreign investment grow more quickly than world trade. Indeed between 1950 and the present, world exports have increased six-fold relative to world GDP. Alongside this we see the continual growth of integrated world supply chains for many products, from automobiles to processed foods. This internationalisation of production activity and commerce has always been reflected in an increasing internationalisation of the production and application of new scientific and technological knowledge, particularly between the advanced triad economies, Japan, USA and Western Europe (European Commission, 1998). National R&D efforts are increasingly interdependent, measured flows in the technological balance of payments are increasing as are exports of hi-tech goods, and there is widening cooperation between firms and governments in relation to techno-scientific activity. One important consequence of this is an inter-dependence, albeit little recognised, in the conduct of national STI policies. The continued development of information and communication technologies, and the spread of internet communications will further encourage these trends.

Third, it is important to recognise that these trends follow from the restless nature of capitalist economies and that this property follows inevitably from the central role of knowledge in their operation. Production depends on the use of knowledge but in the very use of knowledge a further change in knowledge is produced thereby opening up new productive opportunities, a never-ending process. Economic development is open-ended because the development of knowledge is open-ended. Capitalism is never in equilibrium, if we mean by that a state of rest, and its development is necessarily uneven in respect to both space and time. Hence the ever-present problems of (uneven) economic development itself, of shifting patterns of comparative advantage and trade patterns, and of incessant structural change within and between economies. Even advanced economies do not develop in a uniform way; rural Wales, Galicia, Southern Italy all speak to the local and uneven nature of development. Indeed, it is one of the distinguishing features of economic development that it produces strong spatial patterns, concentrated around cities and their respective hinterlands.

Before proceeding further some important caveats are in order. It is most important to recognise that science policy differs from technology policy which, in turn, differs from innovation policy as I shall explain below (see Box 4.1). Equally, it is obvious that many other economic policies will have implications for the availability of resources to advance scientific and technological knowledge, for the incentives to do so and for the climate of innovation. A stable policy framework at macro and microeconomic level is of primary importance if innovation is to flourish. Equally, an acceptance of market processes and the rule of contract together with a supportive set of policies in relation to education and skill formation at all levels are required in any knowledge-driven economy (Wignaraja, 1999). In this regard the institutions in relation to property rights, law and public administration matter very greatly. We should also recognise that policies in relation to STI are investment policies in the sense of seeking to raise future levels of GDP per head and to do so in part by enhancing the international competitive ranking of national industries. Consequently they take time to work and they will not be helped by frequent changes in objectives or national commitments. Moreover, the outcomes of such investments are necessarily uncertain and unpredictable, the unintended consequences of policy are part of the process.

Box 4.1 STI policies

Science policy

To manage and fund the *accumulation* of knowledge in relation to natural phenomenon by creation and support of appropriate organisations – research laboratories and universities.

Technology policy

To manage and fund the *accumulation* and *application* of practical knowledge needed for particular productive activities, including transfer of technology from overseas and the transfer of scientific knowledge into wealth creation. Appropriate organisations are research laboratories, universities and firms.

Innovation policy

To encourage the *transfer* of science and technology knowledge into application by ensuring that necessary complementary resources (e.g. capital finance) and knowledge are available, by supporting entrepreneurship and by protecting intellectual property.

Finally it is clear that STI policies play an important but a secondary role in the development process. Rarely is it the case that the objective should be to develop innovation capabilities at the world frontier. More often the problem is one of catching up which does not usually need an indigenous capability to advance frontier science and technology. But nor is the solution one of passively copying the technologies of more advanced nations. Indigenous capabilities are needed to transform and modify to suit local conditions, capabilities which can at later stages underpin the attempt to 'forge ahead', to gain technological independence. Imitation is an active, creative process, it involves adaptation not adoption. Thus, for most developing economies the problem remains one of inward and adaptive technology transfer. Seen in this light one may avoid the danger of expecting too much from science and technology, particularly at the early stages of development. One can then enquire more carefully as to the proper role of government in this area and the appropriateness of different strategic views.

Let me begin by observing that development entails dissatisfaction with the status quo, economies in equilibrium, by definition, do not develop. That development involves ongoing structural change in the absolute and relative importance of different economic activities and that it is premised upon an ever more extensive division of labour, will be accepted without question. In modern capitalism these attributes appear in an extreme form. Capitalist economies are restless economies, there are always reasons to challenge established economic positions, and the primary reason for this lies in the knowledge generating system which is characteristic of capitalism. There is immense micro diversity in the sources of new knowledge. There is a highly developed division of labour in relation to this production of knowledge viewed either in terms of 'disciplines' and 'sub-disciplines' or in terms of knowledge generating institutions. Now systems based on this division of labour also depend upon coordination, and in capitalism this involves a blend of interacting market and non-market institutional forms. Market institutions provide the incentives for change and they also make possible adaptation to new opportunities. I shall say a great deal more about their relative importance and interdependence below, but here it suffices to summarise the developmental system, which capitalism is, as a complex system. It is a system in which the apparent anarchy of individual attempts at innovation is coordinated into the patterns of economic change that have characterised the past two hundred and fifty years of the world economy. Indeed, it is this combination of micro creativity and institutional coordination which leads many modern scholars to recognise capitalism as an evolutionary economic system (Nelson and Winter, 1984; Mokyr, 1990). New knowledge opens up opportunities for new activities that in turn lead to further knowledge in a self-reinforcing, autocatalytic process. As Frank Knight put it so accurately, in societies premised on the division of labour and the role of markets, economic development is a 'self-exciting' process. He might have added that the pace of 'self-excitation' depends crucially on the institutional structure of the economy. Since the growth of and application of new knowledge is vital to this evolutionary process we need to make some careful distinctions between science, technology, and what I shall call managerial or administrative knowledge. This leads us directly to the insight that there this is much more to technology than scientific knowledge and that there is much more to innovation than technology and science.

STI: basic concepts

As the central concern of this chapter is the relation between knowledge and economic development we must begin with some clarification of what these terms might mean. No one doubts that the accumulation of practically applicable knowledge is the foundation of the development process in all societies, rich and poor alike. Nor, I hope less confidently, would they doubt that the relationship is very different for societies at different levels of development and that the policy consequences vary accordingly. All economies are knowledge-based economies, they could not be anything else. What distinguishes different economies is the nature of the knowledge that underpins development at different stages and the different ways in which that knowledge is accumulated and applied to practical effect. In particular there are important differences in the strength and depth of their institutional structures for generating and applying knowledge, whether old or new. To express it at its simplest, the relation between knowledge and development is highly complex and this is so because capitalism in its many varieties is also complex. How is this complexity manifested?

Science and technology

A wealth of recent scholarship has established that science and technology are different and mutually reinforcing bodies of knowledge, created within distinctly different communities of practitioners characterised by different institutional contexts and rules of accumulation. They have in common a dependence on imagination and creativity in the solution of problems, and on the cumulative building of knowledge upon knowledge. *But their differences are profound.* In science the focus is upon the law-like status of natural phenomena at all scales of observation. Its natural institutional context is the academic discipline, and its organisational form is the university or the private or public research laboratory. The method of knowledge accumulation is that of conjectures and experiment, of rejection or provisional acceptance of hypotheses. Moreover, the conjectures are not formulated at random but follow cumulatively from the established state of theoretical understanding. The search is for truth and truth depends on the conformance between observations and theory. Science is open, its results are diffused widely within an international culture of publication and its primary reward mechanisms are closely related to priority of publication and the breadth of impact of the discoveries. To this degree science is an international institution, following commonly accepted procedures and it increasingly involves international collaboration in its prosecution.

By contrast the world of technology is that of the law-like nature of man-made phenomena and its natural institutional context is the profession and its organisational form is the firm. Conjectures and experimentation are just as important as in science but conjecture builds on practical experience and is far less bounded by theoretical speculation. As science seeks after confirmable truth, so technology seeks after practical effect, and practical effect is embodied, in products and processes, in technique. The natural outputs of technology are designs, artefacts and practices and their modes of operation, and their value is judged not by their intrinsic truthfulness but rather by their practical utility. Not, 'Is it True?' rather, 'Does it Work?', that is the question. As science is 'open' so technology is 'closed', at least relatively, with quite different dissemination cultures and a natural concern for secrecy or, where possible, patent protection. In particular, the development of technology and its reward mechanisms depend upon successful exploitation in the economic and social sphere, and the formation of technological conjectures is strongly shaped by those practical experiences. Indeed, the complexity of technology frequently takes its operation beyond the bounds of theoretical understanding, which is one reason why 'disasters' play such an important role in shaping the development of technologies. Many of our technologies are operated in contexts in which experience is the only guide to operational validity and further development. It often is a case of learning as one goes along, producing, applying and using. Now this has a very important consequence, namely that the development of technology cannot be separated meaningfully from the market process in which it is continually tested to meet commercial or social ends.

However, these differences must not be overdrawn, the dividing line between science and technology is often extremely difficult to draw, as it always has been, for example, in relation to medicine. In truth, the accumulation of knowledge defines a spectrum of activities along which the scientific and the technological merge naturally, one with the other. Modern science and technology are becoming increasingly interdependent. Wherever an understanding of the natural world is relevant to an understanding of the practical world this will be so. Thus, developments in science may open up new opportunities for technology and equally, the converse is true; the demonstration of a technological effect can stimulate the search for the underpinning natural principles. This is as much true of the discovery of say iron or steel as it is of the transistor. This is one reason why a substantial number of firms engage in pure scientific research. Their competitive position depends on an understanding of relevant sciences and so they conduct science to solve their own problems, and, of equal significance, so that they can interact with and draw upon the far wider world of science beyond their own laboratories.

Understanding innovation

The first point to note in answering this question is that innovation involves much more than knowledge of the relevant science and technology. At least since Schumpeter (1911) economists have accepted a distinction between the formulation of a working idea for a product or process (an invention) and the

Box 4.2 Invention, innovation and diffusion

Invention

The conception and *realisation* of a working design for a product or process, or an improvement to a product or process. If sufficiently novel can be patented.

Innovation

The *application* of invention to economic activity, that is to say, the *economic use* of an invention. Normally restricted to the first example of economic use.

Diffusion

The *spread* within the economy of an innovation, the process by which innovations gain economic significance. Diffusion invariably leads to the modification and development of an innovation. Sometimes equated with imitation, the process of *copying* ideas from application to another.

Radical vs Incremental innovations

Radical innovations open up *new design spaces* in the innovation process. Usually involve new principles behind the product or process. Incremental innovations *explore* this design space with 'small' step-changes in working principles or design performance.

---

application of that idea to the economic process (an innovation). They distinguish the wider application of an innovation beyond its originating firm by the term diffusion. (See Box 4.2.)

Innovation requires much more in the way of knowledge than science and technology. It requires a sound judgement of what potential users might demand in a product and what they would be willing to pay. It requires an ability to organise the production process, to acquire the appropriate inputs at economical prices and to manage the new activity. It requires the ability for creative conjecture well beyond that associated with the advancement of science and technology. The concept of entrepreneurship captures this well. The entrepreneurial function is to bring together market opportunities with scientific and technical opportunities. It requires an ability to combine conjectures and knowledge from these different sources, to see in them a new profit opportunity, and to carry this opportunity into practice. Without a capability to combine together these complementary kinds of knowledge innovation does not occur. This is especially so with many new technologies that draw upon information from multiple disciplines and sources. New managerial, organisational and market knowledge is also a highly practical knowledge, like engineering production knowledge it accumulates on a trial and error basis and is only weakly guided by theoretical supposition. It has a much greater claim to be tacit, localised knowledge and it is certainly deeply connected with the market process.

Before proceeding further it will be helpful to summarise some of the important functions of this innovation process. First and foremost is its *unpredictability* arising naturally from the two concepts that define any innovation, change and novelty. Unpredictability implies uncertainty and an inability to predict with any accuracy either the contributing elements in the innovation process or the uses to which innovations are put. Since all innovations are business experiments within a wider process of knowledge discovery, the unexpected plays a more than a usual role. Nevertheless, we know a good deal about the kinds of phenomena that define the innovation process.

Innovations are not best understood as isolated events. Rather they are located in sets of innovation opportunities from which *sequences* of innovations typically emerge in a *cumulative* fashion (Utterback, 1996). Many of the innovations will be *incremental* improvements in current practice, a much smaller number will be the *radical* innovations which open up whole new fields of opportunities. Consequently one of the features one expects of any internationally competitive industry is the ability of the firms within it to sustain a trajectory of innovation, not their ability to make a single innovation. Single innovations give only transient competitive advantages and, often, rivals who understand the significance of maintaining the momentum of innovation overtake the pioneering firms.

In assessing the factors that shape innovation it is convenient to distinguish four elements; *opportunities*, *incentives*, *resources* and managerial *capabilities*. The significance of these categories is that they become the *targets of innovation policies*.

The opportunities depend on the combination of technological and market ideas to identify a new product, process or method of organisation. The incentives depend on the expectation of profits sufficient to compensate for the risks in relation to the capital invested. The resources include not only the elements of formal R&D but also all the complementary assets required to transfer ideas into practice. The capabilities relate to the knowledge skills and organisation of firms involved in the management of the innovation process. Innovation capabilities are a distinctive type of capability, involving the management of knowledge and change, additional to the capabilities in relation to production, investment and interaction identified by Lall (1987), although, clearly, they overlap to a considerable degree. Innovation policies can be defined in relation to all four attributes as we show below.

One of the most important factors governing the generation of innovation opportunities is the fact that much of the relevant knowledge lies outside the firm either in suppliers or customers or research institutions such as universities.

An ability to gain access to and absorb external knowledge into the firm is crucial in a world where even the largest firm cannot accumulate all its knowledge in-house. Innovation is more likely to occur when firms are located within a rich knowledge base and when they have developed the skills to interact with this knowledge base. Consequently, networks play a very important role in the innovation process. Some of these may be concentrated geographically, so-called clusters, others may be distributed nationally and internationally. Again we shall see below that network formation is an important dimension of innovation policy.

There has been considerable debate about whether the stimuli to innovation reflect demand-pull or science-push in the innovation process (Mowery and Rosenberg, 1973). It is now accepted that both views are mistaken, it is the interaction between push and pull which matters and this interaction is reflected in the multiple kinds of knowledge required in the innovation process. Adding to the stock of scientific knowledge, without making the complementary investments in supporting technological, managerial and market knowledge simply leads down the path of rapidly diminishing innovation returns. Some idea of the importance of these complementary activities is given in Table 4.1 based on a selection of OECD countries. It shows that R&D expenditures, on average, account only for one-third of total innovation expenditures, and that on average a quarter of total expenditures are incurred outside of the firm. Equally, market opportunities remain unfilled if the innovation capabilities are missing.

Table 4.1 Breakdown of innovation expenditure (percentage share)

	R&D	Patents and licences	Product design	Market analysis	External spending
Australia	35.1	4.1	—	7.6	—
Belgium	44.7	1.5	11.3	6.6	21.2
Denmark	40.1	5.3	15.8	8.2	9.0
Germany	27.1	3.4	27.8	6.1	29.2
Greece	50.6	6.4	—	13.2	11.7
Ireland	22.2	4.3	22	38.5	20.4
Italy[a]	35.8	1.2	7.4	1.6	47.2
Luxembourg	29.3	8.9	8.4	4.3	26.4
Norway	32.8	4.2	14.2	5.5	17.6
Portugal	22.9	4.1	24.5	5.4	16.8
Spain	36.4	8.0	—	8.8	6.3
The Netherlands	45.6	6.1	7.6	19.8	20.2
United Kingdom	32.6	2.7	28.4	8.9	15.9
Average	33.5	4.6	24	6.6	22.4

Sources: Bosworth et al. (1996), Community Innovation Survey Data, ISTAT (1995), Australian Bureau of Statistics (1994).

Note
a Adjusted according to ISTAT. Data do not total 100 per cent, as 'other expenditures' are not included in the table.

It follows from the above that we cannot treat the categories of invention, innovation and diffusion as a logical, temporal sequence with invention first and diffusion last. The stages interact: knowledge gained in the diffusion process stimulates further invention that stimulates additional innovation in never ending sequences. In the case of most technologies we observe streams of *multiple innovations* which are shaped by the process of application and diffusion, by the interaction between technological possibilities and market opportunities (Bell and Pavitt, 1993). Thus, innovations are in practice *sequences of related improvements*, a stream of developments within a particular technological and market context. This is not really surprising. The growth of knowledge reflects the emergence of particular problems that act as focusing devices to guide enquiry. Some of these problems are internal to the science or technology, others arise from experience in production and use of the particular devices. Either way they give rise to the *cumulative* nature of scientific and technological advance.

To make the best of these distributed forms of knowledge requires that this division of labour be coordinated and that the institutions needed to achieve this are in place. Second, the returns to investment in innovation fall sharply if the complementary sources of knowledge are not properly coordinated. Third, many of the important complementary types of knowledge constitute practical knowledge of market needs, of how to organise production and distribution, with little relation to science and technology as normally understood. Finally, there is an important complementarity between the two principle ways knowledge is acquired in relation to innovation, through the experience of the market process and through formal R&D programmes. *It is the bringing together of these two complex ways of learning that I define as a central problem of STI policy in developing economies.*

By way of summary, a policy for innovation cannot be reduced to a policy for science or even technology and on this misunderstanding has foundered many a promising initiative. Innovation policy is necessarily broader. It must address the availability of complementary assets and knowledge. It must address the supply of skilled labour and the supply of risk capital. It must address the ways in which those with knowledge of science and technology can be brought together with those who have organisational and market knowledge. It must address the incentives to innovate and, most fundamentally of all it must address the capabilities of firms to manufacture with new technology and to market new products.

STI policy: underlying principles

Policies do not exist in a vacuum nor do they emerge at random, they are always grounded in a wider set of beliefs about the world: those beliefs in relation to economic activity have played an important role in shaping the practice of STI policy. Behind these developments are two very different accounts of market economies. One set of principles is defined by the economic theory of competitive equilibrium, it focuses upon the efficiency with which the market system allocates given resources to competing ends. In a perfectly competitive price system, the prevailing prices measure and thus equate the marginal valuations placed on commodities and resources by producers and consumers. Such a system has quite a remarkable efficiency property; neither is it possible to produce more of any commodity without sacrificing some of another commodity, nor is it possible to increase the welfare (utility) of any agent other than by reducing the welfare of some other agent. The intellectual force of this 'Pareto Principle' cannot be underestimated since it underpins many of the ideas in relation to tariff and tax policy in the world economy. However, as we shall see it also leads to strong implications in relation to the efficiency with which knowledge is produced and used in market economies.

A second contrasting set of beliefs is associated with the Austrian and evolutionary schools of economic thought. It was Hayek (1948) who put the problem of knowledge at the heart of his economic analysis and who argued that market systems have developed as solutions to two distinct but interrelated problems. The first is the idiosyncratic individual nature of knowledge, its distribution among all the actors in an economy and the impossibility of any one mind comprehending in total the knowledge of what individuals want and what they can do. Markets and the price mechanism 'solve' this problem of knowledge dispersal. The second and related problem is that of learning and the growth of knowledge. Markets provide a framework for experimentation and the trial and error formulation and testing of business hypotheses, and they provide a means of adaptation in which new events require new knowledge for their solution. In this perspective, development and the growth of knowledge are inseparable adaptive consequences of market institutions, which like all institutions are best evolved as the outcome of a trial and error process. To say that economic growth and development depend on the accumulation of knowledge is also to say that they depend upon a competitive process. The micro diversity of creative behaviours which innovation reflects has its economic impacts through market relations in the competitive process. Successful innovators develop new products and or processes that enable them to attract customers and resources from rivals. In this process growth and profitability are closely linked and the outcomes depend very much on the operation of the prevailing market institutions. We shall see below that this makes competition policy a natural complement of innovation policy (Metcalfe, 1998). Firms can compete in many ways, some of them socially unproductive. Competition policy can guide firms to compete in productive ways, of which competition through innovation is the most beneficial to economic development. The profitability of firms is not only crucial in connection with the ability to expand and attract scarce resources, it is also crucial to their ability to fund investments in knowledge creation and thus maintain a sequence of innovations. The directions in which technology advances depend very much on who is successful in this market process.

This line of thinking has become vital to the development of evolutionary accounts of market activity (Nelson and Winter, 1984; Metcalfe, 1998). Evolution depends on micro diversity of individual behaviours and the market processes that resolve that variety into patterns of economic change. In this view, markets are devices for communicating information about what is available on what terms and firms are devices for deciding what is to be produced and how. Micro diversity is in turn created by acts of innovation and acts of innovation depend upon the idiosyncratic development of knowledge. These processes will not be efficient by the canons of the Pareto Principles, for they inevitably involve elements of failure, of waste. However, they will be creative, *and it is to this creativity that evolutionary economists point in charting the rise of the Western economies* (Rosenberg and Birdzell, 1986).

Thus, the dynamics of capitalism is a reflection of its creativity in generating and applying new knowledge to economic problems that are largely self-generated. Capitalism is not then the particular state of affairs emphasised by equilibrium theory but rather a process of change, a process of discovery with particular properties. It is in all relevant essentials a development system. Both development and the competitive processes depend on the imperfect distribution of knowledge that in turn is a reflection of the division of labour.

These different views on the nature of a market economy lead to two very different justifications for STI policy, namely market failure and system failure. We explore each one in turn.

Market failure

Let us turn first to the problem of knowledge in terms of the competition equilibrium theory of resource allocation. It was Arrow, in a seminal paper (1962) who drew attention to the peculiar economics of the production and use of information. From this has come the principal modern justification for STI policy, the doctrine of *market failure*. Arrow fully recognised the fact that information is not knowledge and that the peculiar economics of information *qua* commodity, have deep implications for the role of the market mechanism in the generation and application of knowledge. What are these peculiarities?

First and foremost, information has the property of *a non-rival good*, the same ideas may be accessed and used any number of times by any number of people. Information is used but it is not used-up. This is true whether it is used to produce goods and services or to produce more knowledge. In this it also has properties akin to a public good and it had long been understood that market systems undervalue the true social worth of public goods. Second, the value of an idea is highly *uncertain* and the economic system lacks the depth of future markets to give the necessary comfort to stimulate investment in information production. Since one cannot foresee the future one cannot know which innovations will emerge or how needs will evolve, there is no basis for writing future contracts to trade commodities not known about. Here the probability calculus does not help, one cannot transform uncertainty into risk when one cannot write down all the options which will define those risks.

Third, and reinforcing the first two points, it is extremely difficult to establish *secure property rights* to protect the producers of ideas. There is a natural tendency to experience spill-over information externalities, which allow individuals to benefit from the knowledge investments of others while avoiding the costs required to make those investments. The 'theft of ideas' thus undermines the incentive to produce ideas. Why sow when others will reap?

Fourth, the production of ideas is subject to significant *indivisibilities* in terms of the investment required to generate that information, and indivisibilities give rise to scale economies in the application of knowledge. One cannot have half an innovation, all the ideas must be present for it to work. The consequences of this are profound. While the non-rival nature of information suggests that it be widely diffused at a nominal communication cost, such a pricing regime would mean that the producers of these ideas could be unable to cover their fixed costs of information production. Marginal cost pricing will not work to efficiently distribute ideas and simultaneously cover the costs of production. As many economists have understood such indivisibility gives rise to increasing returns and monopoly.

Fifth, the production of knowledge creates *asymmetries* in what is known by buyers and sellers, and asymmetries can lead to opportunistic behaviour, adverse selection and moral hazard. It is then difficult to create incentives for each side of the market to behave in an efficient way. Familiar examples of this problem are provided by the markets for commodities with unknown characteristics (the lemons problem) or by insurance markets where the seller of insurance can neither observe nor control the behaviour of the insured. Now the essential point about innovation is that it, of necessity, requires information asymmetries; an innovating firm knows and acts in ways different from its rivals, and the outcomes are always uncertain.

Finally, Arrow pointed to a paradox that strikes at the heart of the idea that efficient market transactions depend upon their property rights. Imagine, he suggests, that you are to sell an item of information. Quite reasonably the purchaser needs to know what the information is before she can place a value on it and decide whether or not to meet the asking price. Thus, for the transaction to occur the information must be divulged in advance. But then, once divulged, why should the purchaser pay. Rather like a market in 'lemons' the transaction process seems to self-destruct.

To the extent that all economic activities require prior and continuing investments in information production and the translation of information into knowledge, the conclusion that necessarily follows from the above is that no activities can be organised in a Pareto efficient fashion (Stiglitz, 1997). Capitalist economies will at best be imperfectly competitive in the sense made clear by Edward Chamberlin. The normal mode of organisation in an ideas-based economy is monopolistic, and prices have to stand above production and distribution costs by at least the degree necessary to cover and reward the costs incurred in generating the underpinning knowledge as well as the normal costs of production and distribution.

If these are the principles behind market failure what are the consequences for STI policy? Consider first basic science as a type of knowledge whose areas of potential economic application are highly uncertain, and highly diffuse (Nelson, 1959). Private firms, it is argued, will not invest in producing pure science, a market solution will not work. Consequently pure science has to be funded by the state and be prosecuted in non-commercial institutions. This is broadly what we observe, basic science and basic technology is the preserve of universities and dedicated public laboratories. Moreover, the institutions of science are particularly favourable to its having the impact one would hope from the production of a non-rival, quasi-public good. Scientific awards are allocated according to priority of publication, and publication in international journals is a device to disseminate that information at a minimal, marginal communication cost. This is just as true for work in basic technological research.

Consider next the related problems of spill-over externalities and property rights in ideas. The solution here is the patent system and the copyright system. In return for public disclosure of relevant information, a patent holder is given a limited term monopoly right to use or license the information within a particular domain of application (the scope of the patent). Information is placed in the public domain and the inventor can extract a reward for her efforts. Notice though that the reward is not necessarily linked to the cost of inventing nor does it typically capture more than a fraction of the wider social value of an invention. Neither are the patents the only way to protect intellectual property, secrecy or a rapid rate of innovation, nor the complexity of the invention, often, more than effective barriers to imitation. Thus, the required degree of patent protection and the use of patents varies very greatly from sector to sector. In pharmaceuticals they are vital elements in innovation, in the engineering industries they are not.

These institutional devices, patents and the public funding of open science, are remarkable in themselves and they reflect the sense in which the Nelson and Arrow arguments are exactly right. However, to link this to a general presumption of market failure in relation to innovation is simply a mistake. *Rather it is clear that many of the alleged sources of market failure are, in fact, essential for the market process to work at all.*

Here it is important to recognise again that information is not knowledge and that economic activity depends directly on the latter, not the former. This has nothing to do with the less than satisfactory tendency to equate information with a flow and knowledge with a stock. Rather it reflects the much deeper point that a flow of information will reflect a certain state of knowledge in the sender and may generate quite a different state of knowledge in the recipient. It is not knowledge that flows between them but a message embodying the intended information. *While information can be public, knowledge is not naturally so.* What one learns from particular information depends on one's prior state of knowledge and this is necessarily *idiosyncratic and individual*. Thus, the ability to interpret information messages is not to be taken for granted nor is it a costless process. To understand information one must make the necessary investments in background knowledge. This takes time and resource, and, since both are scarce, we cannot invest in everything, it follows that the emergence of specialised knowledge is a consequence of and reinforces the division of labour in society. As Rosenberg (1990) has indicated this is the reason firms in high technology activities make major investments in basic science and technology, not only to develop their knowledge internally but also to interpret the flow of external information and hold intelligent conversation with its producers in universities. That knowledge is non-rival we can agree; that it is publicly accessible at a negligible price once it is produced is a far more doubtful proposition.

Here we can see an important weakness of the Arrovian framework. Since information may be disseminated readily at negligible cost, it treats knowledge in the same way, as if it were part of the atmosphere, or, as others put it, readily available off the shelf. This assumption is very far from the reality: the substantial costs of turning information into knowledge means that knowledge is not readily available to all. *Knowledge is sticky and it does not flow like water to find a uniform distribution. If it did flow uniformly, it would be difficult to explain why the development paths of countries are so different.* This is true of basic science and technology just as it is of more applied knowledge. Scientists and technologists are necessarily specialists in what they know, they often have very limited abilities to claim expertise outside of their competence. Even within disciplines, access to knowledge is subject to substantial barriers, barriers which become greater the further one's knowledge is behind the frontier. If one wanted further proof of the significance of this distinction between information and knowledge one need look no further than the current complaints about information overload in modern society: more information than can possibly be translated into useful knowledge.

It does not follow from the above discussion that there cannot be *workable* markets in information. These have always existed. Books, newspapers, compact disks are all devices that embody non-rival public information in rival physical goods and make market transactions possible. Moreover, the provision of scientific and technical knowledge on a commercial basis has for at least two centuries been the basis for viable business activities. Consulting chemists and engineers, contract R&D companies, and more recently, management consultants are each examples of the market provision of specialist information. The Arrow Paradox does not destroy this market. Contractual arrangements are readily devised and problem-solving capability readily becomes a matter of reputation and trust. These markets may not be Pareto efficient, the information providers may act as limited monopolists but that this is surely better than having no information markets at all.

Thus the thrust of the argument is that market failure can only be part of the rationale for policy. Moreover, it is clear that uncertainty and information asymmetries are necessary for the market process to work in knowledge-based and innovation-driven economies. It is perverse, consequently, to identify them as sources of market failure.

Consider the problem of uncertainty, which all agree is the essential characteristic of the innovation process. It cannot be avoided and to suggest that this stands in the way of a fully articulated set of futures markets is simply irrelevant for the real competitive process. To eliminate uncertainty one would need to eliminate innovation – scarcely a sensible policy stance. This is also the case with asymmetries in information and knowledge. Far from being a nuisance they are, in fact, essential to the innovation process: innovation is exactly the process of creating and trading an information and knowledge difference between rival firms. These particular kinds of asymmetry cannot be labelled market failures if the market process cannot operate without them. Here we see the source of the difficulty. Market failure has been judged by the standards of equilibrium resource allocation. Innovation, however, resides in the world of market process not market equilibrium. It is essential to that competition process and indeed it is the combination of innovation and a competitive process that delivers economic development.

Consider next property rights. What is important here is the comparative weakness of the patent system: it protects against pure imitation but it does not protect against rival invention based on different principles, and rightly so. Capitalism depends for its development on the principal that every economic position is open to challenge. Thus, while patents can be important, it is equally necessary that their scope not be drawn so widely as to make it too difficult to invent around an established idea. As with many domains of policy, difficult trade-offs have to be identified and exploited.

System failure

We turn now to the second kind of rationale for STI policy. We have seen above, that a central feature of the innovation process is the division of labour in the production of innovation related knowledge. That innovation is not a relay race proceeding sequentially from science to market but that it is more like a basketball game in which all players contribute their different skills at different points in time. This leads us directly to the idea of *innovation systems and to systems failure as the rationale for STI policy.*

The central idea is straightforward. As innovation system is a set of interacting organisations charged with the production, communication and storing of all the elements of specialised knowledge required in the innovation process (Freeman, 1987; Lundvall, 1992; Nelson, 1993; Carlsson, 1995; Edquist, 1997). As systems are formed from components and interactions between those components we can think of innovation systems failures in two ways. *An STI systems failure arises whenever access to needed knowledge is prevented either because the appropriate organisation to produce or give access to that knowledge is missing, or because the linkages to communicate ideas between the respective organisations are missing or operate defectively.* Then STI policy becomes a problem of institutional design, a problem in building the appropriate social capabilities to realise the potential for development (Abramovitz, 1989). It is this aspect of STI policy that is of particular relevance to developing economies.

Firms are obviously key players, directly, and indirectly through their roles as users of technology and suppliers of technology in the innovation process. So are universities and public and private research laboratories, professional societies and consulting firms. Indeed in any knowledge-based economy there is a rich network of organisations that contribute to innovation. Some of them are national in domain of influence some of them are specific to particular sectors of economic activity.

Now what matters for the operation of the STI system is how these different organisations interact, how the knowledge generated in one part is communicated to another part where it is combined in the process of producing yet new knowledge. The system becomes a framework for compound learning, what one organisation can learn depends in the learning ability of the other organisations in the system. In this way the system provides for the collaborative activity necessary to produce innovations from the combination of different hands of knowledge. Scholars recognise these institutional arrangements as components of the social capital or social capabilities of an economy (Edquist, 1997; Fountain, 1998). Relationships based on reputation trust and reciprocity enable the benefits to be gained from multiple sources of learning, from group problem solving and from working together for mutual gain. These relationships are based on transactions within networks and within markets and they provide the basis for the collaborative and cooperative development of innovation capability. They are an appropriate response to the increasing technological diversity and complexity of the innovation process; a reflection of the need to combine multiple kinds of knowledge created by multiple organisations. The division of labour in knowledge production reduces the society-wide cost of knowledge generation, social capabilities enable this division of labour to be coordinated in the innovation process.

Now this coordination process is not easily achieved. Specialisation of purpose can result in incompatibilities in incentives and difficulties in the communication of knowledge. Knowledge is sticky, and there have to be receptive capabilities as well as transmission capabilities, or, putting it differently, intelligent users as well as intelligent producers of knowledge. This requires investment throughout the innovation system. It may also require the creation of specific bridging organisations to create, for example, the interface between firms and universities or public research laboratories.

From this system failure perspective, the role of policy in the innovation process is clear. It is the embedded nature of firms in a wider network of knowledge producing organisations which matters. While the market failure perspective focuses on lack of incentives to invest in innovation in the single firm, the systems failure perspective points to the creation of opportunities and capabilities in a cooperative fashion. Firms remain the key actors in the innovation process but their knowledge generating capabilities are greatly enhanced by their being embedded in a wider matrix of knowledge generating organisations.

It is clear that in the past two decades with Europe, and to a lesser extent the USA, the balance of policy has shifted markedly in favour of the systems perspective. Indeed a recent OECD report (1999) defines a new agenda for innovation policy focused upon the development of what I called above social capabilities. This recognises the importance of an innovation culture, the need to promote networks and clusters of the relevant organisations and the opening up of the science base to new patterns of entrepreneurship.

A framework for STI policy choices

We have drawn attention already to the fact of the diversity of conditions between developing economies and the implausibility of applying similar STI policies in all conditions. For low-income countries the principal policy should be one of learning via imitation, achieving inward technology transfer typically through importing the appropriate machinery, product designs and manufacturing procedures and through accepting complementary direct foreign investment. A point will be reached, however, when this passive policy is no longer appropriate and an innovation possibility threshold is passed in one or more sectors. The problem passes to a more active phase of technological learning, in which adaptation replaces adoption to build a capacity for incremental innovation not least to fit technology more closely with national market needs and the resource base. This is more likely in middle-income economies when market processes are working, the public finances are sound, export markets are securely established that there is a well established educational infrastructure and a well distributed and adequate level of economic competence to identify, develop and exploit business opportunities. It is in these circumstances that policy choices in relation to STI arise, and it is this case that I focus upon.

STI policy choices

As with many policies the primary question involves an understanding of the relative roles of the public and private sectors in the innovation process and a strategic assessment of how national activities in agriculture, industry and service activities are to be developed. Once answered, the secondary questions relate to identifying national deficiencies in relation to the opportunities to innovate, the resources to innovate, the incentives to innovate and the capabilities to innovate.

Innovation indicators

To achieve this level of understanding requires that the policy maker have access to *appropriate indicators* of the state of the national innovative effort. A good deal of effort has been put into developing appropriate indicators in the OECD countries and a sample of the most important ones is given in Box 4.3. These are divided into three categories, in relation to inputs, intermediate outputs and final outputs. Input indicators cover R&D activity either in the form of expenditures, employment of qualified scientific and technical personnel or lists of

Box 4.3 Innovation indicators

Inputs

R&D expenditures	a Annual rates
	b Cumulated expenditures net of depreciation
Qualified scientists and engineers	a Working in R&D
	b Working in production and marketing

Intermediate outputs

Counts of scientific papers
Number of patents
Number of expenditure on collaborative innovation projects
Number of public and private R&D laboratories in operation
Citation analyses of patents and papers

Outputs

Productivity growth statistics for firms and sectors
New product launches
Stock market valuation of intangible R&D-related assets
Fraction of sales from products launched in the past 'X' years
Diffusion measures of the use of new technologies

---

projects and programmes. Their chief limitation is in knowing the quality of the inputs, which are necessarily very idiosyncratic in the case of the people involved, and the effective organisation of research teams. At best these issues can be assessed indirectly but should include measures of public and private inputs and measures of engagement with the wider world of science and technology, for example, through attendance at seminars and conferences. Intermediate indicators include patents and scientific and technological papers. It is well recognised that the quality of patents varies enormously and that different industries place very different weights on patent activity, nonetheless, they remain a tolerable measure of inventive activity. Publication-based measures can always be accompanied by citation analysis, due allowance being made for time-lags. Final output indicators are the least well developed, again because of quality problems. These can include lists of innovations, measures of diffusion of technology and measures of new business formation to commercialise innovations. When used with care, *benchmarking* of firms against each other and foreign rivals can provide useful information on performance gaps.

Table 4.2 Government support for industrial technology by type, 1995

	Australia 1993	Canada 1995	Finland 1996	France 1995	Germany 1993	Japan 1995	Mexico 1995	The Netherlands 1995	UK 1995	USA 1995
Fiscal incentives										
Fiscal incentives	38.9	46.9	0.0	8.8	0.0	1.8	0.0	25.0	0.0	6.2
Grants and forgiven loans	14.1	9.7	42.7	14.6	28.0	1.2	2.3	12.0	4.9	15.5
Others	0.0	0.4	2.7	0.0	0.0	1.3	0.0	0.0	0.0	0.0
Total	53.0	57.0	45.4	23.4	28.0	4.3	2.3	37.0	4.9	21.6
Mission-oriented contracts and procurement										
Defence	9.7	4.7	0.0	35.6	19.5	8.3	—	7.4	61.2	58.8
Space	0.2	9.8	7.4	19.4	11.2	7.5	—	12.3	4.5	8.7
Others	0.0	14.8	0.0	4.3	1.8	10.9	—	1.7	7.3	9.4
Total	10.0	29.3	7.4	59.4	32.5	26.6	—	21.3	73.1	76.9
S and T infrastructure										
Technology institutes, etc.	28.8	5.6	34.7	0.9	13.7	21.6	14.3	11.0	2.6	0.5
Academic engineering	0.2	0.0	0.0	0.0	1.6	0.0	—	0.9	6.3	0.0
Others	8.0	8.1	12.5	16.4	24.3	47.5	83.4	29.7	13.2	0.9
Total	37.0	13.7	47.2	17.2	39.5	69.1	97.7	41.7	22.1	1.4
Grand total	100.0	100.0	100.0	100.0	100.0	100.0	100.0	100.0	100.0	100.0

Sources: OECD calculations based on R&D database, PSI database and information supplied by Member countries, March 1998.

Any information system is only a prelude to analysis and action. The next step is to identify relevant technologies, singly or in combination, to decide which firms and other research organisations are to play an innovative role and whether this is to be reflected in the identification of particular innovation projects and programmes. The policies can be general or they can discriminate between sectors, technologies, firms or projects. R&D tax credits, for example, are an entirely general policy, applying in principle to all firms in all sectors. Project-based R&D support is at the other end of the spectrum, being highly specific in its application.

There is, however, a simple and useful way to categorise alternative kinds of policies. This involves distinguishing between policies that take innovation opportunities as already established and needing only to be realised, and policies that are designed to create those innovation opportunities. The two groups are, of course, complementary not mutually exclusive. The first group includes policies that deal with market failures and the second group with policies that deal with system failures. Table 4.2 provides a breakdown based on OECD data of support for industrial technologies. Fiscal incentives fall in our first category and infrastructure policies in the second, while mission oriented policies may fall in either group. The considerable differences between countries in their policy mix is immediately noticed.

Group 1: *policy with given innovation opportunities*

R&D subsidies

The first case to consider is when the innovating firm does not have available the internal financial resources to fund the profitable projects, nor can it raise the money in the capital market for well known reasons in relation to risks, imperfect information and adverse selection. In short it cannot convince potential backers that its hopes are justified, nor can it offer sufficient collateral against the required loans. Its resources are too small relative to the options for innovation that it can identify. Here lies the case for government support, typically in the shape of a specific project grant, a fraction of the project costs (often 50 per cent) is made available to the firm which, if successful, will be repaid in part or in full. Subsidised R&D loans from banks, and R&D tax breaks are alternative ways of achieving the same end, namely to guide more resources into innovation by reducing the marginal cost of R&D activity (Metcalfe, 1994). This type of policy is often of utmost importance in relation to innovation in small firms. Usually the subsidy applies directly to R&D expenditures but it can equally be directed at the employment of R&D personnel.

R&D tax breaks have a number of attractions, not least in that they do not involve government in making micro decisions on particular innovation projects. Most of these schemes involve treating R&D as an allowable expense for tax purposes and granting a tax credit on a fraction of these expenditures. What is to be included as allowable expenditure is not always transparent. A good case can be made, for example, for including market identification and development expenditures under the broad heading of R&D. While there are obvious dangers in the encouragement of creative accounting, there is some evidence that tax incentives can be effective. Although, clearly, their effectiveness depends on the efficiency of the prevailing tax administration.

Public purchasing

A final type of policy is found in public purchasing and market developing policies more generally. Since innovations require indivisible investments in their realisation it follows that their exploitation gives rise to increasing returns. The bigger the market the lower become the average costs of innovation. The same principles lead us to more general policies in support of the demand-side of the innovation process. Public purchasing can have a very effective role in supporting demonstrator projects to establish feasibility to users, and more generally, in providing innovation products for public services such as health and utilities which have to be supplied domestically. Moreover, in relation to metrology, quality assurance services and standards, the government can act as the proxy customer for what are essentially public goods. It is the same argument that holds with respect to basic science and basic technology research. On the demand-side, governments have a positive role to play.

There is a related way that policy can stimulate the innovation process, namely by export promotion policies that create awareness of and incentives to exploit foreign markets. Indeed any policy that increases penetration of foreign markets will help to encourage innovation. As pointed out above, knowledge of market possibilities is an essential component in the definition of innovation opportunities.

Group 2: policies to identify innovation opportunities

However, fiscal incentives of this kind address only one dimension of the innovation problem, namely resources. In many other cases the lack of awareness of opportunities and managerial capabilities will be of far greater importance. It is knowing how and where to innovate is the problem. We can interpret the negative innovation stance as a lack of economic competence in either of two dimensions. Either a firm lacks the design manufacturing and marketing capabilities, so that its projects are not profitable in the prevailing competition situation. Or, alternatively, it lacks the managerial ability to carry out innovation projects. In response to these problems, innovation subsidies of any kind are not the answer, and this is where the innovation systems perspective comes into play.

The crucial point is that the relevant capabilities or knowledge lie outside the firm because we can take it that it is not operating at the world frontier. Then the problem is how to access the necessary ideas, and here collaborative arrangements are potentially of great importance.

Collaborative innovation

We have already pointed to the multiple kinds and sources of knowledge that characterise modern innovation activities, their embeddedness in *distributed innovation processes*. The policy issues here are of two kinds. Is the STI infrastructure sufficiently well developed for current innovation needs, and, are there appropriate academic and other research facilities for the needs of local industry? Second, if the infrastructure is satisfactory are the networking arrangements and incentives in place to support local innovative activities?

For developing economies these are likely to be the key issues. Investments in infrastructure need to be the prime aim of policy. But then the organisations so created need to connect with the rest of the economy in a range of collaborative activities. The following are the more important examples:

- collaborations between firms suppliers and customers to develop new technologies;
- collaborations between firms and local STI institutions;
- collaborations between local STI institutions and overseas universities and laboratories, in part to promote the exchange of research staff;
- collaborations between local firms, STI institutions and foreign multinationals to transfer capabilities in jointly executed projects.

In respect to each of these possibilities *specialist local research organisations* (SROs) have an important role to play by being the bridge between different contributors to innovation. They can act as focal points in the innovation process in a number of ways. By collating, codifying and disseminating knowledge on the industry's technology, thereby raising awareness. By providing technological and innovation management services for firms; by engaging in pre-competitive research projects and supporting the innovation projects of specific firms and, by acting as a bridge between firms and other knowledge-based institutions such as overseas universities. In particular, they can organise collaboration research projects in an industry bringing together the viewpoints of different firms and sharing the costs of innovation. Thus, SROs can act as organisers of innovation networks within supply chains and between firms and other knowledge creating organisations. They can be the most effective institutions to coordinate the division of labour in the innovation process. By encouraging the demand for innovation they justify expenditure on the supply of innovation.

We also find the importance in an innovation systems perspective of policies in relation to training, education and research. When feasible, it is obviously sensible to access the work of overseas innovation systems through secondments or joint research projects with foreign universities and research institutes and by using the R&D facilities of foreign firms that have invested locally.

Technology infrastructure

One of the key lessons of the innovation systems perspective on policy is that the government has responsibility to develop a country's technological infrastructure.

This is not just a question of supporting advanced research and education activities in universities and specialist research organisations. A particularly important aspect of this is contained in the need to support an infrastructure of *metrology, testing and standards* activity. There is no area of productive activity that does not require the use of accurate measurement techniques and this dependence increases as technologies become more advanced and dependent on the interconnection of multiple components and systems. The creation of national standards, metrology and testing services is essential for economic development and this falls on governments in all the advanced countries. Metrology is the classic example of information as a public good and is to be funded by the state though not necessarily managed in all its dimensions by the state. In the UK, for example, public laboratories hold the fundamental standards in relation to measurement while a network of private laboratories is accredited to provide metrology and testing services for industry.

The importance of these issues is difficult to overestimate. Accurate measurement and the ability to meet standards is essential if a country is to compete in international markets, it is essential to the design and development process for new products and processes, and it is essential to the successful conduct of R&D activity at any level. Metrology and related services are central elements in any country's innovation system. Policy must not only establish these services it must ensure that they are coupled with training activity and that procedures are in place for the effective diffusion of this information. It is no accident that the reorganisation of the service in the USA in 1988, with the creation of the National Institute of Standards and Technology, was legislated for in the Technology Competitiveness Act of that year. Competitiveness depends on standards and standards depend on metrology (Tassey, 1992).

Lessons from different countries

Following this rather long discussion of the principles behind innovation policy it will be helpful to turn to three specific cases, the UK, South Korea and Colombia, since they provide insights on policies at different stages of development.

The UK

It will be instructive to begin with the UK since it has a well-established STI infrastructure, dating from the immediate postwar years. Yet, there is a sense that this system does not contribute to innovation in the UK as well as it might and there is a continual search for policies to enhance this innovation system. In the early 1960s, the UK's STI system reached the climax of its first postwar phase. This system is built around a broad division of labour. Universities are funded from the public purse primarily to carry out basic research in science and technology. The policy missions of government departments were supported by publicly funded laboratories (with the bulk of the spending being on defence and nuclear energy). Industry funding was directed at applied research and development in support of innovation, primarily in large firms in a small number of sectors, chemicals, engineering and aerospace. In addition, it had long been recognised (from 1918 in fact) that fragmented industries, predominantly made up of smaller to medium-sized firms, could benefit from cooperative research arrangements (the Industrial Research Associations) and many had been established in the 1920s and 1930s.

The Labour governments of the mid- to late 1960s began a long process of change and reform, setting up the framework of Research Councils and the ill-starred Ministry of Technology. The thrust of the policy was to support key industries (computing, aerospace and nuclear power) and establish a policy of support for innovation in private firms. This framework, which lasted until the mid-1980s, and a Conservative government under Margaret Thatcher, provided innovation grants to the UK companies in support of innovation. Support was project-based and single firm based and required firms to provide 50 per cent of collateral funding for their projects. The rationale for these schemes was pure market-failure, as explained previously, and co-funding was designed to suppress any tendencies to exploit opportunities for moral hazard at the public expense. It is clear that these policies simultaneously provided firms with resources to innovate and increased their incentives to spend more of their own resources on innovation.

In the 1980s, this framework for innovation support came under increasing scrutiny. Despite its commitment to the free market, the realities of innovation in the UK meant that an active role was maintained for government. The stimulus for change was the observed policy of the Japanese government where major investments were being made in the development of new enabling technologies through joint public/private funding. The UK response was the adoption of collaborative research programmes focused on pre-competitive research. In this model, public funding was provided to groups of research collaborators, firms and universities, provided the research was not deemed to be 'near to market'. The idea of generic, enabling knowledge that could be exploited in many different ways by different firms was central to this approach. At first these principles were adopted in a major programme of collaborative research and development in computing and information technology, the Alvey programme that treated the UK industry and university system as a distributed research laboratory. However, the principles were soon extended more widely. The idea of single company support for innovation was progressively abandoned from the late 1980s onward and confined to very limited programmes in relation to innovation in SMEs. By 1993, the last vestiges of single, large company funding support, the Advanced Technology Programmes, had been terminated. This transition is extremely important for it signalled the beginning of a systems failure perspective on STI policy. At the same time the development of a European Community policy on cross-country collaborative research began to shape the thinking of policy makers.

Altogether these developments amounted to a shift towards a systems failure perspective on policy that was endorsed by the 1998 White Paper, '*Our Competitive Future: Building the Knowledge Driven Economy*'. The new policy is driven by three considerations, stimulating competition, developing innovative capabilities by encouraging entrepreneurs and developing skills throughout the workforce, and by encouraging collaboration in the innovation process. This policy is now firmly entrenched, and an indication of the current types of innovation policy instruments, as managed by the Department of Trade and Industry (DTI), is shown in Box 4.4. Of all these schemes only the SMART programme provides innovation subsidies for single company projects. These are directed at small firms and the purpose is to enable them to develop an innovation to the stage where it can attract venture capital support. The LINK programme is the major initiative that funds projects and programmes on a collaborative basis between firms and the science base. A total of seventy-five have been funded since the acceptance of the scheme although the overall scale of funding is limited. These collaborative programmes provide 50 per cent of co-funding of pre-competitive research programmes. Eureka and the Fifth Framework are programmes of collaborative work within Europe.

All of the other schemes are examples of policy directed at creating connections within the innovation system. The ISI (information society initiative) scheme is directed at the encouragement of the widespread adoption and effective use of information technology in small firms. Business Excellence is devoted to networking and the promotion of best practice in industry, and ITS (information technology service) is a scheme to enable UK companies to access foreign technologies. The TCS is a very important scheme that links graduate study in Universities to projects that are of strategic significance to the companies concerned. In 1997, industry committed £35 m to the scheme with 222 new projects involving 356 graduates. Large companies pay 60 per cent of the costs, a figure reduced to 30 per cent for small firms.

Box 4.4 UK innovation policies administered by the DTI

Aims	Scheme
1 Enhance the capabilities of firms	Teaching Company Scheme SMART University for Industry
2 Collaborative innovation	LINK Foresight EUREKA EU Framework
3 Diffusion of best practice	Innovation unit Business Excellence Information society initiative International technology services

However, perhaps the most significant policy development of the 1990s, at least from a system's failure viewpoint, has been the Foresight Programme.

Technology foresight

There is no more appropriate indication of the switch in policy from matters of resources and incentives to matters of opportunities and capabilities than the adoption of a Technology Foresight Programme by the UK Government and indeed other governments (de Laat and Laredo, 1998). Foresight activities have been defined as:

> a systematic means of assessing those scientific and technological developments which could have a strong impact on industrial competitiveness, wealth creation and the quality of life.
>
> (Georghiou, 1996)

and they appear to have been applied on a most consistent, long-term basis within the Japanese science and technology system (Freeman, 1987). The process involved in conducting a large-scale foresight programme is precisely a matter of bridging and connectivity within a nation's science and technology base and between that base and its areas of application. In particular, the crucial point about foresight proper is its inclusion of knowledge about demand and market developments in its activity. Foresight activities of this kind are necessarily broadly defined to explore the social and economic constraints and opportunities in relation to the development of scientific and technological knowledge. The process involved the creation of fifteen sectoral panels of 'experts' that consult on a wide basis with the relevant communities in industry, academia and government through regional workshops, a major delphi survey and numerous other activities. Panels covered fields as diverse as transport, biotechnology and the service economy. Each panel produced a report indicating the main forces for change and the policy issues which flow from the analysis as well as identifying the likely constraints on change. It is without question the most extensive consultation of industrial and scientific opinion that has ever occurred in the UK. It is the fact that the development of modern technology is so heterogeneous with respect to its discipline base and institutional context that makes the sounding of opinion in the broadest possible fashion extremely important.

One way to interpret foresight activity is in terms of Weinberg's (1967) careful enunciation of external criteria for the support of science. Despite strong objections from the pure science lobby, the use of external criteria does not imply that pure science is to be transmuted into applied science. Rather what is at stake is the differential focusing of basic scientific work in relation to non-scientific objectives. Here the crucial point is that the principal lasting benefit of the exercise lies in the process of building the science base into the national innovation system. It is what the process does to the formation of commercial and academic *strategies* to promote innovation; to the creation of lasting *networks* between industry, government and the science and technology community; and to the emergence of coherent *visions* within their communities on *complementary developments* in science and technology. By a coherent vision is definitely not meant a consensus view about specific technologies or routes to innovation but rather an understanding of the breadth and interdependence between the uncertain opportunities open to a particular sector.

Thus, the policy aim has the stimulation of the technology support systems of particular groups of firms; and bridging between those formal and informal institutions which interact in a specific technological area for the purpose of generating, diffusing and utilising technology (Carlsson and Stankiewicz, 1991; Carlsson, 1995). The latest example of this can be found in the renewed concern for industry–university links and the encouragement of university spin-off companies. To create effective webs the policy maker must know the relevant communities of scientists and practitioners, and understand the rival technologies. The sequences of innovations that emerge, and the firms which are successful, are the outcomes of the process and are not a specific concern of the policy maker. Winners emerge, they are not pre-chosen and they cannot be predicted in advance.

The basic principles behind all these system-building policies are network formation and the creation of operational innovation systems. Since, many kinds of knowledge play a role in the innovation process, the different providers have to be coordinated appropriately. When basic science and technology are involved then public funding is provided in recognition of the diffuse and uncertain benefits of this work. Responsibility for the final steps to innovation lies firmly with the private sector. Thus, the central policy question has become 'How do firms gain access to the necessary external knowledge to support their internal innovation activities?'.

South Korea

The case of South Korea provides an instructive example of the role of science and technology policy in a regime of rapid industrialisation. After hostilities ended in 1955, Korea was heavily dependent on inflows of military and other aid from the US and it was not until 1961 that rapid economic development began. This experience well illustrates the different stages of development of a policy for technology acquisition. In the first stage there was no demand to develop technology on a stand alone Korean basis, and the central purpose was the acquisition of overseas technology and the internal diffusion of that technology largely through internal labour mobility. Only in the subsequent stages did an indigenous innovation stance emerge. In the 1960s, the focus was on light industry (e.g. Textiles, Plywood), moving on in the 1970s to heavy industry (e.g. Ships, Steel, Construction). By the 1990s, Korea was at the leading edge of the next generation of electronic products on areas such as multi-media and HDTV. Three aspects of the general environment are important to understanding the Korean case: the extremely high levels of educational attainment; the emphasis on strong internal competitive pressure and the role of the large industrial conglomerates (Chaebol) as the vehicle for industrial organisation (Kim, 1997). The later are unique to the Korean experience and placed it on a quite different path of development from close rivals such as Taiwan, where industry was in the hands of a multiplicity of small and medium enterprises (Hobday, 1995).

In the first stage, foreign technology was acquired through capital goods imports and the purchase of turn-key plants, supported by the posting of nationals overseas and a policy of reverse engineering. At this stage neither foreign licensing nor FDI played a significant role. The emphasis was on catching-up in mature technologies, taking advantage of lower national wages to support a vigorous export policy while protecting the home market through high rates of effective protection. As development proceeded successfully in the 1960s and 1970s the natural consequences was a rise in real wages that began to erode Korea's competitive advantage in advanced country markets.

It was at this stage in the 1980s that the need arose to develop indigenous innovation capabilities and this resulted in new approaches to STI including an increasingly important position for the Ministry of Science and Technology. What Korean experience indicates is that there is little point in developing an elaborate supply infrastructure for science and technology if the demand to use the knowledge so generated does not exist. From about 1984 onwards a concerted group of policies were put in place including public procurement policies to encourage the development of domestic technological capabilities, the management of inward direct foreign investment and the establishment of technology transfer and sectorally specialised public R&D institutes. During the 1980s the transition to greater technological independence was achieved through the development of strong OEM relationships with foreign companies in Japan, the USA and Europe. These relations provided a powerful framework within which to learn new capabilities and provide access to established distribution channels in overseas markets. As the 1980s progressed so did the ability of Korean firms to develop their own independent design and development capabilities (Hobday, 1995). In support of these advances, a total of forty-six industrial research cooperatives had been established by 1989. The public banking system provided preferential financing arrangements for technology projects, which were supported by tax credits for R&D and human capital investment, and the provision of accelerated depreciation for R&D facilities. This combination of infrastructure development and technology incentives provided a powerful stimulus to the R&D expenditures of private firms that increased from 12.3 million won in 1975 to 6,903 million won in 1995. More tellingly this represented an increase in the fraction of GDP devoted to R&D from 0.42 to 2.69 per cent over the same period, with over 80 per cent of the total spending accounted for by private firms, compared to some 10 per cent in 1965 (Hobday, 1995). These figures exceed the proportionate expenditure of many more advanced countries. In the UK, for example, the comparative figure for business R&D from the 1980s onwards is less than 1 per cent of GDP. Similarly the number of corporate R&D laboratories increased from 12 to 2,270. Kim (1997) has documented accurately the nature of this remarkable technological transition that transformed Korea from an Agrarian society to an advanced industrial country in three decades. GDP per head rose over this period from $87 per annum in 1962 to $10,000 per annum in 1995. It was a transition from technological imitation, through a more sophisticated stage of reverse engineering and development of relatively mature technologies, to, finally, the development of an indigenous R&D capability including basic research. The strong export orientation and the competition pressure that this produced have clearly been crucial, and so has the underpinning of a well-educated population. But from the point of view of accumulating capabilities the key lessons are three in number. First, the development of a first rate higher education system and research base focused on technology, mathematics and computing as a necessary element in communicating with and learning from the external world of knowledge at the science and technology frontier. A policy encouraging international mobility of national scientists and technologists is an important element in this communication process. Second, in order to develop beyond a role of technology dependence, it is essential to develop a national innovation infrastructure, and in this the State has a crucial role in providing the foundations for the development of firm capabilities in R&D. In Korea's case it took responsibility for the training of researchers and for the establishment of key public research institutions. Third, the supply capabilities of this infrastructure need to be matched by the development of demand-side capabilities and a supporting policy of R&D incentives. Korean firms achieved this through the formation of strong OEM relationships with foreign technological leaders which they used to frame technological learning. They also benefited from subsidised loans to develop new technology and the ability of the large Chaebol to cross-subsidise R&D investments. Clearly the Chaebol were a key element in this particular path of development, indicating the importance of specific idiosyncrasies in any nation's development. What the Korean experience so clearly indicates is the shifting balance of public and private support for innovation in conditions of rapid development.

Colombia

The experience of Colombia provides a fascinating case study of the evolution of STI policy towards a national system of innovation perspective (UNCTAD, 1999). From the mid-1950s to the mid-1970s Colombia experienced positive if modest rates of GDP growth per head based on a policy of domestic protection and export promotion combined with restrictions on inward foreign investment. In 1991, this policy was abandoned dramatically with a programme of tariff cuts in agriculture and manufacturing and a switch to the encouragement of direct foreign investment. By the mid-1990s Colombia had established itself as a middle-income country with a GDP per capital of $2,300. From 1973 to 1994, it enjoyed the fastest growth rate of all the Latin American economies. The 'Apertura' of 1991 is the key event that ushered in new approaches to STI policy.

Colombian officials describe the period 1957–1974 as one with a 'Defensive Technology Policy' in which initiatives supported the pattern of protection. Public funds supported scientific research in the universities but little attempt was made to connect this work with industrial needs. From 1974 onwards, a technology policy began to be developed and this has developed rapidly after the move to trade liberalisation. The overarching aim of policy was to support the competitive development of Colombian economic activity through innovation. The government plan of 1994–1998 sets out the following objectives in support of this:

- to activate the National System of Innovation;
- to strengthen the research, training and technological services infrastructure;
- to support technological innovation in business;
- to generate an entrepreneurial business culture based on creativity, knowledge and a long-term view;
- to encourage innovation at regional level in order to foster balanced social development;
- to provide financial and other incentives to private investment and R&D.

These are ambitious aims and what is important about them is the recognition of the central role of new R&D organisations and the connections between them. As the UNCTAD report points out (UNCTAD, 1999, p. 9) 'the capability to learn and build new competencies depends on how well the parts fit together and on the strength of their connections'.

The central purpose of the new policies is thus the creation of networks of interacting research organisations, many of these newly created. In total twenty-nine Centres for Technological Development (CDTs) have been created to act as 'virtual' structures, whose function is to coordinate the supply with the demand for new technologies while operating with the existing structures of universities, enterprises and test laboratories. Eight of the CDTs are in industry, ten in agriculture and livestock, seven in new technologies and four in the mining–energy sector. Several broad kinds of organisation are involved. Sectoral Technology Centres whose aim is to give a better definition of the technological requirements of business, in industries which are perceived to face strong foreign competition (e.g. plastics, textiles, shoes and papers). Second, there are technology centres focused on new technologies, biotechnology, optics, electronics, software and automation. In addition to these, a number of research and technology incubator units have been formed. The new CDTs are meant to reinforce the links and connections between sectoral and regional technology organisations including those that provide support services in relation to quality and standards, training and technical assistance. They have four functions: to perform R&D, to provide technical extensive services, to coordinate with the internal and overseas R&D communities and to support collaborative innovation between firms in supply chains.

Mauritius

A final case of a very different character is provided by Mauritius, a small late industrialising economy, that has concentrated on the textile industry for its present stage of development. Over 80 per cent of its exports come from the garment industry leaving the economy over-specialised in relation to foreign competition (Lall and Wignaraja, 1998). The development of new capabilities in sectors such as consumer electronics is central to its future but this presents it with major challenges. The economy is too small to fund the necessary R&D programmes to put it at the leading edge of these industries, so it must acquire established mature technology and rely on inherent cost advantages. In turn this requires the requisite level of skills and capabilities, and a technology policy to develop them. Mauritius has tackled these problems in a number of ways, primarily to build its technological infrastructure. Central among these was the creation of the Mauritius Bureau of Standards in 1975. This body is responsible for the development of standards that are recognised internationally, for the provision of metrology and testing services and for the certification of firms for quality assurance purposes. In regard to each of these it performs a major educative role and its services are vital for firms that wish to develop new export markets. Another important policy instrument has been the Technology Diffusion Scheme, this provided grants for firms to buy consultancy services to raise competitiveness and productivity. An evaluation indicates a very high return measured in terms of additional export revenues, indicating the benefits obtained from the intelligent use of relatively simple managerial and market knowledge. Other programmes have been implemented to improve design and product development skills and to carry out limited collaborative R&D projects.

Clearly Mauritius is very different from Korea and Colombia, yet, technology and innovation policy play an important role in its pattern of development. Relevant knowledge is not necessarily or normally hi-tech knowledge.

A brief evaluation

An evaluation of these different national experiences is clearly premature but several observations are in order. The effectiveness of any innovation system depends on an appropriate matching of demand for innovation with its supply. *Technology-push from universities and public laboratories generally does not work and the major issue becomes the innovative stance of domestic firms and their absorption capacities.* If firms cannot identify benefits to product or process innovation then little of substance is likely to happen. Thus, the first function of this new system is not to produce a list of innovations *per se* but rather to build the capacity to absorb new technology. Furthermore, it is important that the infrastructure of STI institutions be appropriate to the stage of development of the economy, and that this infrastructure can adapt as development proceeds. Failure to do so may act as a severe brake on a nation's development.

Lessons for STI policy

In this final section we can draw together some of the lessons that can be derived from the previous analysis. At the outset it is important to reiterate the point that no single policy stance is appropriate for all developing economies, nor will an appropriate policy remain unchanged over time. Nonetheless, some general policy principles can be identified. The focus of these principles is the creation of a working innovation system. They are not concerned primarily with the scale of expenditure on R&D.

The most important principle of all is the need to put innovation first and recognise that the proper role of science and technology is to support innovation. In turn, innovation is concerned with enhancing national productivity and national competitive performance. Many different kinds of knowledge are required for firms to innovate, and knowledge of market possibilities and the organisation of production and distribution are as important as more formal science and technology. Appropriate knowledge need not be sophisticated, formal knowledge. What innovation requires is practically useful knowledge and much of this can only come from experience. Thus, the studies by Best and Forrant (1994) of the Jamaican furniture industry, and by Wignaraja (2002) of the Mauritian textile industry each make the point that 'simple' managerial knowledge is what is needed to make those industries more competitive. To put it more precisely, it is the ability to combine different kinds of knowledge and skill that is the essential factor in the innovation process. However, the ability to combine knowledge applies in two other ways. First, in relation to linking the results of R&D programmes with the knowledge that is generated through practical experience. Second, in linking with the international world of science and technology which is becoming increasingly integrated.

The second principle requires the State to take a strategic view of the sectors where sustainable competitive positions can be established: positions that reflect the domestic resource base and the nature of international competition. This does not require the national efforts to be the best practice in world terms, it only requires that they be better than the prevailing worst practice, that is to say, they are viable at world prices. Thus, a view has to be established of the countries' strategic location behind the technological frontier and with regard to the complexity of the relevant technologies (Bell and Albu, 1999). Again, the matter is a practical one, the choices must reflect the time needed to build the necessary capabilities and the fact that the external situation will also advance over time. Moreover, STI strategy is not a detailed recipe, a plan for individual innovations. It is a framework of commitment within which the sets of innovating actors can work, confident in the durability of the strategic aims. *If a policy is to have a lasting impact it must be adaptable to the changes that are a necessary element of the competitive process.*

The third principle concerns the need to establish an innovation system infrastructure appropriate to these strategic choices. We have seen that innovation systems consist of sets of organisations and their interactions, and that the principal route to interaction is through a wide variety of collaborative arrangements. A key role in this is played by lead actors whose role it is to coordinate the division of labour in the innovation process. In many developing countries this role will be associated with specialist research institutions either partly or totally funded by the State that sit at the centre of the appropriate, sectoral, innovation networks. Among these, those related to metrology and standards have a particularly important function. Their role is defined in relation to the use and the generation of knowledge (Bell and Albu, 1999). Their main activities can be listed as follows:

- The training and formation of a skilled labour force, including management.
- The provision of metrology services.
- The linking together of firms with their suppliers and customers in pursuit of innovation.
- The generation of collaborative R&D projects and programmes.
- The oversight of standards in the various sectors.
- Interaction with foreign universities and relevant technology institutes to keep a watching brief on world developments in a sector.
- Linkage of foreign firms into the national innovation structure.

One way to interpret this innovation system is to view it as a broadcasting system with transmitters and receivers of information, remembering that information is not knowledge. The capability of the receivers matters as much as the capability of the transmitters and barriers to their interaction may come from misaligned incentives and the 'not invented here' attitude. One of the major difficulties facing research institutes on this model has always been that of keeping their R&D and training activities relevant to the needs of the sector. By far the best way to do this is to have industrial partners as collaborators in these activities not as distant and indirect customers for its services. This requires in turn that the firms in question have the capabilities to interact with the SROs.

Within the innovation system, universities and science policy have an important role to play. They have a training function in relation to advanced research and experimental capabilities, they supply bridging knowledge between pure science and its applications, and they provide a natural focus for links with overseas research programmes. But this role remains secondary to the main task of raising the technological and innovative performance of local firms.

Finally, it is important to recognise that innovation is not to be equated solely with hi-tech sectors. New knowledge can, just as well, open up significant innovation opportunities in established sectors. The essential point about innovation is that it is concerned with change and transformation not with newness in a narrow sense.

Concluding remarks

Innovation presents the policy maker with many paradoxes. It is the driving force in the development of capitalism, yet it remains unpredictable in its content and field of application. Its importance reflects the restless, discovery-based nature of capitalism in which the accumulation of knowledge is embedded in market processes. It cannot readily be managed, let alone accounted for. A world of innovation is in policy terms rather uncomfortable. However, in terms of economic development it is clear that innovation policy matters. I have suggested that what matters for STI policy is the creation of an innovation system and an associated innovation culture. Its role is not to innovate at the world frontier but rather to adapt and develop available technologies to meet the needs of local markets and resources. This is challenge enough.

Note

1 This chapter was prepared as a paper for the workshop on Enterprise Competitiveness and Public Policies, Barbados 22–25 November 1999 and revised following that presentation. I am grateful to Jeremy Howells, Ganeshan Wignaraja and to the participants of the Barbados workshop for helpful comments on the first draft.

References

Abramovitz, M. (1989), *Thinking about Growth*, Cambridge University Press, Cambridge.
Arrow, K. (1962), 'Economic Welfare and the Allocation of Resources to Invention', in R. Nelson, (ed.), *The Rate and Direction of Inventive Activity*, NBER, New York.
Bell, M. and Albu, M. (1999), 'Knowledge Systems and Technological Dynamism in Industrial Clusters in Developing Countries', *World Development*, 27, 1715–1734.
Bell, M. and Pavitt, K. (1993), 'Technological Accumulation and Industrial Growth: Contrasts between Developed and Developing Countries', *Industrial and Corporate Change*, 2, 157–170.
Best, M.N. and Forrant, R. (1994), 'Production in Jamaica: Transforming Industrial Enterprises', in P. Lewis (ed.), *Preparing for the Twenty-first Century; Jamaica 30th Anniversary Symposium*, Ian Randle, London.
Carlsson, B. (1995), *Technological Systems and Economic Performance: The Case of Factory Automation*, Kluwer Academic, Dordrecht.
Carlsson, B. and Stankiewicz (1991), 'On the Nature, Function and Composition of Technological Systems', *Journal of Evolutionary Economics*, 1, 93–118.
C. Edquist, (ed.) (1997), *Systems of Innovation: Technologies, Institutions and Organisations*, Pinter, London.
European Commission (1998), *Internationalisation of Research and Technology*, Brussels.
Fountain, J.E. (1998), 'Social Capital: A Key Enabler of Innovation', in L.M. Branscomb and J.H. Keller, *op cit*.
Freeman, C. (1987), *Technology Policy and Economic Performance*, Pinter, London.
Georghiou, L. (1996), 'The United Kingdom Foresight Programme', *Futures*, 28, 359–377.
Hayek, F. (1948), *Individualism and Economic Order*, Chicago University Press, Chicago.
Hobday, M. (1995), *Innovation in East Asia*, Edward Elgar, London.
HMSO White Paper (1998), *Our Competitive Future: Building the Knowledge Driven Economy*, HMSO, London.
Kim, L. (1997), *From Imitation to Innovation: The Dynamics of Korea's Technological Learning*, Harvard Business School Press, Harvard.
deLaat, B. and Laredo, P. (1998), 'Foresight for Research and Technology Policies: From Innovation Studies to Scenario Configuration', in R. Coombs, K. Green, A. Richards and V. Walsh (eds), *Technological Change and Organisation*, Edward Elgar, Cheltenham.

Lall, S. (1987), *Learning to Industrialise: The Acquisition of Technological Capability by Developing Economies*, Macmillan, London.
Lall, S. and Wignaraja, G. (1998), 'Mauritius: Dynamising Export Competitiveness', Commonwealth Secretariat, London.
Lundvall, B.A. (1992), *National Systems of Innovation: Towards a Theory of Innovation and Interactive Learning*, Pinter, London.
Marshall, A. (1919), *Industry and Trade*, Macmillan, London.
Metcalfe, J.S. (1994), 'The Economic Foundations of Technology Policy', in P. Stoneman, (ed.), *Handbook of the Economics of Innovation and Technological Change*, Blackwell, London.
Metcalfe, J.S. (1998), *Evolutionary Economics and Creative Destruction*, Routledge, London.
Mokyr, J. (1990), *The Lever of Riches*, Oxford University Press, Oxford.
Mowery, D.C. and Rosenberg, N. (1979), 'The Influence of Market Demand upon Innovation: A Critical Review of Some Recent Empirical Studies', *Research Policy*, 8, 102–153.
Mowery, D.C. and Rosenberg, N. (1998), *Paths of Innovation*, Cambridge University Press, Cambridge.
Nelson, R.R. (1959), 'The Simple Economics of Basic Scientific Research', *Journal of Political Economy*, 67, 351–364.
Nelson, R.R. (ed.) (1993), *National Innovation Systems: A Comparative Analysis*, Oxford University Press, Oxford.
Nelson, R.R. and Winter, S. (1984), *An Evolutionary Theory of Economic Change*, Belknap Press, Harvard.
OECD (1999), *Managing National Innovation Systems*, OECD, Paris.
Rosenberg, N. (1990), 'Why Firms do Basic Research (with Their Own Money)', *Research Policy*, 19, 165–174.
Rosenberg, N. and Birdzell, L.E. (1986), *How the West Grew Rich*, Basic Books, New York.
Schumpeter, J.A. (1911), *The Theory of Economic Development*, Oxford University Press, (1938 edition), Oxford.
Stiglitz, J.E. (1997), *Whither Socialism*, MIT Press, London.
Tassey, G. (1992), *Technology Infrastructure and Competitive Position*, Kluwer Academic, Dordrecht.
UNCTAD (1999), *The Science, Technology and Innovation Policy Review, Colombia*, United Nations, Geneva.
Utterback, J. (1996), *Mastering the Dynamics of Innovation*, Harvard University Press, Harvard.
Wignaraja, G. (1999), *Tackling National Competitiveness in a Borderless World*, Commonwealth Business Council, London.
Wignaraja, G. (2002), 'Firm Size, Technological Capabilities and Market-Oriented Policies in Mauritius', *Oxford Development Studies*, 30(1), 87–104.
Weinberg, A.M. (1967), *Reflections on Big Science*, Pergamon, London.

5 Industrial clusters and business development services for small and medium-sized enterprises

Eileen Fischer and Rebecca Reuber

Introduction

In most countries, small and medium-sized enterprises (SMEs) make up the majority of businesses and account for the highest proportion of employment. They produce about 25 percent of OECD exports and 35 percent of Asia's exports (OECD, 1997). Those SMEs that are internationally competitive are better able to grow as well as to survive in their domestic markets. In order to become internationally competitive, SMEs must be market-oriented and offer products and services of international quality. These objectives can be particularly difficult to achieve for SMEs in developing countries, and governments and non-profit agencies often need to provide assistance. This chapter identifies principles and best practices from worldwide experience with SME development assistance in the field of non-financial services. In particular, the chapter focuses on the role of industrial clusters and on the ways that public and non-profit institutions can help to promote SMEs by fostering the development of such clusters.

The following section of the chapter describes SMEs and their role in economic development. The section on "A cluster perspective" describes the nature and benefits of industrial clusters. These benefits are illustrated further through four case studies examined in the section on "Case studies of clusters underlying economic development". The section on "Best practice programs for SME development" outlines best practice examples for establishing business development services (BDS) for SMEs, and the section on "Delivering BDS to SMEs" discusses principles and guidelines for BDS provision. The final section summarizes the trade-offs that need to be made in developing and selecting policies and practices for SME promotion.

SMEs and their role in economic development

Definitions of SMEs

The term SME encompasses a heterogeneous group of businesses, ranging from a single artisan working at home and producing handicrafts to sophisticated software product firms selling in specialized global niches. What is, or is not, an SME is usually defined by the number of employees of the firm,[1] and definitions vary between countries. Definitions used in developed nations will often have higher size thresholds than those in less developed countries. For example, in Mauritius, firms with less than ten employees are considered microenterprises and firms with ten to forty-nine employees are considered SMEs (Wignaraja and O'Neil, 1999), while in Japan, firms with less than 300 employees are considered small (Whittaker, 1997). In addition, some developed countries, such as Australia and Canada, differentiate between manufacturing firms and other firms; in Canada, for example, a small business is considered to be a manufacturing firm with fewer than 100 paid employees or any other firm with fewer than fifty paid employees, while a medium-sized business has between 100 and 500 paid employees.

Within the general SME category a number of sub-groups can be identified: self-employed persons with no employees, microenterprises with fewer than ten employees; small firms with eleven to forty-nine employees, and medium-sized firms with between 50 and 100 employees. In most economies, self-employed persons with no employees will constitute the majority of SMEs, and a very small portion will be medium-sized firms. However, self-employed persons and microenterprises are often under-represented in official statistics because they may be excluded from registration requirements or able to avoid compliance with such requirements. Failure to recognize their prevalence within the population of SMEs can lead to a distorted characterization of SMEs.

Characteristics of SMEs in developing countries[2]

Labor force characteristics

Given that most SMEs are one person businesses, the largest single employment category is working proprietors. This group makes up more than half the SME work force in most developing countries; their family members, who tend to be unpaid but active in the enterprise, make up roughly another quarter. The remaining portion of the work force is split between hired workers and trainees or apprentices.

Sectors of activity

SMEs tend to be engaged in retailing, trading or manufacturing. While it is a common perception that the majority of SMEs will fall in the first category, the proportion of SME activity that takes place in the retail sector varies considerably between countries, and between rural and urban regions within countries. Retailing tends to dominate in urban regions, while manufacturing can occur in either rural or urban centers. Differences across countries in the proportion of SMEs engaged in manufacturing goods are a result of differing endowments of raw material, tastes and consumption patterns of domestic consumers and the level of development of export markets.

Sex of owner

When sole-proprietorships and microenterprises are taken into account, it becomes apparent that the majority of SMEs are owned and operated by women. SMEs headed by women are more likely than those headed by men to operate from home. And since home-based SMEs are most likely to be under-represented in official statistics, it is not uncommon that assistance projects are designed without sufficient consideration of the needs of businesses that women tend to start. This is the case because the owners of these small, home-based, businesses are often the ones that feel they are unable to take advantage of any programs, simply because the administrative burdens of programs designed to benefit them are perceived not to be worth the cost.

Efficiency

If efficiency is defined as net returns per hours of labor, the available evidence suggests that there are considerable differences in efficiency between SMEs of different sizes. In general, one person businesses generate the lowest net returns. Some studies show that both small firms and large firms are inefficient relative to medium-scale firms (Little et al., 1987). While the vast majority of SMEs that are founded do not grow larger, the few that do provide valuable social and economic benefits to their communities.

The contributions of SMEs to development

There is a general consensus that the performance of SMEs is important for both the economic and the social development of developing countries (Levy et al., 1999). From an economic perspective, SMEs provide a number of benefits (Advani, 1997; Halberg, 1999; Leidhom and Mead, 1999):

- SMEs, due to their size, can often readily adapt to changing demand patterns, trade patterns and macroeconomic conditions. This increases industrial flexibility.
- SMEs have a reasonable propensity to acquire technological capabilities and develop new products and processes and can thus contribute to national technological development and competitiveness.
- SMEs can be an important vehicle for generating income and employment and so contribute to gross domestic product, economic growth and reductions in unemployment.
- SMEs provide a setting in which assets and skills can be accumulated. This can lead to better economic opportunities for the individuals who acquire the skills, and for the households they help to support.
- SMEs can decrease wage inequality. They do so largely by increasing economic participation among those in the lower half of the income distribution. However, neither SME owners nor their workers are likely to be the poorest of the poor, and so promoting SMEs to achieve equity objectives is not necessarily as effective as direct methods like income transfers.

The social benefits of SMEs, while analytically separable, are closely linked to economic benefits. From a social perspective:

- SMEs can contribute to the development of particular regions, especially when groups of similar businesses can create collective efficiencies.
- SMEs can help to bring about social change. The experience of owning and operating firms can help develop individual feelings of responsibility for and ability to participate in governance.
- SMEs can help to institutionalize democracy and increase social stability. They do so through the creation of structures that reflect people's needs and objectives.

Why support SMEs?

The economic and social contributions of SMEs suggest that it is clearly in the public interest for SMEs to thrive. But this alone would not necessarily mean that governments should actively intervene by creating policies and programs to support SMEs.

The more pressing argument that favors the development of public policies in support of such businesses is that a strong SME sector, and particularly industrial clusters (Humphrey and Schmitz, 1996), cannot emerge without some form of support from the state. This argument rests on the observation that small enterprises suffer disadvantages in markets because of their size. It holds that market failures routinely occur that systematically undermine SMEs, and that many SME requirements are those of a "public good" nature. Accordingly, SME-specific programs and policies need to be established in the following areas (Levy, 1994; Hallberg, 1999):

- *Information.* For markets to allocate resources efficiently, all competitors must have the same relevant information. The high fixed costs of acquiring relevant information on potential opportunities, for instance on foreign buyers or distribution channels in international markets, create disadvantages for small firms.
- *Training.* When a firm invests in training, it does not gain all the benefits of the training because it is not entirely specific to the job and the worker can change jobs. In addition, training is often used to remedy deficiencies in the education system. Thus, because SME owners are concerned that they will not receive benefits equal to training costs, they often under-invest in training. Policies that support education and training as a public good, remove some training costs from individual SMEs.
- *Linkages between firms.* Linkages between firms create positive externalities, or shared benefits among them. For example, large firms often get volume discounts from suppliers, while small firms require insufficient volumes of raw material to qualify for discounts. If small firms can coordinate their

purchases, then they are better able to take advantage of volume discounts. There is likely to be a high fixed cost, however, in setting up an appropriate coordination mechanism, which creates a substitute cost for SMEs, unless it is covered by a public program.
- *Delivery services.* The overall institutional infrastructure should be appropriate to the characteristics and resources of SMEs. In particular, they require information and services that are customized to their particular needs. Again, small firms are systematically disadvantaged compared with large firms because customization is costly.

Approaches to SME promotion policies

The traditional form of SME assistance has been financial, focused on providing credit. Based on the assumption that the high cost of credit was a main constraint for SMEs, credit policies often included credit guarantee schemes and/or subsidized interest rates. These traditional credit programs have not succeeded well in their basic objective of increasing the SME's access to financial resources (Hallberg, 1999, p. 9). They have inhibited the development of sustainable financial institutions, and created distortions in credit markets by discouraging firms from using non-credit forms of financing. The overall approach reflects an attitude that governments (or other assistance providers) are donors and SMEs the needy recipients of charity. An alternative approach, one that assumes that services can be provided so as to make SMEs competitive and independent of support, is gaining attention. Under this approach, services and policies are directed toward addressing market failures that create disadvantages for SMEs in accessing markets, so that they may become competitive (Hallberg, 1999).

A type of SME promotion policy that fits the latter perspective is one which supports the creation and development of industrial clusters. Support directed at groups of enterprises that form, or could form, industrial clusters has two advantages from the point of view of a policy-maker trying to stimulate competitiveness by overcoming market failures that disadvantage SMEs:

- groups of SMEs acting collectively can overcome some of the limitations that individual enterprises have in acquiring either services or market information;
- groups of SMEs that are in the same region and industry can form rivalries that promote competition and innovation.

Clusters have attracted a great deal of interest and appear to have promise for stimulating the growth of healthy SMEs in developing economies. Accordingly, clusters and cluster-development policies and practices will be emphasized in this chapter.

A cluster perspective

Interest in clusters has grown with the realization that some of the richest regions in both developed nations such as Italy and Germany, and in developing countries such as Brazil, share a particular characteristic: they are home to groups of interlinked firms that tend to collaborate technologically and/or strategically. These groups of firms are often referred to as clusters. Examples from developed nations, such as the textile industry in northern Italy, shipbuilding in Glasgow, and computers and software in Silicon Valley, are well known (for some examples, see Swann *et al.*, 1998). Less celebrated are the many clusters that exist in the developing world, some of which are listed in Table 5.1.

Table 5.1 Examples of clusters in less developed countries

Location	Industry
Argentina	
Rafaela	Metalworking, machinery
Brazil	
Americana	Weaving
Petropolis	Knitting
Sinos Valley	Footwear
Ghana	
Kumasi	Metalworking and mechanical engineering
India	
Agra	Footwear
Bangalore	Informatics
Ludhiana	Light engineering and textiles
Okhla	Garments
Tiruppur	Knitwear
Indonesia	
Tegalwangi	Rattan furniture
Kenya	
Nakuru	Carpentry
Korea	
Daegu	Textiles
Mexico	
Aguascalientes	Children's garments, uniforms
Leon and Guadalajara	Footwear
Monterrey	Garments
Nuacálpan	Women's garments
Tehuacán	Garments
Tijuana	Furniture
Peru	
CBK, Lima	Machine tools
Trujillo	Footwear
Singapore	
Technology corridor	Informatics, biology, microelectronics
Taiwan	
Hsinchu	Microelectronics, biotechnology

General characteristics of clusters

This section describes the stereotypical characteristics of clusters. In reality, any given cluster is likely to vary somewhat from this ideal type. Clusters are characterized by:

- A *predominance of SMEs*. One reason for this is that small firms are simply more numerous than large firms. However, small firms are also capable of a flexibility and capacity for innovation that is essential for taking advantage of the momentum that clustering can create (Humphrey and Schmitz, 1996). This does not mean that large firms are rarely found in clusters. In fact, they are often present and integral to cluster dynamics.
- *Geographical proximity*. A concentration of SMEs within a reasonably close geographic region.
- *Sectoral specialization*. A concentration of SMEs within the same industry.
- *Shared social, cultural or political characteristics*. At one extreme there are clusters where the individuals who run and work in the firms share a strong, homogenous socio-cultural identity and strong family ties. In some cases, it is primarily an ethnic identity or a common local value system that is shared. In other cases, it is largely a political identity that is shared among members of firms (Advani, 1997).
- *Inter-firm collaboration and competition*. There is a high density of inter-firm linkages, including horizontal linkages (between competing firms) and vertical linkages (between suppliers, manufacturers and distributors). Collaboration among firms facilitates collective action and collective learning. At the same time, competition leads to innovation (Humphrey and Schmitz, 1996).

Clusters can be organized in quite varied ways. Vertical relationships range from large firms managing a division of labor among small firms to frequently changing permutations of small firms complementing each other. Horizontal relationships may involve cooperation, such as sharing tools, and/or collaboration, such as sharing orders. It is worth noting that a new form of cluster is emerging in some developing countries, consisting of a major multinational corporation affiliated with numerous local suppliers, for example, Intel and its suppliers in Costa Rica (Spar, 1998).

Rationale for promoting industrial clusters

The reason that geographically concentrated industrial clusters appear to be of such importance in emerging and developed markets is that they both foster inter-firm learning and take advantage of social capital. These two advantages of clustering are discussed in turn.

Inter-firm learning

Clusters assist inter-firm learning in three key ways. First, SMEs can learn from their transactions with buyers and suppliers. For instance Rabelloti (1995) found wholesalers in Mexico's Leon and Guadaljara shoe districts helping manufacturers improve product quality by monitoring production and offering advice on organizational matters. The cluster can also develop industry-specific knowledge in diverse functional areas of business, such as procurement, financing, marketing, law, accounting, R&D and distribution. Since there is a large number of firms in the industry, it is less risky for outsiders (such as investors or buyers) to invest in acquiring this specialized knowledge about the cluster.

Second, SMEs can learn from interactions with similar firms. The tacit knowledge[3] of how to do things in an industry can really only be learned by embeddedness in a community of practice (Brown et al., 1996). This diffusion of tacit knowledge makes clusters more than mere aggregations of firms. Even where the ever-increasing power and bandwidth of communication networks makes it simple to transmit huge amounts of information over long distances, the geographic concentration of activity continues to have value (Swann and Prevezer, 1998). This is because only formalized, codifiable knowledge can be transmitted through communications technology. The acquisition of tacit knowledge that depends upon observation and interaction still requires geographic proximity.

Third, the clustering of firms with a similar industry focus creates an "information rich" environment with specialized labour pools. These specialized workers help to diffuse knowledge both because they tend to communicate with one another and because they may work, over time, for more than one firm in the region. Thus, industry trends and technical innovations can be communicated quickly between firms in a cluster. In addition, inter-firm learning is increased when individual SMEs act cooperatively; for instance to bid on contracts that a single firm would be too small to obtain on its own.

Social capital

Social capital is defined as the "norms and social relations embedded in social structures ... that enable people to coordinate action to achieve desired goals" (World Bank, 1999). The inter-personal linkages between members of different SMEs, coupled with the socio-cultural similarities and family or community connections that are typically shared among people involved in a geographic cluster, tend to ensure that social capital exists within the clusters. This social capital facilitates cooperation, collective action and sharing of information to an extent that would be unlikely if impersonal, arms-length relationships were all that existed between members of a group of firms (Rabellotti, 1995; Advani, 1997).

Case studies of clusters underlying economic development

In order to provide a greater understanding of cluster dynamics, four case studies are presented in this section. The first two case studies describe spontaneously emerging clusters, that is, clusters that developed independently of specific policy initiatives. The first, the shoemaking industry in Brazil's Sinos Valley, is a low-technology cluster, chosen because of its phenomenal growth rate based on export markets. The second, the machining industry in Tokyo's Ota Ward, is a high-technology cluster, chosen because of its innovative capabilities and responsiveness to the changing economic climate in Japan. The remaining two case studies describe clusters that developed as a result of specific policy initiatives. The third case study is that of the woodworking industry in Brazil's Ceará State, which is a low-technology cluster developed through demand-driven support. The fourth, the computer industry in Taiwan, is a high-technology cluster that was developed through supply-driven support. This section of the chapter concludes by summarizing what we know from empirical studies of clusters in developing countries.

The shoemaking industry in Brazil's Sinos Valley[4]

Brazil's Sinos Valley is an example of a spontaneously emerging cluster in the relatively low-technology shoemaking industry. By the end of the 1960, there was an established cluster of over 400 shoe firms located in Brazil's Sinos Valley. The population was largely of German descent, indeed from one particular region in Germany, with an agricultural background and a tradition of self-employment rather than working for others. The cluster mainly consisted of small firms. In 1971, roughly one-third of the shoe firms had ten employees or fewer and 85 percent had 100 employees or fewer serving the domestic market.

During the 1970s and 1980s, the cluster grew substantially. It has been estimated that by 1991, the number of shoe firms grew to somewhere between 1,500 and 4,000, with an increasing proportion of large firms. Employment from the shoe industry grew from 27,000 workers in 1970 to 150,000 workers in 1990. Two weekly papers and four bimonthly technical magazines, specializing in the shoe trade, were established and added to the legitimacy of the Sinos Valley as a center of knowledge and information about the industry. There were two reasons for this growth. First, foreign buyers, particularly from the United States, were attracted to the geographical and sectoral concentration of firms: with so many possible firms to do business with, it was less risky to invest in learning about the region. Second, the concentration of shoe firms was large enough to support collective action among the firms in proactively seeking out foreign markets, through participation in trade shows and joint advertising.

During this time period, the cluster was characterized by many vertical linkages among firms due to a high degree of inter-firm division of labor. Shoe production was carried out through a number of stages, and few firms in the Sinos Valley were involved in each stage. It was most common to subcontract operations that required highly specialized equipment and operations that were simple and labor intensive.

The cluster was also characterized by a high degree of horizontal integration. A wide variety of specialized services for shoe firms, such as design, technical and financial consultants, were developed. There was a high degree of inter-firm cooperation, such as the sharing of knowledge, orders and machinery. There were also more major collaborative efforts such as the establishment of self-help institutions. Indeed, between 1963 and 1972 local producers in the area, together with public sector support, set up a major shoe fair and established four technical schools to develop required labor skills. Despite this culture of cooperation, there was also intense competition between firms at all stages in the local value chain, which bolstered the competitiveness of the region as a whole.

The development of export markets was led by intermediaries, the export agents, who were at first mainly American but who later included many Brazilians. These export agents were key to the growth of the cluster. They understood both the US retail market and the Brazilian production environment. They negotiated deals between the two groups, set quality and delivery standards, and arranged transportation and payment.

From the late 1980s and into the 1990s, lower cost producers from China entered the US market and gained considerable market share from Brazilian shoe makers. In order to respond to new market requirements, the shoe makers in the Sinos Valley had to adjust to smaller orders, tighter delivery times and higher quality requirements. These changes favored the smaller firms which were more specialized and flexibly organized. They also favored flexibility and collaboration. Unfortunately, the period of economic growth through the 1970s and 1980s, was accompanied by the development of a more stratified class structure. This, in turn, resulted in the diminishing of the initial shared socio-cultural ties. There were labor difficulties and the segmentation of industry associations into those serving large firm interests and those serving small firm interests. Collective action became more difficult, which was dysfunctional for the cluster at a time when collaboration within and between firms could help to overcome market challenges.

The challenge from lower cost producers also came at a time when the economic environment in Brazil was unfavorable. There was extremely high inflation, followed by an exchange rate pegged to the US dollar which decreased or even eliminated the profitability of existing contracts. Accordingly, even though there were substantial improvements in quality, flexibility and order processing, export volumes in 1997 had not increased from those in 1990, and profitability had fallen. It appears that the upgrading of market development did not keep pace with technological upgrading. This was at least partly due to the existence of the export agents: they were very interested in the upgrading of the shoe maker's production capabilities, but retained the marketing and design functions themselves. It was also partly due to collective inaction among the shoe makers: despite the establishment by the key associations of a joint "Shoes from Brazil programme," the most influential large producers did not participate.[5] Thus, there was little opportunity for marketing know-how to develop within the industry in the Sinos Valley.

The machining industry in Tokyo's Ota Ward[6]

Ota Ward, a district in south Tokyo with a population of roughly 650,000, is an example of a spontaneously emergent, high technology cluster based around machining industries. As late as 1920 there were few machine factories: workers in the district were primarily engaged in fishing and farming, with some small agriculture-related industries. After the Kanto Earthquake of 1923 that destroyed other parts of Tokyo, there was a movement of people and factories into the district. The Manchurian Incident in 1931 and then the Second World War contributed to even greater production because of the demand for heavy industry and armaments. One quarter of Japan's key military factories were in Ota. This concentration of industry led to heavy bombing. In the period 1942–1945 virtually all the factories in the district were destroyed.

After the war, however, conditions were favorable for the development of small machining factories: both land and labor were inexpensive, local workers had a reputation of being able to handle complex work in small lot sizes, and there was a culture which valued independence and business ownership. Over time, Ota Ward gained a reputation for technological preeminence. From 1960 to 1983, the number of factories in Ota Ward increased from 4,987 to 9,190, with a decrease to 7,860 by 1990. The percentage of *small* factories actually increased over this period as well: the percent of factories with fewer than ten employees increased from 42 to 80 percent.

Many original factory owners were second or third sons of farming families, and valued autonomy and independence. Unlike the stereotypical Japanese salary men, who worked for large established firms, the vast majority of founders of Ota Ward machining factories gained experience in small firms, similar to the one they will later found. Work, business ownership and production are an integral part of their identity. Indeed, over two-thirds of the smaller firms have factories and housing on the same premises.

The cluster is characterized by a high degree of vertical linkages among firms. In all size categories of firms, suppliers are less widely dispersed geographically than are customers. It is usual for previous employers to act as incubators for their employees who have started their own firm, by providing subcontracting work, as well as technical and financial assistance. The industry is characterized by much subcontracting, and smaller firms are likely to derive a higher proportion of their income from it.

The practice of subcontracting has led to a high degree of specialization. Two-thirds of the factories with fewer than ten employees, and one-third of the factories with 10–29 employees, carry out only one production process. This specialization has allowed these small firms to focus their technological upgrading to design their own customized machines. Some small firms have a product or process specialization, others specialize in a "will do or machine anything" service, which makes them particularly able to adapt to changing circumstances, and a third group of firms do both. This combination of specialization and flexibility within individual firms provides the district with a high degree of flexibility, durability and adaptability.

In addition to the prominence of vertical linkages, horizontal relationships among firms are also important to understand how the district operates. These relationships between firms are cemented by a shared sense of community, based on living together in the district, a common occupational identity and shared interests. The relationship is known as "confrere trading," between similarly sized firms which are not product makers. It involves the sharing of orders, tools, workers and managerial and technical information. These horizontal relationships are important in lessening small factories' dependence on orders from a few large customers, where margins are becoming squeezed. Small factories are beginning to switch to low volume, high value-added work in new growth areas such as medical and environmental equipment.

The horizontal relationships between firms are increased through the large number of small firm associations: 600 organizations are listed in the Ota Ward Registrar of Trade Organizations. The most common type of association (roughly 85 percent of the total) is the sub-sector specific cooperative business association, the majority of which have fewer than thirty members. Cooperative business associations carry out collective activities for their members, such as collecting and providing information, joint purchases, training and education, joint advertising, capital loans and credit, provision of insurance and provision of joint facilities such as conference rooms.

Although the machining industry of Ota Ward remains vibrant, there are now substantial threats to its future viability. The first is that the industrial development of the district, with its noise and pollution, make it increasingly an undesirable residential location. Larger factories are moving out of the district and owners and workers are less interested in living there. The skilled craftsmen who founded the firms are retiring: in 1994, 43 percent of Ota Ward's owners were aged sixty or more and younger workers, with a higher degree of education are less interested in the long hours required to start a firm, and in working for small firms and receiving lower wages. Between 1983 and 1993, the number of factories in Ota Ward declined by 25 percent. Many owners have not identified a potential successor, and as they become closer to retirement, there is less incentive for them to undertake the costly innovation and technological upgrading which is needed as manufacturing processes become more automated.

The woodworking industry in Brazil's Ceará State[7]

The woodworking industry in Brazil's Ceará State is an example of a relatively low-technology cluster developed as a result of public policy initiatives. In 1987, the district of São João do Aruaru (SJA) had a population of 9,000 and four small sawmills with three employees each. Five years later, in 1992, the woodworking industry employed 1,000 people and there were forty-two sawmills, employing an average of nine workers each. Despite the fact that a government procurement program was the catalyst for this growth, by 1992 private markets accounted for 70 percent of sales from the district.

The procurement program was launched by the state government in 1987 as part of emergency measures for drought relief. The program required that material and tools for reconstruction projects, as well as customary purchases (e.g. school furniture, grain silos, electricity poles, repair and reconstruction of public buildings), be explicitly targeted to small producers in the drought-stricken area. Two key agencies responsible for the program were the State Department of Industry and Commerce (SIC) and a technical assistance agency, the Brazilian Small Enterprise Assistance Service (SEBRAE). The program worked so well that it became permanent when the drought ended, and, indeed in the period 1989–1991, the total amount contracted was US$ 15 million, which was 30 percent of the total state expenditure for goods and services.

The procurement process involved several entities. The buyer, such as the state department of education or agriculture, contracted with SIC, which in turn contracted with SEBRAE to provide technical assistance to the small firms for a 5 percent commission. Rather than contracting with individual firms, SEBRAE then contracted with existing small firm associations. The buyer advanced 50 percent of the contract to provide the small producers with the working capital required to complete the order.

The key inter-firm linkages were those between the contracting agencies and the small firm associations. The reason for this is that strict warranties underlay the contracts. Each desk, for example, had a metal plate with the producer's name and the number of the contract. If it proved to be defective, it was returned to the producer for repair or replacement; if the producer was no longer in business, the local association was contractually responsible. This collective responsibility for production led to peer pressure for quality and productivity improvements. During this period, the furniture-makers' association became a major civic institution which offered other services to members: organizing group purchases of timber, the sharing of equipment and occupational safety measures. The sawmill association, faced initially with a high volume of defective products, organized night classes to upgrade the skills of sawmill workers.

A key to the success of the program is this independent relationship between parties in the vertical value chain. Buyers were able to buy from elsewhere and were independent of the support agency, which had to prove that the firms in the region could deliver quality goods on time. The support agency had a financial incentive to obtain contracts and to work with the associations in the district to fulfill them satisfactorily. The local associations took the costs of monitoring and the problem of adverse selection away from the public agencies. Individual producers received working capital up-front, but as an advance, much like a large private customer would provide a small supplier, rather than as a government subsidy.

As the economic activity in the region grew, and the producers gained a reputation for high quality woodworking, firms started to expand vertically, and across other sectors into private markets. Sawmills started to assemble and repair sawmill equipment, and then equipment for food-processing industries, such as sugarcane and cheese. Furniture makers were increasingly able to obtain orders from other state governments and private buyers, branching into products such as furniture for summer homes and hotels.

The biggest challenge for the program is how to replicate its success elsewhere. Enormous publicity resulted in the situation where all districts wanted to participate in a similar program. The pressure to expand the program made it politically difficult to choose only a few potential winners, and so the program was dispersed across a number of districts, which is contrary to a focused cluster-based model.

The computer industry in Taiwan[8]

The computer industry in Taiwan is an example of a high-technology cluster developed as a result of public policy initiatives. Taiwan emerged from the Second World War with a weak industrial base that was capable of exporting only sugar, rice and bananas. The Chinese who came from the mainland included experienced professionals and business people, who had valuable expertise to offer when the government took over the businesses established by the Japanese. In 1952, publicly owned enterprises dominated the economy. Due to the limited domestic market, the government adopted an export-promotion policy, which favored industrial development. From 1953 to 1988, the agricultural sector's share in the gross domestic product decreased from 38 percent to 6 percent, and by 1987, the publicly owned sector accounted for only 10 percent of the total value added in the manufacturing sector. Private enterprises are overwhelmingly SME firms, defined as having less than US$ 1 million in paid in capital. The 700,000 SMEs in Taiwan account for 70 percent of employment, 55 percent of gross national product and 62 percent of manufactured export sales.

Due to the low level of capital available, labour-intensive industries predominate and SMEs engage in little R&D. To increase domestic technological capabilities, the government has set up research institutions for technology development, and currently funds roughly half of Taiwanese R&D. Two key organizations are the Industrial Technology Research Institute (ITRI), dealing with hardware, and the Institute for the Information Industry (III), dealing with software. ITRI is responsible for developing new information technology and transferring it to private firms. It does so by publishing technical reports and disseminating information through conferences, by entering into licensing agreements and by establishing spin-off companies to produce and market the end products. The equity in these spin-off companies is normally shared among ITRI itself, domestic entrepreneurs and foreign investors. During the 1980s, six integrated-circuit companies were established through this spin-off company model.

The Institute for the Information Industry is responsible for developing and transferring software technology, supplying technical and market information to the domestic information technology industry, promoting the use of computer-related technologies and training and educating computer professionals. One of the ways it achieves these objectives is to enter into alliances with foreign partners.

This is consistent with the Taiwanese government's history of seeking foreign direct investment, with technology transfer goals. In areas such as the software industry, where domestic firms are weak, foreign firms are required to contribute to the development of local capabilities.

An example of such a partnership is the establishment of Neotech Development Corporation (NDC), jointly established in 1983 by III and IBM. NDC provided exclusive software development and design services to IBM, while IBM paid for NDC's salaries and R&D equipment. IBM engineers spent time in Taiwan directing the research, while NDC engineers visited IBM plants around the world to acquire the necessary expertise. The collaboration was so successful that a second joint initiative was launched in 1988 to develop database and artificial intelligence software.

Not only do SMEs in Taiwan engage in relatively little R&D on their own, due to limited capital and skill bases, they also tend not to have their own marketing channels, for the same reason. Instead, they rely heavily on marketing channels developed by foreign firms, often owned by overseas Chinese. In order to overcome both their technological and marketing constraints, and fueled by large foreign exchange reserves, Taiwanese firms are starting to merge with foreign firms. Mergers can provide the Taiwanese firm with technology, marketing channels and a brand name. The government is supportive of such mergers, through policies such as providing low-interest loans or investment capital.

A challenge for Taiwanese in the computer industries is the upgrading of technology. For small firms, much of the upgrading relates to technology that can easily be imitated and is not protected by patents. Therefore, especially when capital is scarce, there is little incentive for individual firms to innovate.

What can we learn from empirical studies of clusters?

Seven major conclusions can be gathered from the four case studies presented here, as well as from other published empirical studies of clusters in developing countries. They are itemized below.

1. Clusters are characterized by a high density of vertical and horizontal linkages, and so a high degree of trust is necessary for effective functioning. Socio-cultural ties and a common identity, such as kinship, occupational affiliation, residence or ethnicity, provide a social regulatory mechanism that facilitates economic transactions (Schmitz, 1995a).
2. In some sectors, there is segmentation which is important to understanding how the cluster functions. For example, the knitwear industry of Tiruppur, India is segmented into two market-based groups, the larger firms which have contracts with foreign buyers and the smaller firms which produce different types of products for the domestic market (Cawthorne, 1995). In Sub-Saharan Africa, the largest businesses tend to be owned and managed by resident expatriate minorities and this restricts their integration into the local economy (Dessing, 1990).

3   A substantial number of external economies are associated with clusters. Suppliers, trading agents, intermediaries and financiers are attracted to a geographical and sectoral concentration of firms, because with so many potential firms to do business with, it is less risky to learn about the region (Humphrey and Schmitz, 1996). As a collectivity, the cluster generates more experiential knowledge about the industry than do individual firms, and can engage in joint activities, such as promotion, transportation and training, which lowers the cost per firm of carrying out these functions. As well, having a large number of similar firms in close proximity means that orders, equipment and labor can be shared.
4   Local industry associations, or self-help organizations, developed within and by the cluster provide mechanisms for individual small firms to obtain information, expertise and resources that they would not otherwise be able to afford.
5   It is often more difficult for the cluster to acquire marketing expertise and marketing channels, particularly for export markets, than it is to upgrade technology. Firms within the cluster tend to specialize in particular production processes, and rely on neighboring firms to specialize in related processes. Owners can upgrade technology through imitation, without traveling far from the cluster. Marketing expertise and channel development, however, are often carried out by agents outside the cluster, who have greater access to foreign markets. It can be prohibitively expensive for small firms to replicate these activities.
6   It is possible to develop sectors through public policy initiatives, although it remains an open question as to how sustainable these sectors are, and whether particular success stories can be replicated elsewhere. Moreover, different sectors face different challenges, because of their different histories, socio-cultural underpinnings and sectoral bases.
7   Clusters are dynamic entities, and a cluster that functions productively in one decade may function less effectively in the next decade. Clusters are affected both by changing dynamics within the cluster and by the changing global economic environment. Rather than being considered a static model of a local economy, clusters should be considered dynamic trajectories (Humphrey, 1995).

Best practice programs for SME development

So far, the chapter has emphasized the advantages of a vibrant SME sector. There are two interrelated development tasks that can be used to establish and maintain such a sector. The first is the upgrading of firm-specific capabilities, and the second is the promotion of SME clusters, so that inter-firm cooperation is possible. Both tasks are facilitated when there is an economic infrastructure rich in BDS. This section of the chapter describes five programs which constitute best practice in the delivery of BDS. The first group of best practices described here are targeted at developing firm-level capabilities, while the second group are targeted at developing cluster capabilities.

Developing firm-level capabilities

Training programs overwhelmingly constitute the most numerous programs aimed at developing firm-level capabilities. Accordingly, this section outlines two training programs: CEFE (Competency-based Economics, Formation of Enterprise), a training program which has been widely adapted and used around the world, and a voucher payment program which is attracting interest. The third firm-level program outlined here assists firms in subcontracting or locating potential subcontractors.

Training programs

Competency-based Economies, Formation of Enterprise[9] is a comprehensive set of training instruments, based on action learning and experiential learning methods, and developed by GTZ, a development agency of the German government. It was first developed in the 1980s in Nepal and has been introduced into more than sixty countries. Its objective is to improve the entrepreneurial performance of individuals through guiding self-analysis, stimulating enterprising behavior and building up business competencies.

CEFE is highly adaptable to particular needs and economic environments, because each program is designed through an interactive, participatory, four-phase process:

1. *The appreciation workshop* is a course for support agency staff. The objectives of the course are to familiarize agency staff with the nature of the CEFE program and to familiarize CEFE program designers with the cultural norms and economic environment of the particular situation and the support mechanisms that already exist for the target group.
2. *The needs assessment* customizes the training material for the target group. The target group might be identified by characteristics such as sex, ethnicity, educational background, technology, sector, location, business size, business stage and so on.
3. *Identifying complementary support* for the participants is important to minimize the resources needed for the program. CEFE programs prefer to function through existing support institutions that business owners already use. This saves resources on the part of the program providers, as well as the participants who are already familiar with existing institutions. Designers of CEFE programs look for host agencies that are close to the target group and market-oriented and entrepreneurial.
4. *Screening candidates* allows the program to focus on those with the greatest potential for economic impact on their community. CEFE believes there is no one winning formula to pick candidates, and looks at a variety of characteristics, such as general competency, knowledge of their economic environment, motivation, access to resources and existing support systems. CEFE recommends that a fee be charged, the magnitude of which is determined by the economic circumstances of the target group and the demand for the program, since charging a fee for the course introduces a favorable self-selection bias.

An overall evaluation of the program in Asian, Latin American and African countries found that 98 percent of existing entrepreneurs with 1–2 employees and 100 percent of entrepreneurs with 6–10 employees claim that their income and turnover increased by at least 30 percent after going through CEFE training. In addition, an average of four new jobs per participant in CEFE training were created. Analyses indicate that CEFE appears to have more of an impact for participants with lower educational background and income.

The training voucher scheme in Paraguay[10]

A voucher training program was launched in Paraguay in 1995 by the Inter-American Development Bank (IDB). The program was designed to accompany a micro-enterprise lending program and to overcome problems with previous credit-linked training programs. Even though such training programs were obligatory, in the past business owners had tried to avoid them, sending family members or subordinates if they had to. Participants from different sectors and with different backgrounds were grouped together, trainers were lenders with little business knowledge, there was no assessment of entrepreneurs' needs, and programs could not cover costs because business owners were not interested in paying for them.

In the voucher program, both micro-entrepreneurs and the training institutions are screened to assure a certain level of capability. Each owner receives up to six training vouchers that have a fixed monetary value (US$ 20 in Paraguay) and can be used at any participating training institution by the owner or the firm's employees. Each month, the voucher distribution center lists how many vouchers went to each institution, so that business owners find out which courses are the most popular. Regularly, the agency makes surprise visits to, and evaluations of, affiliated institutions, which is important to minimize fraud.

There were two phases of the voucher program in Paraguay: May 1995 through May 1997, and February through December 1998. During the first two-year period, over 19,000 vouchers were redeemed, while during the second eleven-month period, over 29,000 vouchers were redeemed. During both periods, the number of people paying for training programs on their own (without vouchers) increased. The number of training institutions grew to sixty, although a group of about twenty were the most active. They adapted their programs to suit the needs of their clientele: shortening courses, increasing course variety, teaching at night and on the weekends, improving teaching methods, and emphasizing hands-on practical sessions. The average price per course dropped over the two time periods, although the price of courses varied by the topic, for example, electronics and baking courses were more expensive than general business management. The most popular training institutions were small, private, for-profit businesses, and those with a better reputation were able to charge premium prices.

Similar voucher programs have been started in Kenya, El Salvador and Ecuador, although there are a couple of issues still unresolved about the Paraguayan program. One area of debate is whether the program should help training institutions to upgrade training quality. There were no incentives for product development and institutions were expected to invest in R&D. This worked better for established training firms than it did for new entrants to the market, which tended to invest in physical assets (e.g. desks and video projectors) more than training content and delivery. A second area of debate is whether an exit strategy (for example, the gradual reduction of the value of vouchers) should be formulated up-front. If there is a clear exit strategy, it might be difficult to sustain the quality of training programs. If there is no clear exit strategy, it might be politically difficult to end the subsidy.

Subcontracting and partnership exchange (SPX)[11]

The SPX program was developed by UNIDO as part of its objective to increase the networking of SMEs among themselves and with large manufacturing firms. SPXs are organized as non-profit industrial associations, which are centers for technical information, promotion, and match-making among main contractors, suppliers and subcontractors. Using a customized computer system, an SPX can register the manufacturing capacities and capabilities of individual companies; classify these companies by product, sector, manufacturing processes and equipment; and search the data base by multiple criteria to match inquiries from buyers and contractors.

Since 1984, roughly fifty-four SPXs have been established in thirty countries, with forty-five remaining (a survival rate of 85 percent). A regional Latin American network among Latin America's thirty-two SPXs has been created, which provides each SPX with a greater geographical scope, and has supported the establishment of joint subcontracting exhibitions in the region. There are now a total number of 15,588 companies registered; there are higher volumes in nations with a higher level of economic development, because there are greater subcontracting opportunities. In addition, subcontracting is more important in particular sectors, such as the mechanical, electrical and electronics industries, metalworking, plastic, rubber, textile, leather and industrial services (repair and maintenance, testing and quality control, financial accounting, transportation, packaging etc.).

Experience with the program indicates that the best financing model for SPXs involves a variety of sources, each of which contribute at least 10 percent of the budget: the government, professional and industry associations, the individual businesses who participate in the program, revenue-generating services to members and non-members. Such services include participation in supply fairs, training, marketing research, technology and quality (ISO9000) audits, and legal assistance with contracts. It is not recommended that SPXs receive a commission on contracts awarded; such a scheme might bias the SPX towards larger members and would require firms to disclose information they might prefer not to disclose.

An SPX that is performing well after an initial three-year start-up period is likely to have visited and registered 500 firms, been involved in 200 successful consultations per year, concluded 50–100 national contracts (worth roughly US$ 3.8 million) and 25–50 international contracts (worth roughly US$ 2.5 million) per year. A survey in 1997 indicated that, on average, two-thirds of registered companies that year had obtained contracts through their SPX.

Developing cluster-level capabilities

Two programs aimed at developing cluster-level capabilities have already been discussed in a previous section: the procurement program underlying the woodworking cluster in Brazil and the technology transfer program underlying the computer industry cluster in Taiwan. The remainder of this section outlines two other types of cluster-level programs: the first focuses on the development of BDS centers and the second focuses on developing horizontal linkages among firms.

Developing BDS centers[12]

Swisscontact's approach to BDS is premised on the assumption that the poorly functioning BDS sector in many developing nations stems from inadequate service from existing BDS providers and weak demand from SMEs. Accordingly, the focus is to alleviate this supply–demand mismatch by developing BDS institutions that are entrepreneurial, demand-led and for-profit. This approach has two levels of services: (a) Swisscontact provides financial support, and sometime technical support, to business centers; and (b) business centers offer services to their SME clients.

Swisscontact enters into a formal contract, with a limited time frame, with the operator of a business center. Each party makes an equal contribution for investment (e.g. equipment leasing) and preoperational costs. There are performance-based financial incentives, based on, for example, revenue, income or contribution margin. In addition, Swisscontact might provide technical support, such as skills training, to the business center. Swisscontact does not prescribe which SME services the business centers should provide; it is their responsibility to be responsive to demand from their SME clients. Accordingly, there is a wide variation in the services offered by different business centers.

Such a BDS centers program began operating in Java, Indonesia in 1996–1997 and by mid-1998 included eight business centers. Initial investments are roughly $50,000, with annual financial support for two to three years (based on profits) of roughly $35,000. Each business center has a permanent staff of five or six employees, all with at least one university degree. Product offerings differ across the centers, and include technical training, management training, consulting and aid in accessing credit. A total of 300 SMEs have been served by the eight centers, at a cost to Swisscontact of $600–$700 per SME.

A challenge for the business centers is to differentiate themselves so they can compete effectively in environments where there is a predominance of public institutions offering heavily subsidized or free business services. One of the business centers has done so by developing partnerships with larger institutions, such as universities, which pay the full cost of the services to the business center, but then subsidize the participants. Such partnerships increase financial viability in the short term, but may hinder the sustainability of the business center in the long term if it becomes overly dependent on large partners or government support, and does not invest in market research and product development.

Network development[13]

The objectives of the UNIDO network development program in Nicaragua were to develop horizontal linkages among SMEs. The philosophy behind the program is that new market opportunities come about only through firms' abilities to respond to demand in a timely and high-quality manner, which requires long-term improvements in production capability and organization. The program is in its second three-year phase with a budget of US$ 1.3 million. Since the program's start in 1995, twenty networks of firms have been developed in Nicaragua by a team of seven national consultants, and the program is training more national consultants and starting to become involved in vertical linkages, such as supplier upgrading.

The national consultants, known as "network brokers," initially need to upgrade and specialize their knowledge and skills in areas such as export markets and network practices. They are encouraged to share experiences with each other and to visit other clusters as a source of new ideas and best practices. Accordingly, each network is active in inter-institutional networking, and shares activities with other local BDS providers.

The role of the network brokers is to identify groups of SMEs with similar characteristics and growth constraints, and help them to establish common projects. These common projects involve a division of tasks among the firms and group saving schemes. Each network meets weekly to discuss problems and develop a common workplan. The workplan articulates agreed-upon objectives, and how the achievement of the objectives will be evaluated in the short term, medium term and long term. The program assessment techniques are primarily qualitative, based on these previously established evaluation criteria. Once the network is functioning, the network brokers encourage the group towards continuous improvement and more difficult objectives.

An example of a successful network is Ecohamaca, a network of eleven small handicraft firms competing in the town of Masaya. With the establishment of the network, they were able to standardize production to increase joint production volume to a level required for exporting, and to improve product design and quality. They also use environmentally friendly materials, which are more attractive to the export markets in Europe and North America. They are currently exporting an average of 3,000 hammocks per month. The group is now diversifying its product line (e.g. hanging chairs and cribs), participating in joint purchases of raw materials at lower cost, increasing joint marketing activities (e.g. designing an Internet presence) and is starting to market a "Made in Masaya" brand for the entire cluster. In addition, the network has hired an administrator to perform managerial duties, such as identifying appropriate formal training opportunities for workers, researching new technological and funding opportunities and strengthening the marketing strategy.

Delivering BDS TO SMEs

The examples described in the previous section, as well as decades of experience in development assistance, indicates that there are certain principles that should be followed in developing such an infrastructure (ILO, 1997). It is particularly important to recognize factors which limit SMEs' use of such services, so that means of overcoming these factors are identified and implemented.

A starting point for this discussion is to note the broad range of forms that BDS can take. The term can include training, counseling, the provision of market information, the development of commercial entities, technology development and transfer and facilitation of business linkages, to identify only some of the more common forms. The following discussion is, of necessity, general in that it highlights principles that should guide the design and delivery of BDS of any kind.

The next part of this section outlines seven key principles for effective design and delivery of BDS: demand-driven, self-sustaining, customizable, participatory, network-oriented and evaluative. This is followed by a discussion of the factors limiting the use of BDS and ways that BDS outcomes can be measured.

Principles of effective BDS design and delivery

Design demand-driven programs

The experience of BDS providers to date has consistently shown that the organizations providing BDS *must* think and act like businesses themselves. They must be driven by demand, which means they:

- treat those who receive services like customers, not recipients of charitable assistance;
- treat the services they provide as products and ensure that they meet the needs of clients;
- maintain close contact with clients so that their understanding of changing client needs and how to meet them is up-to-date (ILO, 1997).

In general, donors and governments have been ineffective as service providers (Gibson, 1997) because they do not have systems and values like the SMEs they serve. It is better to select local partner organizations (often SMEs themselves) which do reflect local systems and values. For example, the training voucher scheme focuses on developing local training institutions and Swisscontact's BDS Centers program encourages centers to tailor services to their local clientele.

Donors or governments must give these partner organizations the flexibility they need to pursue these objectives (ILO, 1997).[14]

It is in keeping with an emphasis on designing programs based on demand to adopt a specialization-based approach to service delivery. It will rarely be the case that any one partner can or should provide a full range of services. Every service delivery organization will have certain strengths and limitations corresponding to meeting specific needs, not least of which is a capacity limitation. Service providers can have simpler, more demand-responsive relationships with clients if they are focused on doing a few things well. This does not mean innovation by service providers should be discouraged. It only means that a diffusion of energies across a wide range of activities should not be encouraged (ILO, 1997).

Design self-sustaining programs

Two observations support the argument that BDS should be designed to be self-sustaining. First, if services are demand based, they should be of sufficient value to the SMEs who benefit from them that they are willing to pay something for them in the short run (as is recommended in the CEFE and SPX programs), and to pay for their full costs in the long run. Second, it is clear that governments and donors cannot sustain support for a full range of BDS services in perpetuity. Taking these two points together, there is an emerging consensus that private markets must be encouraged to provide BDS (Hallberg, 1999).

This means that the priority should be given to providing those services that can, eventually, become commercialized and privatized. Any interventions which are not designed with a view to finite government or donor involvement need to have a clear rationale, and a plan for long-term support (ILO, 1997). In the long run, it is desirable that only public goods be subsidized (Hallberg, 1999).

Design customizable programs

Different kinds of SMEs need different BDS. Firms performing different functions within a sector (e.g. raw material suppliers vs equipment repair services vs manufacturers) and firms in different industries (e.g. textile manufacturers vs metalworkers vs electronics firms) will have quite distinct needs for services. Public policy agendas often favor making services widely available, or providing services for particular socio-economic groups (ILO, 1997). However, unless tailored services are designed for sectoral or industry groups with distinct sets of needs, the resulting service offerings are likely to suit no group particularly well. An important aspect of the CEFE program, for example, is to customize training material and personnel, as well as program content and teaching methods, so that it is sensitive to characteristics of the target group, such as sex, ethnicity and educational background. Similarly, Swisscontact's BDS program is based on the individual business centers' defining and targeting particular customer segments. Especially when cluster formation is to be encouraged, it is essential that services be tailored to the particular needs and characteristics of specific industries in specific regions at specific points in their development.

Adopt a participatory approach

Involving SME owners and personnel in the design of services can have three distinct kinds of benefits. First, in both developed and developing countries, there tends to be low awareness and even lower usage rates among SME owners of available services (e.g. Wignaraja and O'Neil, 1999; Orser et al., 1999)[15] and involving owners in program design can increase both. Second, it helps ensure that the needs of the client are well understood, and that the services developed for them are appropriate. Third, it can lead to a greater sense of ownership and buy-in among those who have some responsibility for ensuring that the assistance works as intended: service delivery staff, SMEs who are clients, and other participating organizations such as business associations (Gibson, 1997).

Adopting a participatory approach may sometimes mean that the types of services to be provided are not fully specified in advance. Some programs, such as CEFE and the UNIDO network development program, include participation as a key component of program development, while others, such as the training voucher scheme, are flexible enough to enable emerging requirements to be met. Some reasonable limits will need to exist, but it makes little sense to invite participation so that needs can be met, and yet to be inflexible about gathering information about, and meeting, those needs (ILO, 1997).

Stimulate competition

Any policies designed should stimulate SMEs to be competitive and avoid sheltering them from competition. Policies giving SMEs preferential treatment in bidding for particular kinds of government contracts in India and in the United States, for example, have been criticized for creating protected niches where suppliers have limited incentive to compete based on cost or quality, and where public sector customers know they must buy from SMEs for "social" reasons (Tendler and Amorim, 1996). In other jurisdictions, such as the Ceará region in Brazil, SMEs were given assistance in meeting contract specifications for public purchasing agencies, but no guaranteed assurance they would be the favored bidders for these contracts. This kind of incentive helped them produce competitive offerings, and launched them into private markets rather than leading them to depend on ongoing assistance or preferential treatment (Tendler and Amorim, 1996). A further example of encouraging competitiveness is the training voucher program, which publicizes the most popular courses.

Develop networks among service providers

Frequently there are multiple donors and levels of government providing or funding BDS. To avoid imbalance in the system, awareness of other actors is essential, and some coordination may be useful. This means that those involved in funding services should communicate to ensure that conflicting delivery mechanisms are not undermining some of the objectives outlined above (ILO, 1997). One way to do this is to incorporate an investigation of current institutions in the program design process, like the CEFE program does; another is to specifically facilitate networks among program developers, like the UNIDO network development program does.

Competition among service providers is quite likely to occur, and can be useful. But some collaboration is also likely to be beneficial. At a minimum, this can mean service providers refer clients to competitors, but more sophisticated forms of cooperation are also possible. For instance, subcontracting among service providers with differing areas of expertise or focus can sometimes be efficient. Business service providers need to define and to focus on their areas of core competence.

Build-in cost/benefits analyses

While BDS providers routinely have budgets and maintain accounts, they are less likely to use cost analysis to improve business management and to help assess the relative effectiveness of various offerings. The capacity to analyze costs and to monitor performance objectives needs to be built into service design so that service providers are able to conduct cost/benefit analyses of their offerings (ILO, 1997). Although costs should not be considered the only indicator of performance, failure to include cost assessment inhibits an understanding of the efficiency and effectiveness of different practices. For example, Swisscontacts BDS program keeps track of the costs per business center, as well as the cost per SME served.

Factors limiting the effective use of BDS

Excessive administrative burdens

Administrative burdens are often a significant barrier to SMEs in accessing services meant to assist them (Levy, 1994). For instance, completing even well-designed registration requirements can be an obstacle for smaller firms due to their limited management and monetary resources. While large firms can afford to employ specialized staff, micro-enterprises are hard pressed to find the time to deal with voluminous and often cumbersome paperwork, much less to clarify ambiguities about when and to whom particular policies or regulations apply. This problem is compounded when there are "hidden costs" such as extra payments required by officials, or unmanageable delays in dealing with bureaucracies (Meier and Pilgrim, 1994). Too often, the limited resources of the smaller firms simply become over-taxed, and so services designed for SMEs are under-utilized.

Ineffective market segmentation and targeting

Service providers are sometime perplexed to find that the services they have designed for and promoted to particular kinds of SMEs are not used by many of them. This kind of problem may arise when the basis for segmentation and the groups targeted are not well chosen in the first place. A segmentation basis is the principle or criterion by which a whole population of potential service recipients is divided into groups, with the idea that the sub-groups identified have similar kinds of needs. If the basis for segmenting groups of SMEs do not lead to the targeting of groups of organizations that are homogenous with respect to their service requirements, then the service designed for them will work for only a limited subset of the group. Frustration can arise among those SMEs who are being encouraged to take advantage of a service that is not well suited to their needs. Worse, there may be SMEs that would benefit from the service that are not in the group that has been targeted.

In practice, a given segmentation criterion is relevant in some settings and not in others. For example, segmenting groups of SMEs based on the sex of the business owner is a common practice in both developed and developing countries, but it is arguably more appropriate in some cases than others (Orser et al., 1999). For services that will assist in the formation of clusters, it may be that sex of the business owner (like any number of other possible demographic criteria) is *not* particularly relevant. Rather, segmenting based on industry, and targeting particular kinds of services to particular industries may allow for the most effective design and delivery of services. In other cases, however, it may indeed be necessary to stimulate clusters by segmenting on some demographic basis such as social class or ethnicity and to target particular kinds of services to the certain demographic subgroups.

Ineffective communications

When appropriate services have been designed for appropriate target groups, there are sometimes still problems in "getting the word out." In both developed and developing countries, studies have found low awareness of services targeted to SMEs (Meier and Pilgrim, 1994; Orser et al., 1999). These findings are not surprising given that it is common for all resources available to be devoted to the services themselves, with no resources reserved for communicating their existence.

To use fewer scarce resources, social networks can be used effectively to communicate information about services to firms in a cluster. Leaders in local trade or industry associations can often pass on information to a wide array of members. Community leaders may be able to identify opinion leaders among the SMEs in the cluster. If boards of trade or commerce exist, they might be useful for getting the needed information into the right hands as well. The key point here is that what will work depends on the specific social dynamics of the particular industry in the particular region.

Measuring the outcomes of BDS

It is difficult to prescribe in general terms how to measure the outcomes of BDS due to the wide range of services offered. A starting point for assessing the efficiency and

Table 5.2 Four kinds of performance criteria for BDS

Performance criterion	Description	Examples of indicators
Outreach	The quantitative scale of the service's impact	Number of SMEs that participated, number of networks
Efficiency	The rate and cost at which services are turned into measurable "outputs"	Training cost per firm, cost of vouchers redeemed, time to develop a viable network
Effectiveness	The extent to which the objectives of the program have been met	Sales growth per participating SME, number of contracts obtained, growth in exports of a network
Sustainability	(1) The extent to which a service can be financed through client fees (2) The extent to which the outcomes are durable	Number of SMEs paying for their own training, a network that hires an administrator, survival rate of start-up SMEs

effectiveness of a service is identifying the objectives underlying the service. However, it is sometimes the case that the purpose of a particular kind of intervention and/or the specific client group has not been defined as clearly as they should be (Gibson, 1997).

The box in Table 5.2 shows four kinds of performance criteria that can be used to assess whether a service has met its goals once those have been defined (ILO, 1997). These criteria, taken together, reflect a balance of different concerns related to qualitative impact, quantitative scale and efficiency. In practice, few BDS providers measure performance using all these criteria. Focusing on only one or two, however, can lead to a misuse of resources because high performance on one criterion may come at the expense of performance on another. For example, providing minimal assistance to large numbers of firms can yield high outreach performance, but low impact (ILO, 1997). This suggests that efforts must be made to measure performance based on as many of the criteria as possible.

How can these criteria be made tangible? Inventories of measures for each can be developed, but caution is required in applying such inventories (ILO, 1997). Generic lists of measures need to be supplemented with appropriate local indicators in most cases, since no general list is likely to serve the purposes of every service in every setting. Further, there are qualitative dimensions to good practice that no list of measures can capture and reducing evaluation to a pass/fail on a series of indicators must be resisted (Gibson, 1997). Measures of the *process* of the delivery service itself (as opposed to its outcomes) are likely to be situation-specific and not included in a general list, but these process indicators of operational effectiveness are often vital for purposes of monitoring progress (ILO, 1997).

Finally, and of particular relevance when measuring the effectiveness of services targeted at clusters rather than at individual SMEs, it should be recognized that the criteria above are much easier to operationalize when the objectives involve individual firms than when they involve groups of firms within a region. While using measures of effectiveness at the level of individual firms may be appropriate as *part* of an effectiveness assessment for services aimed at cluster development, they are not sufficient, and may conflict.

Cost-based effectiveness measures require specific mention here, because they are sometimes, inappropriately, avoided altogether and sometimes over-valued. There are obvious limitations to any individual measures. Even the most seemingly objective measures are subject to multiple interpretations. Cost-based measures related to operational efficiency can and should be developed and used, as long as their susceptibility to manipulation is never neglected, and their limited linkage to the full range of policy goals is kept in view (Gibson, 1997).

Benchmarking is another topic that must be briefly addressed. If standardized performance criteria and indicators of performance are developed and used, it becomes possible to compare performance across services and, potentially, to evaluate services relative to one another. In practice, the application of benchmarking to the appraisal of BDS is limited for a number of reasons (ILO, 1997). First, completely standardized approaches to measurement are not appropriate: good practice requires services to be focused and specific for particular groups in particular settings. Second, any comparisons based on costs are difficult because of different costs of living in different settings. Third, effectiveness measures are among the most important and the least comparable when they are done well. Fourth, unless data from a significant number of projects is available for any standardized measures that are used, it is unclear that the limited data which does exist will form a sound basis for comparison.

Summary and conclusions: maintaining a balance

Balancing general needs and cluster-specific needs

Business-related policies and services have frequently been developed and delivered in order to benefit "all" SMEs. Public policy-makers are naturally wary of initiatives that would appear to direct resources to one group unless they can offset this by offering comparable resources to all, or at least to many. There are some priorities in policy-based interventions that systematically favor particular groups or activities, for example, increasing gender equity, promoting environmentally sensitive development, and improving working and employment conditions (ILO, 1997). None of these, however, favor developing specific policies or offering targeted programs to foster the development of a cluster.

Yet, if policies and BDS supportive of clusters are *not* created, it is unlikely that clusters can thrive (Humphrey and Schmitz, 1996). Policy-makers therefore sometimes need to make the choice to help foster a cluster in a particular region. Occasionally, they may have to choose between supporting some clusters and not others. These choices need to be based on an assessment of whether a cluster has the potential to develop in a way that will have long-term social and/or economic benefits. Not all clusters have equally high potential. A number of factors can be used to identify those that have greatest potential.

The strength of the socio-cultural milieu. A strong system of shared values and a developed network of social ties between people within a region will make it possible for SMEs in the district to be able to act cooperatively, share information and gain technical know-how from one another.

Strong, existing associations. Associations such as local trade associations, industry associations or chambers of commerce, provide a mechanism through which geographically concentrated SMEs can recognize and advance their collective interests. Programs can encourage the creation of such associations, but it is more promising if a basis exists upon which to build.

The existence of both horizontally comparable and vertically linked firms. Such existence within a region suggests that the potential for synergies from clustering are present. The value of horizontally comparable firms, typically a group of manufacturers or artisans, lies in their potential to learn from one another, to stimulate one another to innovate and to act cooperatively for purchasing inputs, co-producing outputs, or lobbying for collective interests. The value of the vertically linked firms, in particular traders or agents who sell to end customers, is that they are closest to the market and can therefore provide a range of useful information to producers. Without the "downstream" firms, the traders or wholesalers, there is less momentum compelling producers to respond quickly to changing market conditions. Wholesalers and traders are often catalysts in developing linkages between producing firms and in stimulating them to work together to compete, probably because they are the first to suffer from downturns in markets (Advani, 1997).

Comparative advantage. Cluster development will be more successful if it is limited to those industries in which the region has a comparative advantage for an industry of that kind relative to other regions or countries. While it may seem desirable to promote clusters based on the need for economic development in a region, the cluster is unlikely to be successful in the long run unless the region has relative advantages in terms of either cost or other inputs.

The presence of these factors cannot ensure that an effective industrial cluster can be developed. However, without at least some of them in place, it is unlikely that any amount or type of intervention will be successful.

Balancing traditional and market-based approaches to services

Traditional assistance to SMEs has been based on a donor/recipient model. This model has produced services that are provided at little or no cost to SMEs, but that are "supply driven." There is an emerging consensus that the services provided have been too general, of variable quality, and not delivered with an awareness of cost control (Hallberg, 1999).

Increasing recognition is being given to a more market-driven model. This approach stresses the development of markets and networks, and de-emphasizes the role of governments and donors in any direct provision of services or benefits for SMEs. Intervention, under this market-driven model, needs to invest in public goods and to address the market failures that create cost disadvantages for SMEs, that restrict their access to markets, or that prevent the development of markets for services that SMEs need. It needs also to reduce regulations and policies that create fixed costs which systematically disadvantage SMEs (Hallberg, 1999).

It is unlikely, however, that any country will fully cease to subsidize SMEs. Nevertheless, subsidization can be offset by encouraging the development of private markets for SME assistance and by treating the subsidized assistance in as business-like a manner as possible. There should be a determination of whether the market-developing potential of any policy or program offsets its market-distorting impact.

Balancing short-term and long-term cluster dynamics

Nothing lasts forever, and those designing policies and programs to support industrial clusters know that this adage will apply to most clusters as well. It has long been recognized that there is a lack of stability in industrial locations. Economic history is full of many examples of industries that have emerged in particular regions, grown and then eventually declined or disappeared.

Even in the short term, when clusters emerge and create a collective capacity to compete, adapt and innovate "it is important not to expect an island of unity and solidarity" (Schmitz, 1995a, p. 534). As collective efficiency develops, some enterprises prosper and others decline. Relationships between both horizontally linked and vertically linked firms can range from exploitative to collaborative. Negative relationships can lead to conflict, and some of the synergies of clusters can be lost.

In the long term, many possible dynamics are possible. Even Italy's celebrated clusters have not all uniformly continued to prosper. Restructuring has been common, with an increase in the average size of the enterprise in the cluster and more differentiation by size. Cooperation and competition among equals has in some cases given way to a hierarchical structure with the firms that have grown large subcontracting out work to the smaller ones (Humphrey and Schmitz, 1996).

It appears that the likelihood that clusters will innovate and evolve rather than dissipate and dissolve is influenced greatly by the nature and strength of the underlying socio-political networks. In those that have innovated and evolved, the underlying networks of groups and associations have enabled clusters to restructure and thrive. In others, fragmentation and under-investment have occurred when a socio-political basis for joint action has eroded and no other basis for cooperative behavior has replaced it (Locke, 1995). It is ironic to note that to some extent, the success and growth of the cluster leads to a breakdown in the social cohesion which allowed it to perform well in the first place. With growth comes increasing differentiation among firms, and a greater participation in the system by "outsiders." Both can erode shared understandings rooted in social systems, and undermine trust based on those understandings.

Factors exogenous to the cluster also influence its trajectory of behavior. A number of distinct factors can be identified, though in practice these factors are very much interrelated:

- Increasing trade liberalization can both alter the accessibility of local markets to non-traditional competitors, and increase the potential attractiveness of foreign markets. It can also mean that the division of labor for production and commercialization spans continents rather than being contained within national boundaries. When this occurs, clusters may be "left out" of the chain of production for mass-produced goods and be forced to seek out a new market niche, or individual firms in the cluster may enter into production relationships with larger firms located elsewhere and form new networks outside the clusters (Humphrey, 1995).
- The diffusion of new paradigms of industrial production, such as "just-in-time" production or "total quality management," can increase competitive intensity from nations where these processes have already gained ground, and create pressure to reorganize production systems within clusters (Humphrey, 1995).
- Technological changes in an industry can "destroy competences," or make obsolete the competence that individual firms or clusters have. The SMEs that are typically present in clusters may be better able than large firms to cope with such technological changes in that they are better at improvising responses and less committed to routines (Swann and Prevezer, 1998). There are no assurances, however, that this coping potential will be realized.

When the collective impact of these factors leads to rapidly increasing competitive intensity within an industry, the fate of clusters can be affected.

Taking into account both socio-cultural dynamics and global competitive dynamics, it is possible to distinguish analytically between four possible trajectories for cluster development. Table 5.3 illustrates these four possible outcomes. It suggests that when the bases for cooperative action are strong or increasing and the competitive intensity of the industry is stable or increasing slowly, clusters can benefit from the synergies of agglomeration. When those same bases for cooperative action exist, but competitive intensity is increasing rapidly, it is still possible for clusters to prosper collectively, but it is likely that restructuring of the production process will occur, that some of the firms in the cluster will benefit more than others, and that new networks outside the cluster will develop. Under the scenario where the bases for cooperative action have become weak or are weakening, and competition is increasing, it becomes difficult for the cluster to continue to benefit from the synergies of co-location. Even in the situation where competitive intensity is not increasing rapidly, if the bases for collective or cooperative action within the cluster have eroded, it is likely that some firms will begin to develop networks outside the cluster and prosper, and others will simply become non-competitive.

Table 5.3 Contingent development trajectories for industrial clusters

Competitive intensity in global industry	Bases for cooperative action in cluster	
	Weak or decreasing	Strong or increasing
Constant or increasing slowly	Survival with low to moderate collective efficiencies likely	Increasing innovation and heightened competitive position
Increasing rapidly	Fragmentation likely	Restructuring likely; ongoing competitiveness possible

It must be borne in mind when considering this analysis however, that neither the survival nor the demise of clusters is inevitable: trajectories need not be "straight line" ones. As Schmitz (1995a) illustrates, competitive factors can both lead to the temporary dissolution of elements of clusters *and* serve as shocks or crises that stimulate them eventually to higher levels of cooperation. He also observes that the foundations on which this cooperation is built may evolve as the cluster does. Operating in world markets can both break down ties based on shared socio-cultural characteristics and result in new ties based on a conscious investment in inter-firm relationships.

As clusters evolve, so too will the roles that public policies and programs should play. Occasionally, government or donor interventions may be required depending upon a cluster's ability to sustain a basis for cooperative action. Only by focusing attention on the trajectories of particular clusters over time can policy-makers hope to assist them to serve as engines of an economy.

Notes

1 Some countries define SMEs on the basis of the value of sales or the value of assets. The differing definitions can make cross-country comparisons difficult.
2 This discussion of the characteristics of SMEs draws heavily on Leidholm and Mead (1999), who synthesize the most recent available results of surveys of over 65,000 SMEs in Botswana, Kenya, Swaziland, Zimbabwe, the Dominican Republic, Guinea, Jamaica, Lesotho, Niger, Nigeria and South Africa. The businesses they report on include all enterprises engaged in non-primary activities with at least 50 percent of output sold for consumption outside the home, and employing up to fifty workers.
3 Tacit knowledge is practical knowledge that is usually not directly expressed or taught, but rather is acquired through experience (Sternberg and Caruso, 1985).
4 The material on the shoe makers of Brazil's Sinos Valley is taken from Schmitz (1995b, 1998).
5 This program involved all the stakeholders (shoe manufacturers, suppliers, association officials and consultants from the university) but was not a success because the largest firms had integrated vertically and became less interested in collective action. It was therefore the association of suppliers, who were concerned about the impact of foreign competition and exchange rates for their customers, that led the next round of cluster wide initiatives (Schmitz, 1998).

6 The material on machining firms in Japan's Ota Ward is taken from Whittaker (1997).
7 The material on wood making firms in Brazil's Ceara state is taken from Tendler and Amorim (1996).
8 The material on the electronics industry in Taiwan is taken from Hou and Gee (1993); Lall (1999).
9 The material on CEFE is taken from CEFE (1995); Kolshorn (1999).
10 The material on the training voucher scheme is taken from Goldmark (1999).
11 The material on the SPX program is taken from de Crombrugghe and Montes (1999); ILO (1997).
12 The material on Swisscontact's approach to BDS is taken from Hitchins and Gibson (1999).
13 The material on UNIDO's network development program is taken from Ceglie and Dini (1999).
14 The Swiss Agency of Development Cooperation is an example of a donor that has pursued just such an approach in Ecuador and Peru. It has invited business service providing companies to bid on contracts to provide assistance to SMEs. Payment for the contracts is based on achievement of performance objectives. There are incentives for good performance, defined as developing improved services offering SMEs, and institutions that fail to perform receive no more contracts. The evidence thus far available suggests this system has led to the development of a greater variety of services offered to SMEs (Gibson, 1997).
15 In the Wignaraja and O'Neil (1999) study, for example, the authors found that when asked about five institutions offering support services in Mauritius, SME's self-reported awareness levels ranged between 65 and 73 percent. The portion of SMEs who said they had actually used the services ranged between 14 and 23 percent of those surveyed.

References

Advani, A. (1997), "Industrial clusters: a support system for small and medium sized enterprises," Private Sector Development, *World Bank Occasional Paper No. 32*, Washington, DC: World Bank.

Brown, J.S., Collins, A. and Duguid, P. (1996), "Situated cognition and the culture of learning," in H. McLellan (ed.), *Situated Learning Perspectives*, Englewood Cliffs, NJ: Educational Technology Publications.

Cawthorne, P.M. (1995), "Of networks and markets: the rise and rise of a South Indian town, the example of Tiruppur's cotton knitwear industry," *World Development* 23(1): 43–56.

CEFE (1995), CEFE International web site. Available online at http://www.gtz.de/cefe/products/214.htm (21 October 1999).

Ceglie, G. and Dini, M. (1999), "SME cluster and network development in developing countries: the experience of UNIDO," Paper presented at the International Conference on Building a Modern and Effective Development Services Industry for Small Enterprises, Rio de Janeiro, Brazil, March 2–5. Sponsored by the Committee of Donor Agencies for Small Enterprise Development, Enterprise and Cooperative Development Department – Entrepreneurship and Management Development Branch, International Labour Organization. Available online at http://www.ilo.org/public/english/65entrep/isep/bds/donor/rio.htm (20 October 1999).

de Crombrugghe, A. and Montes, J.C. (1999), "UNIDO global experience on industrial subcontracting and partnership," Paper presented at the International Conference on Building a Modern and Effective Development Services Industry for Small Enterprises, Rio de Janeiro, Brazil, March 2–5. Sponsored by the Committee of Donor Agencies for

Small Enterprise Development, Enterprise and Cooperative Development Department – Entrepreneurship and Management Development Branch, International Labour Organization. Available online at http://www.ilo.org/public/english/65entrep/isep/bds/donor/rio.htm (20 October 1999).

Dessing, M. (1990), "Support for microenterprises: lessons for Sub-Saharan Africa," *World Bank Technical Paper No. 122*, African Technical Department Series, Washington, DC: World Bank.

Gibson A. (1997), "Business development services: core principles and future challenges," *Small Enterprise Development* 8(3): 4–14.

Goldmark, L. (1999), "The voucher training model: what next, after Paraguay," Paper presented at the International Conference on Building a Modern and Effective Development Services Industry for Small Enterprises, Rio de Janeiro, Brazil, March 2–5. Sponsored by the Committee of Donor Agencies for Small Enterprise Development, Enterprise and Cooperative Development Department – Entrepreneurship and Management Development Branch, International Labour Organization. Available online at http://www.ilo.org/public/english/65entrep/isep/bds/donor/rio.htm (20 October 1999).

Hallberg, K. (1999), "Small and medium scale enterprises: a framework for intervention." Available online at http://wbln0018.worldbank.org/networks/fpsi/rmfsme.nsf (20 October 1999).

Hitchins, R. and Gibson, A. (1999), "Swisscontact: the business center approach in Indonesia and the Philippines," Paper presented at the International Conference on Building a Modern and Effective Development Services Industry for Small Enterprises, Rio de Janeiro, Brazil, March 2–5. Sponsored by the Committee of Donor Agencies for Small Enterprise Development, Enterprise and Cooperative Development Department – Entrepreneurship and Management Development Branch, International Labour Organization. Available online at http://www.ilo.org/public/english/65entrep/isep/bds/donor/rio.htm (20 October 1999).

Hou, C.-M. and Gee, S. (1993), "National systems supporting technical advances in industry: the case of Taiwan," in R.R. Nelson (ed.), *National Innovation Systems: A Comparative Analysis*, Oxford: Oxford University Press.

Humphrey, J. (1995), " Industrial reorganization in developing countries: from models to trajectories," *World Development* 23(1): 149–162.

Humphrey, J. and Schmitz, H. (1996), "The triple C approach to local industrial policy," *World Development* 24(12): 1859–1877.

ILO (1997), "Business development services for SMEs: preliminary guideline for donor-funded interventions," Available online at http://www.ilo.org/public/english/65entrep/isep/bds/donor/guide.htm (20 October 1999).

Kolshorn, R. (1999), "Inside-out; the case of CEFE," Paper presented at the International Conference on Building a Modern and Effective Development Services Industry for Small Enterprises, Rio de Janeiro, Brazil, March 2–5. Sponsored by the Committee of Donor Agencies for Small Enterprise Development, Enterprise and Cooperative Development Department – Entrepreneurship and Management Development Branch, International Labour Organization. Available online at http://www.ilo.org/public/english/65entrep/isep/bds/donor/rio.htm (20 October 1999).

Lall, S. (1999), "Opening up – and shutting down? Synthesis, policies and conclusions," in S. Lall (ed.), *The Technological Response to Import Liberalization in SubSaharan Africa*, London: Macmillan Press Ltd.

Levy, B. (1994), "Technical and marketing support systems for successful small and medium-size enterprises in four countries," *Policy Research Working Paper No. 1400*, Washington, DC: Policy Research Department, Finance and Private Sector Development Division, World Bank.

Levy, B., Berry, A. and Nugent, J. (1999), "Supporting the export activities of small and medium enterprise (SME)," in B. Levy, A. Berry, and J.B. Nugent (eds), *Fulfilling the Export Potential of Small and Medium Firms*, Boston, MA: Kluwer Academic Publishers.

Liedholm, C. and Mead, D.C. (1999), *Small Enterprises and Economic Development: The Dynamics of Micro and Small Enterprises*, London: Routledge.

Little, I.M., Mazumdar, D. and Page, J.M. (1987), *Small Manufacturing Enterprises: A Comparative Analysis of India and Other Economies*, New York: Oxford University Press.

Locke, R. (1995), *Remaking the Italian Economy*, Ithaca, NY: Cornell University Press.

Meier, R. and Pilgrim, M. (1994), "Policy-induced constraints on small enterprise development in Asian developing countries," *Small Enterprise Development* 5(2): 32–38.

OECD (1997), *Globalisation and Small and Medium Enterprises (SMEs). Volume 1: Synthesis Report*, Paris: Organisation for Economic Co-operation and Development.

Orser, B., Fischer, E., Hooper, S., Reuber, A.R., Riding, A. (1999), *Beyond Borders: Canadian Businesswomen in International Trade*, Ottawa, Canada: Department of Foreign Affairs and International Trade.

Rabellotti, R. (1995), "Is there an industrial district model? Footwear districts in Italy and Mexico compared," *World Development* 23(1): 29–41.

Schmitz, H. (1995a), "Collective efficiency: growth path for small-scale industry," *Journal of Development Studies* 31(4): 529–566.

Schmitz, H. (1995b), "Small shoemakers and Fordist giants: tale of a supercluster," *World Development* 23(1): 9–28.

Schmitz, H. (1998), "Responding to global competitive pressure: local co-operation and upgrading in the Sinos Valley, Brazil," *Working Paper No. 82*, Brighton: Institute of Development Studies, University of Sussex.

Spar, D. (1998), "Attracting high technology investment: Intel's Costa Rican plant," *FIAS Occasional Paper 11*, Washington, DC: IFC/FIAS.

Sternberg, R.J. and Caruso, D.R. (1986), "Practical modes of knowing," in E. Eisner (ed.), *Learning and Teaching the Ways of Knowing*, Chicago: University of Chicago Press.

Swann, G.M.P. and Prevezer, M. (1998), "Introduction: conclusions," in G.M.P. Swann, M. Prevezer and D. Stout (eds), *The Dynamics of Industrial Clustering: International Comparisons in Computing and Biotechnology*, Oxford: Oxford University Press.

Swann, G.M.P., Prevezer, M. and Stout, D. (1998), *The Dynamics of Industrial Clustering: International Comparisons in Computing and Biotechnology*, Oxford: Oxford University Press.

Tendler, J. and Amorim, M.A. (1996), "Small firms and their helpers: lessons on demand," *World Development* 24(3): 407–426.

Whittaker, D.H. (1997), *Small Firms in the Japanese Economy*, Cambridge: Cambridge University Press.

Wignaraja, G. and O'Neil, S. (1999), "SME exports and public policies in Mauritius," *Commonwealth Trade and Enterprise Paper No. 1*, London: Commonwealth Secretariat.

World Bank (1999), "Social capital initiative home page," Available online at http://www.worldbank.org/poverty/scapital/index.htm (19 July 1999).

6 Government policies towards foreign direct investment

Dirk Willem te Velde[1]

Introduction

Governments in developing countries are increasingly looking for best-practice policies towards Foreign Direct Investment (FDI).[2] Renewed confidence in the positive benefits of FDI has led many countries that were restricting FDI in the 1960s–1980s to be more open towards FDI in the 1990s (Safarian, 1999) and beyond. Governments are liberalising FDI regimes as they associate FDI with positive effects for economic development in their countries (e.g. Borensztein *et al.*, 1998; Lall, 2000a). Of course, in actual practice objectives to attract FDI differ by country (e.g. technology, market access, growth and poverty alleviation) and the effects of FDI may not always be desired (neglect of local capabilities, environmental damages, inequality between individuals or regions).

Increased liberalisation and technological advances have led to a rapid growth in FDI flows over the last three decades. FDI increased as a ratio of domestic investment and GDP in many countries (UNCTAD, 2000). However, while some countries attracted large FDI flows, others were less successful, even though they had liberalised FDI regimes. Intensified competition for FDI (Oman, 2000) has led many organisations to look for benchmarks of policies towards attracting FDI (see e.g. IPAP (2000) in the Asia-Europe meetings; CBI (1999) in the case of African countries). Countries are forced to be more open towards FDI in the emerging environment (including WTO rules and the importance of technology transfer) where it is difficult to build up an industrial capacity behind closed doors.

Whilst for some countries there is concern about the quantity of flows, there is a shift in other countries towards the quality of FDI. The term quality usually refers to high value-added FDI and/or to FDI with positive linkages and spillovers effects for the domestic economy. Countries that have had successful development based on FDI need to continue to upgrade FDI, either by encouraging existing multinational affiliates to develop into strategic independents, or by targeting higher value-added FDI. With WTO rules limiting domestic policy options we will look at what policies a government can still use.

Relying on high quality FDI does not guarantee (and sometimes prevents) the improvement of local capabilities. We review whether FDI has positive spillovers for the local economy in terms of growth and productivity. Theoretical developments

(Cohen and Levinthal, 1989; Blomstrom et al., 2000b) and empirical evidence (e.g. Borensztein et al., 1998) show that the development of local capabilities is crucial in benefiting from FDI. The encouragement of linkages between local suppliers and transnational corporations (TNCs) may also be important in developing local firms, for example, through a linkage programme or in a cluster development strategy.

Governments wanting to use FDI as part of achieving a development objective will therefore have to think of policies towards *attracting* FDI, *upgrading* FDI and encouraging *linkages* between TNCs and local firms. Some governments want FDI more than others and may try harder accordingly. Governments can base their FDI promotion strategy on industrial policies (promotion, incentives, etc.) and/or on macroeconomic policies (skills, infrastructure, etc.) taking into account external factors which are only partly under their control (natural resource endowments, international agreements, etc.). This chapter helps to classify such policies into more concrete building blocks. The literature frequently regards Ireland and Singapore as examples of efficient FDI attractors but the countries have been dealt with by separate strands in the literature. Here we find interesting comparisons.

The structure of this chapter is as follows. The next section discusses the benefits of FDI in theory and practice. We then discuss the need to see policies towards FDI as part of a development strategy to achieve pre-defined objectives, before we classify policies into different categories. We subsequently discuss the experience of Ireland and Singapore on the basis of this classification. We conclude the chapter with appropriate policies towards FDI.

The benefits of FDI in theory and practice

TNCs are different from local firms, as TNCs need to overcome the extra costs of operating under different circumstances in another country. The difference is termed an ownership advantage (Dunning, 1993) as shown in tangible (technology) or intangible (brandnames) assets. The studies reviewed in Dunning (1993) and Markusen (1995) show that TNCs are more productive, pay higher wages and are more export intensive than local firms. The distinctiveness and superiority of TNCs can in principle offer benefits to developing countries. FDI possesses a bundle of desirable assets (UNCTAD, 1999; Lall, 2000a), including long-term external finance, new technologies, skills and management and market access, which a national government would like to tap. However, in practice the benefits in terms of economic development are by no means automatic or free, suggesting a role of complementary policy. FDI can also lead to less desirable or undesirable outcomes such as rising inequality between (groups of) individuals (e.g. Tsai, 1995; te Velde, 2000) or regions, direct or indirect crowding-out of local capabilities or an erosion of the tax base or labour and environmental standards (Oman, 2000).

Data on the presence of FDI alone do not indicate whether FDI has been successful for economic development (see e.g. Kalotay, 2000). In order to identify countries that have been successful in attracting FDI conducive to economic development, it seems necessary to rely at least partly on anecdotal and/or econometric evidence on the impact of FDI. For instance the evidence shows that countries such as Ireland (Barry and Bradley, 1997) and Singapore (Lall, 2000c) have benefited from FDI, for example, in terms of growing exports. However, most of the anecdotal evidence does not provide evidence as to whether FDI causes growth (at national or firm level) and hence we need to turn to econometric evidence, despite problems related to the construction of a strategic counterfactual.

Table 6.1 provides a survey of the econometric evidence on the effects of FDI on growth and productivity, and this evidence is growing rapidly. There are three types of evidence. First, there are country studies that examine the effects of foreign presence on the average level of productivity in a sector (e.g. Caves, 1974; Globerman, 1979) or on productivity in the local industry (e.g. Blomstrom and Persson, 1983). These studies test whether there are intra-industry spillovers from FDI. Other sectoral studies (e.g. Pain and Hubert, 2000) find evidence for inter-industry spillovers.

Second, there are country or cross-country studies exploiting (panel) data of countries over time. These studies generally find that a measure of FDI flows is positively related with per capita GDP growth or productivity, although some (Borensztein *et al.*, 1998; Xu, 2000) stress the importance of minimal level of human capital in order to absorb spillovers. Balasubramanyam *et al.* (1996) stress that countries with outward-oriented FDI policies have greater benefits from FDI. However, such country-level studies (and the earlier sectoral type of studies) suffer from the selectivity problem that foreign firms usually locate in the more productive segment of the sector or the national economy, and hence the results cannot be used to assess whether FDI improves national welfare due to composition or genuine spillover effects.

The third type of evidence (microeconometric) examines whether foreign firms are more productive than domestic firms and second whether foreign firms have positive spillover effects on domestic firms. The evidence (e.g. Haddad and Harrison, 1993; Aitken and Harrison, 1999; Djankov and Hoekman, 2000) finds that the productivity level of foreign firms (recipients of FDI) is higher than in domestic firms, but that the effects on productivity growth in domestic firms is mixed. As a result of foreign firms, domestic firms in the same sector could be better off as (foreign) competition forces them to upgrade technologies (as in the case of Indonesia, see Blomström and Sjöholm, 1999), they could be worse off through the market-stealing argument (as in Venezuela, see Aitken and Harrison, 1999), or they could not learn at all as the productivity gap is too large to learn anything (as in Mexico, see Blomström, 1986). In Morocco, Venezuela and the Czech Republic, the presence of foreign firms lowers productivity *growth* in (purely) domestic firms.

The overall effect of FDI on national welfare in the host economy is perhaps weakly positive, depending on whether the superiority of foreign firms compensates for the loss of profits (through repatriation) and for the potentially slower

Table 6.1 Selected empirical studies on spillovers from FDI

Study	Country/level of analysis	Effects of FDI on growth and productivity
Sectoral studies		
Blomstrom and Persson (1983)	Mexico, 215 manufacturing industries, 1970	Positive and significant intra-industry spillovers from FDI on labour productivity progress of domestic firms
Caves (1974)	Australia, sectoral level, manufacturing, 1966	Positive correlation FDI presence and productivity in sector
Globerman (1979)	Canada, sectoral level, 1972	Positive correlation FDI presence and productivity in sector
Kim and Hwang (1998)	Korea, six manufacturing sectors, 1974–1996	Positive but insignificant correlation inward FDI flows and TFP growth
Pain and Hubert (2000)	United Kingdom, sectoral level	Inter and intra-industry spillovers from FDI on technical progress of domestic firms
Macro studies		
Balasubramanyam et al. (1996)	46 developing countries, cross-section over 1970–1995	Beneficial effect of FDI on real GDP is greater in export promotion (EP) countries than in import substitution (IS) countries
Barrell and Pain (1997)	UK and West Germany, 1972–1995	Positive and significant effect of FDI on (labour augmenting) technical progress in West Germany and UK manufacturing
Barrell and te Velde (1999)	Ireland and UK, national level, 1975–1998	Positive and significant effect of FDI on (labour augmenting) technical progress
Barrell and te Velde (2000)	East Germany, national and industrial level, 1991–1998	Positive and significant effect of FDI on overall and industrial (labour augmenting) technical progress
Borensztein et al. (1998)	69 countries, national level	Insignificant positive correlation between FDI and per capita GDP growth overall, but significant positive effects in countries with sufficient human capital
De Mello (1999)	Panel of 16 developed and 17 developing countries over 1970–1990	FDI has positive effect on real GDP growth, but on productivity growth only in developed countries
Xu (2000)	40 countries, national level	Positive technology transfer in developed countries, but not in developing countries and depends on minimum level of human capital

(*Continued*)

Table 6.1 (Continued)

Study	Country/level of analysis	Effects of FDI on growth and productivity
Firm or plant level studies		
Aitken and Harrison (1999)	Venezuela, around 10,000 manufacturing firms, 1976–1989	Positive and insignificant effects on productivity growth for recipient firms, and negative productivity spillovers on productivity growth in domestic firms
Aitken *et al.* (1996)	Mexico (1990), Venezuela (1987) and US (1987), manufacturing establishments	Positive spillovers on wages in domestic establishments in the US, but not in Mexico and Venezuela
Blomstrom (1986)	Mexico, 1970s	Positive correlation FDI presence and productivity in sector, but firms with large technology gaps do not learn
Blomstrom and Sjoholm (1999)	Indonesia, manufacturing firms, 1991	Positive effect on productivity of domestic firms, but only for 'non-exporters'
Blomstrom *et al.* (2000a), chapter 11	Uruguay, 159 manufacturing plants, 1988	Positive and significant spillover effects on domestic firms only with a small technology gap
Djankov and Hoekman (2000)	Czech Republic, 513 industrial firms, 1992–1996	Positive and significant effects on productivity growth of recipient firms, but negative and significant effects on purely domestic firms
Haddad and Harrison (1993)	Morocco, manufacturing firms, 1985–1989	Positive spillovers on level of productivity in domestic firms in low-tech sectors, but negative spillover effects on growth of productivity in domestic firms in high-tech firms

productivity growth in domestic firms. The microevidence calls into question the widespread use of incentives (fiscal and financial) for foreign firms often justified on the basis of correcting a market failure that the social rate of return on multinational investment for the national economy is larger than the private rate of return.

However, this conclusion would be premature. For instance, Blomstrom *et al.* (2000a) and Borensztein *et al.* (1998) find that productivity spillovers of FDI are increasing with the level of policy-created, absorptive capacity (skills, R&D, infrastructure etc.), suggesting an important role for policy in shaping the ultimate effect of FDI. Hence, in situations of good policy (here defined as raising the absorptive capacity of the local economy) spillovers are more likely to occur than in situations of bad policy. This would lead to the conclusion that incentives may still be justified from a national welfare perspective if combined with 'good policy'. Hence, it is important to define which policies maximise benefits from FDI.

FDI policy as part of a strategy

A crucial question that should be at the forefront of economic policy makers concerned with FDI is how FDI can link into a country's development strategy. Depending on the pre-conditions (presence of local capabilities; endowments of production factors such as labour, natural resources and capital; small or large economy, etc.) and ideologies regarding the degree of state intervention, a government can define a development strategy and objectives, for example, in terms of economic growth, improved governance, industrial and technological capability or poverty alleviation. It can then decide whether FDI is an efficient and effective way of achieving this and what type of FDI is needed, if it had such choice.

A country has to ensure that using FDI is more efficient and effective than pursuing a different strategy without using FDI or with FDI to a lesser degree. A country can use FDI as part of its long-term development strategy, but provided it has a strong government, it can also rely on local capabilities (as in Korea and Taiwan, see Lall, 1996) and encourage technology inflows, for example, by reducing import duties on machinery and equipment. Implementation of policies towards FDI requires financial resources, either through up-front grants, promotion activities and institutional reform or through tax concessions. Rodrik (2000) argues that 'opening up' is not a simple matter of revising tariff codes and removing barriers to foreign investment, but requires institutional reform, which needs financial, bureaucratic and political resources. While institutional reforms aimed at maximising trade and capital flows may produce broader benefits, they are not necessarily the most effective way to enhance development. In an ideal world, all costs associated with policies towards FDI would be weighed up against the benefits of attracting FDI and the cost-effectiveness of fostering development using FDI needs to be compared with the cost-effectiveness of other development strategies.

There are a variety of strategies based on FDI implying different degrees of interventionist policies. A host country needs to address two information-related market failures with regard to FDI (Moran, 1998; Lall, 2000a). First, policies need to address information failures in the investment process. In general, all lumpy investment, foreign or domestic, suffers from uncertainty. But foreign investors are naturally at a disadvantage with regard to information on the host country and they prefer to wait until other investors have tested the grounds. To speed up the investment process, host governments may want to intervene by offering modest grants or other incentives, such as information, to potential foreign investors. Second, policies need to address the divergence in interest between mobile foreign investors and the host economy. A country can intervene in the market for skills and technologies, where market failures are most likely and gear the development of technologies and skills more towards the needs of TNCs. Related to this, a country may want to intervene in capturing possible externalities (e.g. technology spillovers) associated with TNCs, but used in the firm's decision.

There is, however, less agreement on the optimal degree of intervention. Lall (1995) defines four different approaches: (1) Passive open-door policy with limited policy interventions and no industrial policy, (2) Open-door policy with selected interventions to improve supply conditions, (3) Strategic targeting of FDI and (4) Restrictive policy. Altenburg (2000) argues that while options (1) and (4) are not sufficient to exploit opportunities for technological learning, the optimum for many low-income countries will be near the second approach and only if local capabilities develop, a more strategic and targeted approach may produce better results.

A choice for certain FDI policies needs to be followed by adequate implementation. Governments that can follow FDI policies consistently and respond in a flexible way to demands by potential investors are most successful. The Economic Development Board (EDB) Singapore is generally seen as providing all the necessary information and permits within a limited period (Lall, 2000b). The example of Intel in Costa Rica also shows that the flexibility of decision making (offering quick provision of infrastructure and training courses) positively affected the decision to locate (Spar, 1998). The implementation of policies in a consistent way over time is also important. Uncertainty deters investment, and hence consistent implementation of financial and fiscal incentives by Development Agency (IDA) Ireland (Ruane and Gorg, 1999) and consistent skill upgrading (Fitz Gerald, 2000) have signalled a long-term commitment to improve the business climate for TNCs in Ireland.

The importance of certain policies may also vary with the type of FDI. While greenfield investment is likely to stimulate competition, Mergers and Acquisitions (M&A) are likely to lead to more concentrated sectors (UNCTAD, 2000). Hence, there are different effects on competition, and the formulation and effective implementation of a competition policy deserves priority.

Finally, in the emerging context of multilateral trade negotiations, new rules brought into or extended by the WTO following the Uruguay Round, such as Trade-Related Investment Measures (TRIMs), Trade-related aspects of Intellectual Property Rights (TRIPs) and the Subsidies and Countervailing Measures (SCM) agreement, limit domestic policy options to conduct an FDI strategy. Other agreements such as General Agreement on Trade in Services (GATS) are *plurilateral* and contain commitments that lock-in the status quo. This begs the question how much scope still exists in the implementation of domestic policies and an industrial policy aimed at FDI in particular, putting aside the question whether you want to treat TNCs differently at all.

There are three categories of policies in this respect: those that specifically target FDI and are still possible at least to some extent (e.g. employment requirements, incentives subject to limits in SCM), those that target TNCs but are disallowed or will be disallowed in the near future (TRIMs such as local content, trade balancing) and those that target all firms but have special effects on TNCs due to their characteristics (trade and R&D related measures, provision of information on investment opportunities). Hence, there is still some scope (e.g. categories (1) and (3)) for an industrial policy to target TNCs, although multilateral agreements do limit the number of options (category (2)). The latter may not be that serious as TRIMs cover just 2–6 per cent of total FDI in practice (Brewer and Young, 1997). In addition, TRIMs and SCM deal with goods but not services. It is more a question of whether a government desires or is able to conduct a proactive FDI policy.

Classification of government policies towards FDI

Overviews of FDI policies usually address either entry conditions or after-care services or TNC–SME linkages, but not all at the same time. While the inclusion of all such elements inevitably entails a loss of detail, there is a gain in regarding policies towards FDI as integrated and running through the entire economic policy spectrum. Here, we provide a framework for identifying what type of policies a government can use (Table 6.2). We distinguish between policies and other factors affecting the locational decision of foreign investors (row 1), policies and factors affecting established foreign investors (row 2) and policies and factors affecting domestic firms (row 3). Spelling out policies in rows 2 and 3 is important as technology transfer affects affiliates as well as other domestic firms. In order to assess the importance of particular policies we cross-classify the type of policy by degree of domestic control. There are specific industrial policies (column 1), macroeconomic policies (column 2) and other policies and factors beyond the control of domestic economic policy makers (column 3). This classification, a three by three matrix, imposes some structure on the myriad of policies and factors affecting FDI. The framework can help governments in developing countries to formulate an integrated FDI policy within an FDI-based development strategy.

Factors affecting potential foreign investors (row 1)

There are many reasons why foreign investors decide to invest in a particular location. Following Dunning's OLI paradigm (Dunning, 1993), TNCs invest abroad because they have an ownership (O), locational (L) and internalisation (I) advantage. In our analysis we assume that TNCs have an ownership advantage (tangible and intangible assets) and that they want to internalise this advantage rather than, for example, license other firms. We, thus, analyse the policy framework, domestic as well as international, and the economic factors that affect the locational advantages of TNCs.

The determinants of FDI in developing countries can be divided into industrial policies specifically relating to FDI, and more general macroeconomic policies. *Industrial policies* have been important determinants of foreign investment (cell 1, 1; Table 6.2). Governments, often in the form of investment agencies, have promoted FDI and targeted TNCs abroad at the national, sectoral or even

Table 6.2 Policies and factors affecting inward FDI

Economic policies largely under domestic control		Other policies and factors
Industrial policies	Macroeconomic policies	

Affecting potential foreign investors ('determinants')

Cell 1, 1	Cell 1, 2	Cell 1, 3
– Financial and fiscal incentives and bargaining – Efficient administrative procedures and rules on ownership – Promotion, targeting and image building – Developing key sectors (agglomeration and clustering) – Developing export platforms (EPZs)	– Availability of infrastructure and a skilled workforce and good labour relations – Sound macroeconomic performance and prospects – Privatisation opportunities – Development of financial market and debt position – No impediments to trade of goods and services	– Global economic integration and transportation costs – International, regional and bilateral treaties, including BITs and WTO – Insurance (ICSID, MIGA, ECGD, OPIC) and political risk ratings – Location near large and wealthy markets – Availability of natural resources – Historical ties and language-use – Absence of corruption – Financial conditions in home countries

Affecting established foreign investors ('upgrading')

Cell 2, 1	Cell 2, 2	Cell 2, 3
– Taxation – Performance requirements (abolished in most cases under TRIMs etc.) – Interaction with research institutions and other firms – Encouragement of R&D – Training of employees	– Labour market policy – Trade policies, export promotion and infrastructure – Competition policy – Development of financial market	– Regional and international investment treaties – Global economic integration – Civil society

Affecting the response of domestic firms ('linkages')

Cell 3, 1	Cell 3, 2	Cell 3, 3
– Encouragement of linkagaes with TNCs – Encouraging technological capabilities (R&D) – Encouraging human resources (training) – Supply-side management	– Education and skill generation – Labour mobility – Competition policy – Export promotion	– Global economic integration

firm level by providing general information, advertising, undertaking matchmaking activities and sector promotion, organising site visits, supporting feasibility studies and project proposals and other activities. Wells and Wint (1990) show that FDI promotion is significantly and positively related with FDI inflows, though less so in developing countries. FDI promotion addresses a market failure related to imperfect information on the investors' as well as on the host government's side (Moran, 1998). However, promotion can be expensive and expenses vary considerably by country.

Almost all developing countries have one or more investment promotion agencies (IPAs) responsible for dealing with TNCs (Wells and Wint, 1990; Wells, 1999). There are different forms of organisation, ownership and funding arrangements, and can be grouped into (1) government organisations, (2) more autonomous, quasi-governmental organisations, and (3) private organisations. These organisations can provide four different type of services: (1) image building, (2) investment generating, (3) investor services and (4) policy advocacy.

While government organisations often lack the skills and experience to facilitate FDI and used to screen and approve investments, quasi-autonomous organisations (IDA Ireland, EDB Singapore) are often more in touch with the business side. With sufficient power (high level, single ministry) such quasi-autonomous organisations can help to get business approvals (e.g. building permits) and advocate with other government departments to provide university graduates, etc. This is frequently called a one-stop shop, where one agency deals with screening, approval of and obtaining permits for foreign investors.

Governments have also offered special financial and fiscal incentives to TNCs by offering discretionary grants (sometimes related to performance) and tax holidays or special tax rates on business profits in host countries and on dividends payments to home countries (bilateral tax treaties usually determine *total* taxes paid in home countries), see, for example, UNCTAD (1996). Hines (1996), reviewing a number of studies, finds that taxation significantly influences FDI, corporate borrowing, transfer pricing, dividend and royalty payments, R&D activity, exports, bribe payments and location choices. Experience, however, seems to suggest that incentives are most effective for foot-loose, export-oriented investment, in countries or regions that are similar to neighbouring countries or regions and in places where other aspects of the business climate are already favourable (UNCTAD, 1996; Bergsman, 1999; Oman, 2000). Tax incentives play a potentially decisive role once the fundamentals are sufficient.

Less evidence is available on the effects of up-front grants on FDI attraction. Grants implicitly lower effective tax rates. They provide governments some discretionary powers during the negotiating process, which could foster corruption if not monitored properly. They can also be used as part of industrial policy to stimulate certain sectors.

Other industrial policies towards FDI include the formulation of administrative procedures and rules on ownership. Administrative procedures can form a significant barrier to FDI, especially in developing countries (Emery et al., 2000). Governments are sometimes unaware that certain regulations can be streamlined without losing its regulatory powers. This can be the result of the past when government often relied on screening of new investment projects rather than on facilitating. Competition amongst governments on administrative procedures (especially if they put FDI at a disadvantage) towards investment should eventually lead to the convergence of procedures towards best-practice, so that investors do not distinguish between countries on the basis of administrative procedures.

Governments can also impose restrictions on ownership varying from outright bans to a maximum percentage of equity owned in joint ventures. An outright ban outside certain sensitive sectors has not proven to be helpful (with the major exceptions of Korea and Taiwan, where the indigenous capacities have prospered instead), as this restricts the potential benefits from FDI completely. However, there is little evidence on whether restrictions on equity shares (minority stakes) are in the interest of host economies (see Blomstrom and Sjoholm, 1999). On the one hand, local participation may enhance technology transfer, but on the other hand, imposed joint ventures may lead to less upgrading in affiliates as parents could keep secret their (in)tangible asset.

Another industrial policy towards potential FDI is that of developing key sectors (e.g. electronics, e-commerce, pharmaceuticals) through clustering. Focusing industrial development policies on clusters of activities is practiced by only a few IPAs. A cluster is a geographic concentration of interconnected firms, specialised suppliers, service providers, firms in related industries, and associated institutions in particular fields that compete but also cooperate (Porter, 1998). Firms with the *same* functions compete, stimulating further productivity growth, while firms cooperate *along* the value-added chain. The fundamental concept that holds together a cluster of firms is that of the 'value chain' that links downstream to upstream industrial activities. Clusters are believed to be of high linkage potential for domestic firms, and can also lead to agglomeration economies: firms benefit from other firms in the same cluster, for example, through knowledge transfer and the availability of particular supplier services. An incumbent firm in particular clusters therefore has an advantage to signal information to potential investors (Braunerhjelm *et al.*, 2000). Moran (1998) relates this to the bandwagon effect. If one star multinational decides to locate, this provides a positive signal to other potential investors and improves the image of a sector/country in general. A policy question is how and how much effort should be devoted to attract the first few firms.

Despite the impact of selective industrial policies, *macroeconomic policies* that shape the underlying fundamentals of cost-competitiveness are also important in attracting mobile FDI (cell 1, 2). Lall (2000b) argues that FDI location decisions will increasingly depend on economic factors and not on temporary policy interventions. Macroeconomic policies should be sound and deliver a skilled workforce, an adequate infrastructure and could signal commitments to privatisation. FDI increasingly takes the form of M&As, which are fuelled by privatisation.

An increasing number of surveys show that the lack of availability of skills and physical infrastructure is amongst the major impediments to investing in African countries (UNCTAD, 2000; Businessmap, 2000). Whilst TNCs spread similar techniques across the world, and technical progress has been mainly skill-biased (see Berman *et al.*, 1998; Berman and Machin, 2000) countries increasingly diverge in availability of skills through diverging trends in education. Strategic asset-seeking TNCs have less to look for in countries with fewer skills and education, partly because they cannot use the techniques that they want.

Macroeconomic policies can also guide labour market policy. Competitive wages are likely to lead to higher inflows of efficiency-seeking FDI (e.g. Wheeler and Mody, 1992). Further, good labour relations prevent labour disputes, which Jun and Singh (1996) found to be important impediments to FDI inflows, especially in countries that receive relatively little FDI.

The importance of an adequate infrastructure system to attract FDI has also been underlined in many studies. For instance, in a much cited study of determinants of US FDI abroad for forty-two countries in manufacturing and in electronics in particular, Wheeler and Mody (1992) found that 'infrastructure quality clearly dominates for developing countries', while specialised support services were a better determinant in developed countries who already have an adequate infrastruture. Tax incentives were not significant as determinants. This contrasts with other findings (e.g. Hines, 1996) and can in part be explained by the use of different specifications.

With regard to other macroeconomic policies, TNCs are helped by the development of financial markets, although this is not crucial. If countries have a weak financial market it becomes more difficult to raise funds locally, but funds (especially in large firms) can be channelled from parents to affiliates.

Impediments to trade in goods and services in the form of tariff and non-tariff trade barriers have encouraged market-seeking and tariff-hopping FDI (Jun and Singh, 1996).[3] In order to attract export-intensive FDI it is important to signal commitments to liberalise trade policy by concluding regional and global treaties. As a trade policy instrument, the establishment of free-trade zones or export-processing zones in countries with a stable economic environment and commitment to trade liberalisation has helped to attract export-intensive FDI (Madani, 1999). Export-processing zones (EPZs) are often defined as fenced-in industrial estates offering free trade conditions and a liberal regulatory framework for firms exporting a minimum share of output. EPZs have also been criticised for poor labour standards.

Finally, there are *other policies and factors*, affecting locational decisions of TNCs, but which are beyond the direct influence of host country policy makers (cell 1, 3). They include global economic integration and transportation costs that have an impact on where TNCs source inputs as well as the conclusion of international agreements. Imposition of international agreements such as the WTO agreement on TRIMs, TRIPs (Maskus, 2000) and GATS should encourage FDI, though less is known about their ultimate developmental impact on host economies (see below). The explosion in bilateral investment treaties during the 1990s (from 385 in the end of the 1980s to 1,857 at the end of the 1990s, see UNCTAD, 2000) is also likely to further encourage FDI. Blonigen and Davies (2000) find that bilateral tax treaties relating to US inward and outward investment for sixty-five countries over 1966–1992 raise FDI activity. Each additional year of a bilateral treaty raises outward US FDI activity in the partner country by 7 per cent (affiliate sales), 9 per cent (stocks) or 6 per cent (flows).

Insurance against political risk facing FDI in developing countries, whilst likely to facilitate FDI, may not be as important as economic variables in driving TNCs' locational decisions. While some argue that African countries generally are rated more risky than warranted by economic fundamentals (Collier and Pattillo, 2000), and hence there is a potential role for political risk insurers, others suggest that economic variables are the primary determinants of risk ratings, and political variables merely reinforce the picture sketched by economic variables (Haque et al., 1998). The latter would imply that host countries should really think about improving *economic* conditions.

Corruption of officials poses another problem during the locational process. It reduces inward FDI and affects the choice of entry, that is, joint venture vs wholly owned affiliate. Smarzynska and Wei (2000) using firm-level data in Eastern Europe and the former Soviet Union find that corruption makes bureaucracy less transparent, thereby reducing the probability to invest and raising the value of a joint venture. Hines (1995) shows that corrupt countries had lower US FDI flows, equivalent to 6 per cent annual declines in host country GDP in forty-one countries, over 1977–1982.

Further, apart from 'being a large market' (an extremely powerful explanatory variable of market seeking FDI inflows!), geographical proximity to a large and wealthy country is also likely to boost FDI although by itself this is insufficient. Take the examples of East Germany and Ireland (Barrell and te Velde, 1999) and Mexico. Only after investment policies had been changed and trade policies were liberalised, did these countries attract significant levels of FDI. Further, the presence of natural resources can also attract FDI as do historical ties and links. Many countries benefit from the fact that the working language is English. Finally, favourable conditions in home country capital markets (low real interest rates, etc.) and unfavourable investment opportunities at home are likely to enhance FDI.

Factors affecting established foreign investors (row 2)

Factors affecting established investors (row 2) can be crucial in determining whether TNCs decide to exploit the static comparative advantage (e.g. low-wage workers, enclaves, natural resources) of their affiliates with little incentive to raise productivity and quality of products, or whether they decide to upgrade skills and products of their affiliates affecting their dynamic comparative advantage with potentially positive effects on capabilities of domestic firms. A significant part of FDI is in the form of expansion of existing operations of affiliates.

As part of their *industrial policy*, governments have offered permanent or temporary tax concessions to TNCs, imposed performance requirements (following 1995 WTO rules these are now disallowed in most cases), encouraged interaction between TNCs, domestic firms and research institutions, encouraged R&D, promoted exports and offered incentives to training of employees within firms (cell 2, 1).

Some corporate tax concessions are for a specific time (tax holidays) but others are permanent (as in many EPZs) thereby implicitly subsidising TNCs. Very often these tax concessions are not the major reason to attract FDI. If they are, TNCs may well leave or form a new company after the tax concession expires as argued in Bergsman (1999). Hence, they may attract less-committed TNCs. Such tax holidays are also not very useful in terms of upgrading.

There are other tax incentives that can be designed in such a way that they affect TNCs more than domestic firms and can be incorporated in a country's industrial policy. The tax system can be used to encourage R&D, or reduce import duties on machinery and equipment. Another tax incentive (in Singapore) is to encourage the production of innovative products through tax concessions to pioneering firms. These tax incentives affect TNCs (trade and R&D intensive) more than domestic firms, and encourage the development of capabilities in TNCs. In the business literature developing technological capabilities through R&D encouragement is generally seen as part of multinational affiliate development. Higher R&D in affiliates reduces the cost of technology transfer from their parents (Teece, 1977). Many countries struggle to maximise the (development of) high technology content of TNC affiliates.

In the past countries imposed performance requirements on TNCs. These included local content, export, and trade balancing requirements (TRIMs). TRIMs will have to be phased out (or already are) in most cases under WTO rules. Here, we consider the effects on TNCs, while we return to the effect on the national economy and local firms below. The performance requirements aim to minimise imports and maximise the use of local sources. Rules such as local content have prevented imports of machinery and equipment, potentially of better quality. Moran (1998) argues that performance requirements in protected local markets lead to less efficient production than allowing foreign firms to set-up operations oriented towards global or regional markets.

The government can encourage the interaction between TNCs, research institutes and domestic firms, for example, through linkage programmes (discussed below). TNCs and local research institutes can both benefit from increased interaction. However, if research institutes are below the standard required, TNCs will develop in-house capabilities.

The government can also design training schemes, where TNCs help to train their employees. Some countries offer incentives or impose a tax levy on firms to finance the training of low-skilled employees (Lall, 1996). The training of employees enables the TNCs to upgrade production in a situation of skill-shortages. Firms have insufficient incentives to train employees as firms may not capture all the benefits of training. Improved capabilities (skills and R&D) in affiliates enhances technology transfer from the parent to their affiliates (Teece, 1977).

Macroeconomic policies, including labour market and trade policies, also affect established foreign investors (cell 2, 2). If TNCs can draw on a pool of skills, this stimulates the upgrading of their affiliates. Labour market policies can be geared to the needs in various ways (we discussed training above). For instance, with a forward-looking government that can predict skill requirements in the future, universities can deliver graduates to TNCs and other firms thereby preventing the so-called cobweb effect. With regard to trade policies, TNCs are generally more open to trade than domestic firms (e.g. UNCTAD, 1999), so that improvements in exporting conditions (tariff liberalisation or general export promotion) are likely to affect TNCs relatively more. Free choice of imports also allows TNCs to import quality goods and services.

An important element in macroeconomic policies is to avoid unfair competition or abuse of market power especially since TNCs are usually larger firms and can dominate sectors through large mergers and acquisitions (UNCTAD, 2000) and due to their characteristics tend to locate in concentrated sectors (Caves, 1996). However, not all countries have an effective competition policy. Competition induces TNCs to compete strategically, thereby providing an incentive (to all firms) for upgrading (Blomstrom et al., 2000b) while at the same time reducing the abuse of market power (Morrissey, 2000). Moran (1998, p. 25) confirms that competition policy is crucial in determining the long-term effects of FDI, arguing that host actions in stimulating or retarding competition wherever foreign investors are located constitute the most important determinant of whether the host benefits or suffers from the presence of foreign firms.

Finally, as mentioned above, the development of local financial markets is important for affiliates to secure loans. With affiliates maturing, local finance becomes important and can help the affiliate to become a strategic independent.

Other factors affecting established TNCs include development of regional and international agreements, forces of global economic integration and civil society (cell 2, 3). The adoption of WTO rules (e.g. TRIMs, state subsidies) limits the power of governments to impose performance requirements on TNCs. TNCs have more freedom to choose their suppliers. The abolition of TRIMs (low-income countries have been given a grace period) can have a static effect by reducing supplies of local companies, but it could also have a dynamic effect on TNCs by improving the quality of inputs. Various regressions in Blomstrom *et al.* (2000a) indicate that the presence of fewer performance requirements raises the payments of royalties and licence fees to US parents in 1982 in a sample of thirty-two developing and developed countries, and hence the abolition of TRIMs may encourage technology inflows.

While the government has less scope to enforce the use of local suppliers, TNCs increasingly specialise in certain stages of the value added thereby intensifying contacts and placing higher demands on partners upstream and downstream (Altenburg, 2000). As part of the process of *global economic integration*, just-in-time techniques and the complexity of relationships between partners require proximity of location. Hence, while the regulatory framework allows foreign affiliates to source from abroad, they may not always choose to do so if local capabilities are up to standard. Having frameworks in place to let TNCs 'upgrade' local firms can be beneficial for the TNCs as well as local suppliers.

TNCs increasingly have to deal with organisations and groups from *civil society*, especially in the extractive industries. This occurs because TNCs are growing in importance, while effective governments defending the needs or expressing concerns of local people are often lacking. For instance, there are concerns about the impact of TNCs on social development. TNCs are increasingly expected to have a code of conduct, a social report or a partnership with civil society (UNCTAD, 1999).

Factors affecting the response of local firms (row 3)

The third and final row of Table 6.2 reviews key policy areas and *other factors determining the response of local firms to the presence of TNCs*. Factors in this row affect whether local firms benefit from foreign firms. Lall (2000a) and Blomstrom *et al.* (2000b) argue that the development of local capabilities and an absorptive capacity is an important factor behind spillover effects from TNCs to local firms. Kalotay (2000) describes the absorptive capacity in Eastern Europe on the basis of survey responses and actual FDI inflows. He defines the concept as a close to ideal situation, under which a host country can both maximise the FDI inflow and derive maximum welfare from it. In line with Blomstrom *et al.* (2000b), we relate the concept to the ability to capture productivity and other spillover effects and hence to Kalotay's term 'FDI absorptive capacity utilisation'. Productivity spillovers are not free or automatic and domestic firms need basic capabilities to absorb spillovers.

The government can play an important role in developing local capabilities in a number of ways. For instance, it can *encourage general R&D* in local firms. Investment in R&D has two purposes (Cohen and Levinthal, 1989): it can raise the innovative capacity, but also the absorptive capacity. Te Velde (2001) finds evidence for the latter in thirty-five US manufacturing sectors over 1977–1994. Sectors with a higher R&D to value-added ratio have higher price and productivity spillovers from FDI than sectors with a lower R&D to value-added ratio. Neary (2000) argues that governments in developing countries should raise the general level of research expertise rather than use targeted R&D subsidies, which would require picking winners.

When FDI plays a significant role in the domestic economy it has become common to set-up linkage programmes. Such programmes are designed to encourage linkages (mainly backward) between TNCs and small to medium sized local companies, which can subsequently lead to spillover effects. Often TNCs are willing to source locally, but do not have the information on local suppliers. Conversely, local firms are not always informed about opportunities or specific requirements by TNCs. Institutionalised linkage programmes combined with human resource training (through government and/or TNCs) can help to overcome this informational related market failure. Such programmes can also identify and fill gaps in the value-added chain of cluster, that is, particular supply services that TNCs demand. Linkage programmes are now more market-oriented than used to be the case with TRIMs often forcing links. Examples of successful linkage programmes have been scarce apart from some Asian examples (Battat *et al.*, 1996), and little is known about the relative importance of these linkage programmes on the local economy.

There are also *macroeconomic policies* that affect local firms (cell 3, 2). One is to encourage labour mobility between TNCs and local firms. Foreign-owned firms provide a good experience for new entrepreneurs who can then start up new firms. Labour mobility is often seen as an important mode through which spillovers occur (Blomstrom et al., 2000b). Governments can also raise the local absorptive capacity through general education. A skilled workforce is not only an important attractor of FDI, as argued before, but it can also facilitate spillovers by creating an absorptive capacity to assimilate new techniques (Xu, 2000; Borensztein et al., 1998 – looking at the combination of recipient and purely domestic firms). Another factor behind the variation in absorptive capacity between countries or regions is the quality and quantity of infrastructure. Easier transport and communication facilitates the operation of TNCs and linkages with TNCs.

Global economic integration will also force many local suppliers out of the market (cell 3, 3). Nevertheless, those firms that survive competition with imports are generally well equipped to supply to TNCs or export abroad.

We summarise the policy options discussed in this section by presenting econometric studies on the effects of policy in Table 6.3.

The experience of Ireland and Singapore

We now discuss the experience of Ireland and Singapore through the lens of Table 6.2. Both countries have been highlighted for using best-practice policies towards attracting FDI. Singapore is often dealt with in the literature on developing countries while Ireland is more often discussed in the literature on FDI in the EU. Here we find interesting similarities between the two case studies.

Ireland

There have been various reviews of Irish FDI policy and the impact of FDI on the Irish economy (see e.g. Barrell and te Velde, 1999; Braunerhjelm et al., 2000; Ruane and Gorg, 1999; O'Connor, 2001), but these have not looked at policy through the lens of Table 6.2. We first discuss the importance of FDI in the Irish economy briefly, and the role played by policy in attracting and upgrading FDI and enhancing linkages between TNCs and local firms.

Economic commentators agree that FDI has played an important role in the economic development of the Irish Republic (see for instance the chapters in the edited volume by Barry, 1999). As we will see below, FDI has helped to transform a largely agricultural society into one of the fastest growing economies in Europe with one of the highest per capita GDP. FDI has created jobs in *new* sectors, raised investment and enhanced overall and local productivity. In 1995, foreign affiliates in Irish manufacturing were responsible for 47.1 per cent of the total number of employees, 76.9 per cent of value added,[4] 52.6 per cent of wages and salaries, 68 per cent of R&D expenditure (in 1993), 82.3 per cent of exports and 77.8 per cent of imports (OECD, 1999). Value added per employee in foreign-owned firms was over 60 per cent higher than in domestic firms, pointing to superior productivity in foreign-owned firms. Barrel and te Velde (1999) estimate the impact of FDI on overall technical progress and find it to be significant and positive.

Table 6.3 Selected econometric evidence of policy effects on attraction and effects of FDI

Study	Type of policy	Effects and other comments
Attracting FDI		
Wells and Wint (1990)	FDI promotion	The presence of a promotional presence in the US raises total inward FDI flows in developing countries by 30 per cent. Using cross-country study in 1985.
Hines (1996), table 1	Tax policy	−0.11 to −1 elasticity of US capital demand/investment/assets abroad to local tax rates or after tax returns. Mostly using cross-section studies.
Blonigen and Davies (2000)	Bilateral investment treaties	Each additional year of a bilateral treaty raises outward US FDI activity (65 countries over 1966–1992) by 7 per cent (affiliate sales), 9 per cent (stocks), or 6 per cent (flows).
Jaspersen *et al.* (2000)	Corruption	Corruption reduces FDI as percentage of GDP.
Smarzynska and Wei (2000)	Corruption	Corruption reduces the probability to invest but raises the possibility of a joint venture.
Hines (1995)	Bribery	Corrupt countries had lower US FDI flows equivalent to 6 per cent annual declines in host country GDP. 41 countries, over 1977–1982.
Maskus (2000), table 3	Intellectual property rights	A 1 per cent increase in degree of patent protection in host economy raises US investment stock by 0.45 per cent.
Morisset (2000)	Trade openness	Trade as percentage of GDP raises FDI business climate and FDI inflows in Africa, but note endogeneity problem.
Wheeler and Mody (1992)	Infrastructure quality	Infrastructure is the most important determinant (with highest elasticity) of US capital expenditure abroad in manufacturing and electronics in 42 countries over 1982–1988.

(*Continued*)

Table 6.3 (Continued)

Study	Type of policy	Effects and other comments
Upgrading FDI		
Blomstrom et al. (2000a), chapter 13, table 13.2	Performance requirements (import, local content and local employment), competition and education	Various regressions indicate that the share of relevant age-groups with third-level education, the presence of fewer performance requirements, and a rising to investment–output ratio (to proxy competition) raise the payments of royalties and licence fees to US parents in 1982. Using 32/33 developing and developed countries.
Hines (1996), table 3	Tax policy	0.1–0.3 elasticity of R&D with respect to royalty withholding taxes.
Teece (1977)	R&D policy	In a sample of 27 cases of technology transfer (of which 20 between TNCs and affiliates), it is shown that higher R&D/sales ratios, manufacturing experience and size significant reduce the costs of technology transfer to affiliates.
Linkages and absorptive capacity		
Borensztein et al. (1998)	Raise human capital	Countries with higher human capital benefit from more FDI than countries with lower human capital.
te Velde (2001)	R&D policy	Sectors with an average R&D to value-added ratio of over 5.7 benefited from FDI spillovers 35 US manufacturing sectors over 1977–1994.
Xu (2000)	Reach minimum level of human capital	Raise average (male) educational attainment to above the range 1.4–2.4 to obtain positive technology transfer effects.

Much of the recent literature that has emerged as a consequence of the visible influence of FDI in Ireland has stressed the importance of policy, industrial (Ruane and Gorg, 1999) and macroeconomic (Fitz Gerald, 2000), as well as other factors (Ruane and Gorg, 2000) such as its location in relation the EU in attracting FDI. There has also been attention to upgrading FDI and linkages between TNCs and local firms (O'Malley, 1998).

Industrial policy (column 1, Table 6.2) towards FDI has been implemented by IDA. Initially a part of the Department of Industry and Commerce in 1949 with powers to issue grants that covered the costs of land and buildings, IDA was established as a separate state agency by the Industrial Development Act 1969 with the responsibility for national industrial development. IDA expanded quickly in terms of staff (230 initially) and location of operation with IDA staff operating worldwide including Japan, Taiwan, South Africa and Australia. IDA targeted aggressively and firm-specifically involving telephone calls, presentations, provision of research, visits and other meetings. The IDA identified electronics and pharmaceuticals companies from the US as offering the best opportunities for Ireland's drive to industrialise through FDI. These sectors now form the basis of industrial clusters (cell 1, 1; Table 6.2). In 1999, 15 per cent of employment in foreign companies (IDA supported) was in phamaceuticals/healthcare and 49 per cent in electronics/engineering, confirming that much FDI in Ireland has been in high-tech industries. Financial services gained in importance and now accounts for 27 per cent.

The IDA was also able to award grants to firms covering part of their initial capital expenditure and these were later coupled to employment generation. Nowadays, the higher the grants the more benefits the Irish economy can reap, but they also need to be consistent with EU rules on state aid (in low-income regions state-aid grants are still allowed). Combined IDA expenditure per job decreased from over IR£ 35,000 in the period 1981–1987 to IR£ 10,000 over 1993–1999 (Forfas, 1999). Total expenditure of IDA Ireland in 1999 amounted to IR£ 160 million, with IR£ 129 million paid in grants and IR£ 21 million paid towards promotion and administration (of which IR£ 5 million directly towards marketing, consultancy, promotion and advertising), see IDA (2000).

Fiscal incentives have been (and still are) perhaps more important in attracting FDI (Ruane and Gorg, 1999). There was a fifteen-year (zero) tax holiday on profits from new *export* profits from the 1950s, which changed into a 10 per cent corporate tax to *all* new firms (compared to around a standard 50 per cent corporate tax rate by that time) from 1982 to be consistent with EU rules. Under further international pressure Ireland is now committed to a 12.5 per cent corporation income tax for all firms from 2003, with some concessions until 2010.

Fiscal incentives helped to stimulate investment in export-intensive manufacturing. Thanks to specific targeting, the IDA was in the position to develop key export-intensive sectors (electronics and pharmaceuticals) leading to bandwagon and agglomeration effects. The IDA plans visits to existing firms as part of their promotion strategy for potential investors. An FDI-friendly image is now apparent after forty years of aggressively promoting FDI.

While specific industrial policies have been very important in attracting FDI, there are also *macroeconomic policies* (cell 1, 2) and *other important factors* (cell 1, 3), without which it would have been difficult to attract FDI. The government has consistently followed a policy of skill-upgrading by providing education (FitzGerald, 2000). The availability of skills further improved recently through net immigration of Irish and other nationals. While the physical infrastructure was initially neglected until the late 1980s, EU structural funds (6 per cent of GDP in early 1990s) have helped to develop the infrastructure since then. IDA Ireland also develops land and industrial parks for foreign investors.

Other important factors have been strong historical ties with the US (through emigration of Irish nationals), which helped to attract US investment, the use of the English as the official language (the only country in the Euro area) and more recently the boom in the US and the electronics sector.

Last but certainly not least, proximity to the EU has been extremely helpful. Geographical proximity to the EU is helpful because of the large size of its market, but it is not sufficient to attract FDI. As Ruane and Gorg (2000) find, Portugal and Greece are also close to the EU, but have been less successful in attracting FDI. Geographical (as opposed to economic) distance becomes less important as transportation costs fall and the 'weightless' economy (software and the like) gains in importance.

Nevertheless, the opening-up of the Irish market has been of crucial importance behind the development of Ireland as an export platform to the EU (Barry and Bradly, 1997). Following the foundation of Ireland in 1922, it followed a closed-door policy as regards trade and investment. By the 1950s the limits of protectionism were seen, culminating in the Anglo-Irish Free Trade Agreement in 1965, EU membership in 1973 and Euro Area membership in 1999. This involved dismantling of (tariff and non-tariff) trade barriers and while local firms suffered as a result from intensified competition through imports, it facilitated the attraction of export-intensive manufacturing investment.

Ireland focused initially (up to early 1990s) more on attracting quality FDI rather than on *upgrading* existing FDI (row 2, Table 6.2). Firms in high-value added sectors were targeted (e.g. through higher grants) more because they added new, high-value added exports, rather than because they could link in with existing (read non-existing) local manufacturing capabilities. Now there is also concern about developing affiliates (as 'strategic independents'), focusing on raising the level of R&D in foreign (and domestic) firms. While business R&D as a percentage of GDP has been rising on OECD data from 0.7 in 1981 to 1.4 per cent in 1997, it is low internationally

While attracting export-intensive TNCs ensures fewer fears of crowding-out of domestic operations, there was considerable concern that the economic distance between local and foreign firms was too great to lead to significant spillovers and *linkages* (row 3, Table 6.2). Indeed, many economic commentators pointed to the lack of linkages in the 1980s. This forced a policy response and three state agencies (IDA, a marketing agency – Coras Trachtala (CTT), and a science and technology agency – EOLAS) formed the National Linkage Programme (NLP) aimed at improving organisational and marketing skills as well as quality and productivity to bring it up to the standard required by TNCs. TNCs helped to upgrade local suppliers by providing technical know-how. Partly as a result of the NLP, but also because TNCs were present in the market for a longer time, Irish raw material purchases rose between 1988 and 1998, from 15.4 per cent to 21 per cent in non-food manufacturing and from 13.2 to 22.8 per cent in electronics.

Ruane and Gorg (1998) examined purchases of raw materials (as a measure of backward linkages) in the electronics sectors using a panel of firms over 1982–1995. They confirmed earlier evidence that foreign firms have lower backward linkages than domestic firms, and found that larger and expanding firms have lower linkages but also that backward linkages improve over time.

While the IDA was involved in the NLP in the beginning, it was recognised after the 1992 Culliton report that TNCs and local firms required different attention. CTT, EOLAS and the part of IDA responsible for local firms formed Enterprise Ireland. IDA became IDA Ireland, responsible for TNCs. A key strategy for developing local capabilities is to develop sub-supply industries along the value-added chain, not only for TNCs in Ireland but also for exporting, thereby also reducing the dependence on TNCs. Many local companies have reached a critical scale to be able to compete internationally.

The development of the Shannon area in the West of Ireland is a good example of the influence of policy (through a regional agency) on attracting TNCs and benefiting from TNCs (see e.g. Callanan, 2000). The Shannon Development Company (SDC) is the state agency responsible for economic development of the Shannon region, for attracting FDI to Shannon, and the negotiation and provision of investment incentives.

The SDC originated from a regional initiative to revitalise the area surrounding Shannon airport in 1957. This was needed because airplanes flying between the US and Europe that usually landed at Shannon to refuel, no longer had to do this after the development of long-haul jets. A significant initiative was to extend the concept of a duty-free shop to establish the *first* Free Trade Zone in the world that developed an industrial estate with factories (cell 1, 1; Table 6.2) and infrastructure provided. The idea was to import, process and re-export without customs duties or formalities (1958 Customs Free Airport Bill). This should help to attract the air traffic and associated business. The SDC offered land and grants to investors and sent out brochures to the US. Significantly, it also built factories, and after one investor took one, the ball was finally rolling – an example of the bandwagon effect. While fiscal incentives (low corporation tax) have been important for Shannon as they have been to the rest of the country, of additional importance has been a deferral system of taxes on imports until they leave the Customs Free Zone, leading to opportunities such as packaging. Nowadays taxes in the Free Zone are equalised (or will become soon) with the rest of the country, with the Shannon Free Zone used as a marketing tool.

The importance of the Shannon area was evident from the fact that around 25 per cent of Irish manufactured exports came from the area by the late 1960s. However, statistics also show that it was a risky undertaking with costs (grant

188 *Dirk Willem te Velde*

payment and infrastructure etc.) exceeding the benefits (adjusted value of net exports) in the period 1960–1964. Callanan (2000) approximated the benefit/cost ratio as 0.4 for 1960–1964, 2.4 for 1965–1969, 6.1 for 1970–1974, 11.8 for 1975–1979, 9.2 for 1980–1984 and 21.7 for 1985–1989. It therefore took at least 5–10 years before the regional agency was profitable for the Shannon region.

Singapore

The development path of Singapore is truly remarkable. Singapore developed from a struggling low-income colony in 1960 to a modern and 'developed' high-tech country. GDP growth rates have continued to be 10 per cent on average over the past four decades. At the same time, the accumulated stock of FDI as a percentage of GDP has risen from 5.3 per cent in 1965, 17.1 per cent in 1970, 51.8 per cent in 1980, 87.2 per cent in 1990 and 98.4 per cent in 1998 (see Yeung, 2001b), the highest in the South-East Asian region. The share of non-manufacturing FDI has been rising from 46.7 in 1980 to 63.4 in 1997. In 1997/1998, foreign firms employed 50.5 per cent of workers in manufacturing, 29.1 per cent in trade and 25.7 per cent in finance. There is clearly a story to tell on how these changes came about and while there are several accounts of the role that FDI has played (e.g. Lim and Pang, 1991; Lall, 2000c), we focus on Singapore through the lens of Table 6.2. The Singapore story is one of strong leadership, proactive industrial strategy, a consistent and favourable FDI policy, continued industrial upgrading and also of risk taking, but not one of rich natural resources or geographical proximity to large economic markets.

An outward-looking approach based on FDI was inevitable. Singapore became independent after a two-year stint with Malaysia failed in 1965. Singapore, though traditionally an important trading port, was now isolated from its hinterland, as Indonesia (Confrontation policy) refused to import goods and Malaysia wanted to cut out the middle-man Singapore in its trading activities (e.g. rubber). This made an import-substitution strategy virtually impossible for Singapore, and unlike in other developing countries this was never an ideology. Singapore also lacked natural resources and an entrepreneurial business elite (Honk Kong did have an influx of Chinese entrepreneurs) and there was a time lag before domestic entrepreneurs would be sufficiently capable. Further, there was the impending withdrawal of the British armed forces, which contributed an estimated 20 per cent to the economy. Singapore had no policy option but to industrialise and because of a lack of indigenous capabilities, the industrial strategy had to rely on TNCs bringing their expertise and technologies.

An industrial strategy was designed under the capable and authoritarian leadership of Lee Kuan Yew (Prime-Minister from 1959 until 1990) and Goh Keng Swee (economics minister), and was partly based on a 1960 UNDP study, prepared by Albert Winsemius (served as economic adviser until 1984), on the future of Singapore. Winsemius recommended the establishment of an EDB (founded in 1961) to be responsible for industrialisation of Singapore. The EDB got a budget of around US$ 25 million (over 4 per cent of GDP), a hundred times more than its predecessor, the 1957 Industrial Promotion Board. Winsemius also recommended that EDB be a one-stop agency (cell 1, 1; Table 6.2), sorting out all investor's requirements, and focusing on ship repair, metal engineering, chemicals and electrical equipment and appliances.

The EDB has acted proactively (developing sites, seeking promotion) and responded to market forces ever since it began operations. The EDB's aim was to promote industries (mainly foreign after 1965) in Singapore and began to build up offices abroad. It had four divisions: investment promotion, finance, projects and technical consultant service and industrial facilities. It was set up as an autonomous government agency, which could set its own wages, had a board comprising business and other agencies, and had an international advisory board comprising executives of major foreign companies located in Singapore, and hence the EDB was 'in contact' with business. While in the initial stages the notion of a one-stop centre was helpful to attract FDI, the operations became more complex over time and resulted in the specialisation towards FDI promotion while other activities were left to other agencies: for example, finance into the Development Bank of Singapore (1968), technical and project consultant service into the Productivity and Standards Board (1968) and industrial facilities into the Jurong Town Corporation (JTC) (1968) to name a few. The EDB has maintained close links ever since and still acts as a one-stop service.

The EDB decided to spend a significant share of allocated funds on the development of the Jurong Industrial Estate (cell 1, 1; Table 6.2). An uncultivated piece of land was quickly transformed into an industrial estate with adequate infrastructure and factories and a new port was built. However, the estate was unsuccessful in the early years and with only twelve pioneering firms in 1961, it had a slow start (activity remained sluggish until 1965). The EDB had invested vast sums in joint ventures, some of which had failed. Nevertheless, there have never been real doubts about the FDI-led industrialisation industry as Singapore was forced to rely on TNCs (including foreign staff), an unusual vision at the time compared with other developing countries' views that TNCs only exploit developing countries. Another reason was that Lee Kuan Yew distrusted local Chinese entrepreneurs who, he thought, had too much association with mainland communists. The government therefore also relied on state-owned enterprises.

The real breakthrough came when a star multinational at the time (Texas instruments), decided to set-up a plant to assemble semiconductors of US$ 6 million. The contract was won by EDB in four months and due to providing facilities ahead of demand it was able to start production fifty days after the decision to invest was made.

A fully prepared industrial estate reduces an investor's search and transaction costs. JTC, spun-off from the EDB in 1968, has since maintained responsibility for preparing industrial sites. By leasing and renting industrial sites it was able to more than pay to conduct a proactive stance and prepare sites ahead of demand. Over time the JTC has begun to spread activity over Singapore, with wafer fabrication in dust and vibration-free area, employment-intensive activities where people live and pollution-intensive industries in the West of the Island away from people.

The industrial strategy proved to be successful by the late 1960s and early 1970s and was able to reduce the unemployment rate fairly quickly. Whilst employment generation was a major focus of policy in the 1960s and early 1970s, this shifted to capital-intensive projects in the 1980s, and knowledge-intensive sectors in the 1990s (upgrading, row 2, Table 6.2). The incentive structure is complex and has developed over time. No effort is made to go over all the incentives here, but just a few important ones will be mentioned. A significant incentive was the Pioneer Industries Ordinance of 1959, with firms exempted (or significantly reduced) from the 40 per cent corporation tax for a fixed period of time provided that firms developed 'new' products (the share of manufacturing output by firms with pioneer status increased from 7 per cent in 1961 to 51.1 per cent in 1971 and 69 per cent in 1996). There were many other tax incentives, among them the Economic Expansion Incentives, reducing the corporation tax for approved firms to 4 per cent. The minimum level of capital or sales required for approval was quickly raised in 1970 after realising that Singapore needed capital-intensive rather than employment-intensive firms. Over time wages rose, especially in the period 1985–1986, and Singapore realised that it could only survive by upgrading FDI (cell 2, 1; Table 6.2) and the work force (cell 2, 2; Table 6.2)[5] to be able to compete with neighbouring low-cost locations. The EDB began to target knowledge-intensive industries that could pay higher wages. To tackle the emerging skill shortages, firms were encouraged to recruit foreign workers. Recently the EDB has begun to attract foreign universities. The EDB's regionalisation programme encourages firms to set-up skill intensive regional headquarters in Singapore, with labour and land-intensive production processes transferred abroad (see e.g. Yeung et al., 2001a).

The period 1985–1986 was Singapore's first postwar recession. The recession changed labour relations and initiated or accelerated new schemes to link local firms with TNCs (row 3, Table 6.2). Singapore could only cope with rising wages if local firms developed capabilities (technical and human resources) and if TNCs continued to upgrade (using R&D incentives, incentives to set-up high skilled head quarters and encouraging joint research institutes through government funding). The EDB also sought to upgrade local industries through the establishment of a Local Industry Upgrading Programme (LIUP) in 1986, under which TNCs were encouraged to enter into long-term supply contracts with local firms, leading to upgrading. Local firms benefited most in the electronics sector by supplying maintenance services, components and equipment to the semiconductor TNCs. Initiatives such as LIUP also embed FDI more in the host economy, with mutual benefit and dependence. Other initiatives from the EDB for local firms included the Local Enterprise Finance Scheme, which was transferred to the Productivity and Standards Board (PSB) in 1996.

As part of a number of relevant skill-upgrading schemes (see Lall, 1996), the PSB is responsible for the Skill Development Fund (SDF). Set up in 1979, it first imposed a 4 per cent levy on the payroll on employers for every worker earning less than a pre-determined amount. It is an efficient way to enhance within-firm skill upgrading of unskilled workers. Firms themselves do not have sufficient incentives to do so. After the 1985 crises, the levy was reduced to 1 per cent, but it still plays an important role in the upgrading of skills.

More recently, the EDB has followed a cluster approach, targeting firms around the electronics/semi-conductor, petrochemicals and engineering industries. The cluster approach is an instrument of industrial policy which attracts FDI (cell 1, 1; Table 6.2), but which also leads to enhanced linkages and spillovers (cell 3, 1; Table 6.2). The EDB's cluster-oriented approach seeks to determine which value chains dominate and where gaps can be identified and potentially filled. Government policy can avoid what is essentially a market failure, and can support services (The EDB began a S$ 1 billion Cluster Development Programme in 1994, and has recently tripled in size) or prepare infrastructure for joint use. The JTC has prepared special wafer fabrication parks and a reclaimed Jurong Island (a S$ 6 billion project) for the petrochemical cluster. By also investing in R&D centres, the government further enhances the value of the cluster and with it the locational advantages (the dynamisation of the L factor in the OLI paradigm was originally underestimated).

While the above indicates a strong role for industrial policy, macroeconomic policies have also played a role, albeit in support of TNCs. Infrastructure has been built with regard to the needs of TNCs. Trade policies have always been very liberal compared to other countries, with very low tariffs and thanks to an increase in ISO certificates also low non-tariff trade barriers. Besides training, general education has also been important (Lall, 1996).

However, there are also 'external' factors (column 3, Table 6.2), which have shaped policies towards FDI or have been important in attracting FDI, and which may make the case of Singapore less general in its application to other countries. Singapore is a city-state with a relatively authoritarian state that can formulate policies without much resistance from either other levels of government, or from civil society. The Peoples' Action Party (PAP) has won every single general election since 1959 and the mandate and legitimacy of the PAP government has become virtually unquestionable. This makes the country almost unique, and enables the government to be technocratic in implementing an FDI strategy. Further, Singapore never runs government deficits, which is helpful to find capital for (profitable) investment (in part financed out of a high statutory pension levy). Perhaps another factor for attracting FDI is that despite a multi-ethnic society, the working language is English. Further, the location in the time zone enabled financial services to fill the gap between the US and Europe during the 24-hour day.

Comparing Singapore and Ireland

There are some interesting comparisons to be made between Singapore and Ireland. The main point here is that both countries have had policies in place in most cells of the policy matrix for some time, and hence they have moved towards an integrated and consistent FDI policy. More specifically, both countries:

- had an aggressive investment agency, effectively a one-stop agency with ample political power to swing policies towards foreign investors (cell 1, 1 in Table 6.2);
- began to target firm-specifically in the 1960s, tagged into the globalisation phase of the electronics sector, followed a pro-FDI policy consistently and are now benefiting from agglomeration economies (cell 1, 1 in Table 6.2);
- had a strong proactive industrial policy approach (perhaps not always explicit in policy documents) with fiscal incentives and grants (share of equity investment in Singapore) (cell 1, 1 in Table 6.2);
- did not suffer from direct crowding-out by targeting export-intensive TNCs, but;
- suffered from indirect crowding-out as TNCs pushed up factor prices (land and/or skilled labour) which kick-started a process of upgrading (row 2, Table 6.2);
- have introduced cluster approaches (cell 1, 1 in Table 6.2);
- realised that local capabilities did not develop sufficiently, and put in place linkage programmes between TNCs and local firms (cell 3, 1 in Table 6.2);
- had a supportive macroeconomic environment (column 2, Table 6.2), with consistent skill-upgrading coupled with good labour relations and little or no corruption;
- had favourable external factors, but which were not decisive towards FDI (column 3, Table 6.2);
- experienced a time lag (5–10 years) before proactive policies (e.g. preparing industrial estates) to attract FDI were successful, and hence can be considered risky policies.

Conclusions

We began this chapter by saying that governments in developing countries are increasingly looking for best-practice policies towards FDI based upon the positive effects associated with FDI. Whilst FDI can bring positive effects (market access, technology, finance etc.), it can also bring negative effects. Moreover, the positive effects are not automatic for host countries and depend on policies in place and other factors.

Which policies are important in which countries depends on the specific country characteristics, the objective of the country and the derived FDI strategy. However, there are some common elements. FDI policies are likely to be some combination of policies in each of the categories identified in Table 6.2 and should fit in with a country's development strategy to achieve certain objectives. An important contribution to help formulating FDI policies was to divide policy factors into (1) specific industrial policies and (2) macroeconomic policies and into whether they are used to (a) attract FDI (b) upgrade FDI or (c) enhance linkages and spillovers to domestic firms. There are many policies to enhance the positive effects of FDI, and these do *not* stop with attracting FDI alone, and do certainly *not* stop with incentives. Realising that FDI policies should comprise policies in each of the above categories is a positive step towards enhancing the benefits of FDI.

On the basis of this chapter, including a review of arguably successful FDI policies in Ireland and Singapore, a number of steps for developing countries wanting to attract FDI seem appropriate, but need to be implemented on a country by country basis:

- Determine whether FDI fits in with your country's development strategy and if so, what type of FDI. Is FDI an efficient and effective way at promoting current or future objectives?
- Build up local capabilities (R&D, education etc.) and infrastructure to establish economic fundamentals to attract FDI *and* benefit from FDI.
- Implement a sound and consistent macroeconomic policy.
- Establish an investor-friendly climate by opening up to foreign investment without discriminating against local firms (as now appears to be the case in Ireland and Singapore). Follow this policy consistently over time as it may take time to convince investors and reap the benefits from FDI.
- Target specific firms that fit into your development strategy. This can be coordinated by a true one-stop investment promotion agency, which can oversee promotion, negotiation, facilitation and perhaps policy advocacy (see IDA Ireland and Singapore EDB).
- The facilitation of trade in goods and services is useful for foreign investors, as they are usually more trade intensive (trade agreements and low tariffs in Ireland and Singapore).
- Be flexible enough to change the targeting of FDI to upgrade FDI to other, higher value-added activities when factor costs are rising (e.g. Singapore regionalisation programme).
- Encourage training of employees within TNCs (see e.g. Singapore SDF).
- Encourage linkages between TNCs and local suppliers through linkage programmes (Singapore NLP, Ireland LIUP).

Much more research is needed to formulate appropriate development policies in developing countries and the role that FDI should play in these. For many countries this is likely to involve the adequate formulation of an industrial strategy, a competition policy, a trade policy and a strategy to enhance local capabilities. There are many other outstanding issues relating to who and what gains from FDI within societies, for example, poor vs rich people, rural vs urban, small firms vs large firms etc.

Notes

1 Without implicating I am grateful to John Fitz Gerald, Khalil Hamdani, Sanjaya Lall, Simon Maxwell, Sheila Page, Frances Ruane, Henry Wai-Chung Yeung, Ganeshan

Wignaraja and (other) participants at the conference 'Policies towards Foreign Direct Investment in developing countries. Emerging best-practices and outstanding issues', held at the Overseas Development Institute, London, March 2001, see http://www.odi.org.uk/FDI_conference/FDIhome.html. The UK Department for International Development supports policies, programmes and projects to promote international development. DFID provided funds for this study as part of that objective but the views and opinions expressed are those of the author alone.
2 FDI has the objective of obtaining a *'lasting interest by a resident entity in one economy in an entity resident in an economy other than that of the investor'*, with lasting usually defined as a 10 per cent stake in the entity.
3 We do not discuss the effects of FDI on imports and related policies (see e.g. Buckley, 1996).
4 Data on output and value added are inflated by transfer pricing (Barrell and te Velde, 1999).
5 Upgrading by raising the quality of inputs into multinationals.

References

Aitken, B., Harrison, A. and Lipsey, R. (1996), 'Wages and Foreign Ownership: A Comparative Study of Mexico, Venezuela, and the United States', *Journal of International Economics*, 42, 345–371.

Aitken, B.J. and Harrison, A.E. (1999), 'Do Domestic Firms Benefit from Direct Foreign Investment? Evidence from Venezuela', *American Economic Review*, 89, 605–618.

Altenburg, T. (2000), 'Linkages and Spillovers between Transnational Corporations and Small and Medium-Sized Enterprises in Developing Countries – Opportunities and Policies', *in* UNCTAD, *TNC-SME Linkage for Development. Issues-experiences-best practices*, UNCTAD, Geneva.

Balasubramanyam, V.N., Salisu, M. and Sapsford, D. (1996), 'Foreign Direct Investment and Growth in EP and IS countries', *Economic Journal*, 106, 92–105.

Barrell, R. and Pain, N. (1997), 'Foreign Direct Investment, Technological Change, and Economic growth Within Europe', *Economic Journal*, 107, 1770–1776.

Barrell, R. and te Velde, D.W. (1999), 'Labour Productivity and Convergence Within Europe: East German & Irish Experience', *NIESR Discussion Paper* 157.

Barrell, R. and te Velde, D.W. (2000), 'Catching-up of East German Labour Productivity in the 1990s', *German Economic Review*, 1, 271–297.

Barry, F. (ed.) (1999), *Understanding Ireland's Economic Growth*, Macmillan Press, London.

Barry, F. and Bradley, J. (1997), 'FDI and Trade: The Irish Host-country Experience', *Economic Journal*, 107, 1798–1810.

Battat, J., Frank, I. and Shen, X. (1996), *Suppliers to Multinationals. Linkage Programs to Strengthen Local Companies in Developing Countries*, FIAS Occasional Paper 6.

Bergsman, J. (1999), 'Advice on Taxation and Tax Incentives for Foreign Direct Investment', FIAS.

Berman, E. and Machin, S. (2000), 'Skilled-Biased Technology Transfer: Evidence of Factor-Biased Technological Change in Developing Countries', Boston University, Department of Economics.

Berman, E., Bound, J. and Machin, S. (1998), 'Implications of Skill-Biased Technological Change: International Evidence', *Quarterly Journal of Economics*, 113, 1245–1280.

Blomström, M. (1986), 'Foreign Investment and Productive Efficiency: The Case of Mexico', *Journal of Industrial Economics*, 35, 97–110.

Blomström, M. and Sjöholm, F. (1999), 'Technology Transfer and Spillovers: Does Local Participation with Multinationals Matter?', *European Economic Review*, 43, 915–923.

Blomstrom, M. and Persson, H. (1983), 'Foreign Direct Investment and Spillover Efficiency in an Underdeveloped Economy: Evidence from the Mexican Manufacturing Industry', *World Development*, 11, 493–501.

Blomstrom, M., Kokko, A. and Zejan, M. (2000a), *Foreign Direct Investment. Firm and Host Country Characteristics*, Macmillan Press, London.

Blomstrom, M., Globerman, S. and Kokko, A. (2000b), 'The Determinants of Host Country Spillovers from Foreign Direct Investment', *CEPR Discussion Paper* 2350.

Blonigen, B.A. and Davies, R.B. (2000), 'The Effects of Bilateral Tax Treaties on US FDI Activity', *NBER Working Paper* 7929.

Borensztein, E., De Gregorio, J. and Lee, J-W. (1998), 'How Does Foreign Direct Investment Affect Economic Growth?', *Journal of International Economics*, 45, 115–135.

Braunerhjelm, P., Faini, R., Norman, V. and Seabright, P. (2000), *How the Right Policies can Prevent Polarisation*, CEPR, London.

Brewer, T.L. and Young, S. (1997), 'Investment Incentives and the International Agenda', *The World Economy*, 20, 175–198.

Buckley, P.J. (1996), 'Government Policy Responses to Strategic Rent-seeking Transnational Corporations', *Transnational Corporations*, 5, 1–18.

BusinessMap (2000), *SADC Investor Survey: Complex Terrain*, BusinessMap, Johannesburg.

Callanan, B. (2000), *Ireland's Shannon Story*, Irish Academic Press, Dublin.

Caves, R.E. (1974), 'Multinational Firms, Competition and Productivity in Host-Country Markets', *Economica*, 41(162), 176–193.

Caves, R.E. (1996), *Multinational Enterprise and Economic Analysis* (2nd edition) London, Cambridge University Press.

CBI, Cross-Border Initiative (1999), *Road Map for Investor Facilitation*, Paper for consideration at the Fourth Ministerial Meeting in Mauritius, October 1999.

Cohen, W.M. and Levinthal, D.A. (1989), 'Innovation and Learning: The Two Faces of R&D', *Economic Journal*, 99, 569–596.

Collier, P. and C. Pattillo, (eds) (2000), *Investment and Risk in Africa*, Macmillan Press, London.

De Mello, L.R. (1999), 'Foreign Direct Investment-led Growth: Evidence from Time Series and Panel Data', *Oxford Economic Papers*, 51, 133–151.

Djankov, S. and Hoekman, B. (2000), 'Foreign Investment and Productivity Growth in Czech Enterprises', *World Bank Economic Review*, 14, 49–64.

Dunning, J. (1993), *Multinational Enterprises and the Global Economy*, Addison-Wesley Publishing Company, Harlow, England.

Emery, J.T., Spence, M.T., Wells, L.T. and Buehrer, T. (2000), *Administrative Barriers to Foreign Investment. Reducing Red Tape in Africa*, FIAS Occasional Paper 14.

Fitz Gerald, J. (2000), 'Ireland's Failure – and Belated Convergence', First Draft, September 2000, The Economic and Social Research Institute.

Forfas employment survey (1999), see http:/www.idaireland.com/

Globerman, S. (1979), 'Foreign Direct Investment and "Spillover" Efficiency Benefits in Canadian Manufacturing Industries', *Canadian Journal of Economics*, 12, 42–56.

Haddad, M. and Harrison, A. (1993), 'Are there Positive Spillovers from Direct Foreign Investment? Evidence from Panel Data for Morocco', *Journal of Development Economics*, 42, 51–74.

Haque, N.U., Mark, N. and Mathieson, D.J. (1998), 'The Relative Importance of Political and Economic Variables in Creditworthiness Ratings', *IMF Working Paper* WP/98/46.

Hines, J.R. (1996), 'Tax Policy and the Activities of Multinational Corporations', *NBER Working Paper* 5589.

Hines, J.R. (1995), 'Forbidden Payment: Foreign Bribery and American Business after 1977', *NBER Working Paper* 5266.

IDA Ireland (2000), *Annual Report 1999*, IDA Ireland, Dublin.

IPAP (2000), see http://www.mofa.go.jp/policy/economy/asem/asem3/statement.html

Jaspersen, F.Z., Aylward, A.H. and Knox, A.D. (2000), 'Risk and Private Investment: Africa Compared with Other Developing Areas', in P. Collier and C. Pattillo (eds), Investment and Risk in Africa, Macmillan, London.

Jun, K.W. and Singh, H. (1996), 'The Determinants of Foreign Direct Investment: New Empirical Evidence', *Transnational Corporations*, 5, 67–106.

Kalotay, K. (2000), 'Is the Sky the Limit? The Absorptive Capacity of Central Europe for FDI', *Transnational Corporations*, 9, 137–162.

Kim, J.D. and Hwang, S.-I. (1998), 'The Role of Foreign Direct Investment in Korea's Economic Development: Productivity Effects and Implications for the Currency Crisis', *Korea Institute for International Economic Policy Working Paper* 98–04.

Lall, S. (1995), 'Industrial Strategy and Policies on Foreign Direct Investment in East Asia', *Transnational Corporations*, 4(3), 1–26.

Lall, S. (1996), *Learning from the Asian Tigers*, Macmillan Press, London.

Lall, S. (2000a), 'FDI and Development: Research Issues in the Emerging Context', *Policy Discussion Paper* 20, Centre for International Economic Studies, University of Adelaide.

Lall, S. (2000b), 'Evaluation of Promotion and Incentive Strategies for FDI in Sub-Saharan Africa', First Draft, January 2000, Oxford University.

Lall, S. (2000c), 'Export Performance, Technological Upgrading and Foreign Direct Investment Strategies in the Asian Newly Industrializing Economies. With special reference to Singapur', *ECLAC serie desarrollo productivo* 88, Santiago, Chile.

Lim, Y.C. and Pang, E.F. (1991), *Foreign Direct Investment and Industrialization in Malaysia, Singapore, Taiwan and Thailand*, OECD Development Centre, Paris.

Madani, D. (1999), 'A Review of the Role and Impact of Export Processing Zones', *Policy Working Paper* 2238, World Bank, Washington.

Markusen, J.R. (1995), 'The Boundaries of Multinational Enterprises and the Theory of International trade', *Journal of Economic Perspectives*, 9, 169–189.

Maskus, K.E. (2000), 'Intellectual Property Rights and Foreign Direct Investment', *Policy Discussion Paper* 22, Centre for International Economic Studies, University of Adelaide.

Moran, T.H. (1998), *Foreign Direct Investment and Development*, Institute for International Economics, Washington, DC.

Morisset, J. (2000), 'Foreign Direct Investment in Africa. Policies Also Matter', *Transnational Corporations*, 9(2), 107–126.

Morrissey, O. (2000), 'Investment and Competition Policy in Developing Countries: Implications of and for the WTO', *CREDIT Research Paper*, 00/2.

Neary, J.P. (2000), 'R&D in Developing Countries: What Should Governments Do?', *Centre for Economic Performance Discussion Paper*, July 2000.

O'Connor, T.P. (2001), 'Foreign Direct Investment and Indigenous Industry in Ireland: Review of Evidence', Report to ESRC.

O'Malley, (1998), 'The revival of Irish Indigenous Industry 1987–1997', *Quarterly Economic Commentary*, ESRI.

OECD (1999), *Activities of Foreign Multinationals*, OECD, Paris.

Oman, C. (2000), *Policy Competition for Foreign Direct Investment: A Study of Competition among Governments to Attract FDI*, OECD Development Centre, Paris.

Pain, N. and Hubert, F. (2000), 'Inward Investment and Technical Progress in the United Kingdom Manufacturing Sector', *NIESR Discussion Paper* 175.

Porter, M. (1998), 'Clusters and the New Economics of Competition', *Harvard Business Review*, 76, 77–90.

Rodrik, D. (2000), 'Can Integration into the World Economy Substitute for a Development Strategy?', Note prepared for the World Bank's ABCDE-Europe Conference in Paris, June 26–28, 2000.

Ruane, F. and Gorg, H. (1998), 'Linkages between Multinationals and Indigenous Firms: Evidence for the Electronics Sector in Ireland', Trinity Economic Papers Series, 98/13.

Ruane, F. and Gorg, H. (1999), 'Irish FDI Policy and Investment from the EU', Chapter 3 in R. Barrell and N. Pain (eds), *Investment, Innovation and the Diffusion of Technology in Europe*. Cambridge University Press, pp. 44–67.

Ruane, F. and Gorg, H. (2000), 'European Integration and Peripherality: Lessons from the Irish Experience', *World Economy*, 23, 405–421.

Safarian, A.E. (1999), 'Host Country Policies towards Inward Foreign Investment in the 1950s and 1990s', *Transnational Corporations*, 8 (2).

Smarzynska, B.K. and Wei, S.J. (2000), 'Corruption and Composition of Foreign Direct Investment: Firm-Level Evidence', *NBER Working Paper* 7969.

Spar, D. (1998), *Attracting High Technology Investment: Intel's Costa Rican Plant*, Washington, DC, Foreign Investment Advisory Service, IFC and World Bank, FIAS Occasional Paper 11.

Teece, D.J. (1977), 'Technology Transfer by Multinational Firms: The Resource Cost of Transferring Technological Know-how', *Economic Journal*, 87, 242–261.

Tsai, P.L. (1995), 'Foreign Direct Investment and Income Inequality: Further Evidence', *World Development*, 23, 469–483.

UNCTAD (1996), *World Investment Report 1996*, UNCTAD, Geneva.

UNCTAD (1999), *World Investment Report 1999*, UNCTAD, Geneva.

UNCTAD (2000), *World Investment Report 2000*, UNCTAD, Geneva.

te Velde D.W. (2000), 'Foreign Direct Investment and Unskilled Workers in US manufacturing, 1973–1994, presented UK at Royal Economic Society conference, July 2000, downloadable from http://www.st-andrews.ac.uk/~res2000/papers/pdffiles/wednesday/te%20Velde.pdf

te Velde, D.W. (2001), 'Foreign Direct Investment and Factor Prices in US Manufacturing', *Weltwirtschaftliches Archiv*, 137(4), 662–643.

Wells, L.T. (1999), 'Revisiting Marketing a Country: Promotion as a Tool for Attracting Foreign Investment', *Harvard Graduate School of Business Administration*.

Wells, L.T. and Wint, A.G. (1990), *Marketing a Country: Promotion as a Tool for Attracting Foreign Investment*, Washington, DC, International Financial Corporation.

Wheeler, D. and Mody, A. (1992), 'International Investment Location Decisions: The Case of US firms', *Journal of International Economics*, 33, 57–76.

Xu, B. (2000), 'Multinational Enterprises, Technology Diffusion, and Host Country Productivity', *Journal of Development Economics*, 62, 477–493.

Yeung, H.W-C. (2001a), 'Towards a Regional Strategy: The Role of Regional Headquarters of Foreign Firms in Singapore', *Urban Studies*, 38, 157–183.

Yeung, H.W-C. (2001b), *Enterpreneurship and the Internationalisation of Asian Firms: An Institutional Perspective*, Cheltenham, Edward Elgar.

7 Financial sector policies for enterprise development

Andy W. Mullineux and Victor Murinde

Introduction: general policy considerations

If there were full information about the creditworthiness of enterprises and other economic agents, banks would probably not exist. All finance would be 'direct finance' in the sense that lenders would finance enterprises and households directly, without the need for banks or other financial intermediaries.[1] Likewise, there would probably be no need for money since borrowing and financing would involve transfers of assets between portfolios of wealth.

Moreover, the vision of a world without banks may be contemplated in the context of the full consequences of the ongoing communications and information technology (CIT) revolution. Electronic funds transfer is already replacing cheques and giro-based transactions; it is anticipated that 'electronic wallets' will replace wallets and purses containing bank notes and coins. Although it is normally assumed that 'direct finance' in the form of bonds and equity (shares) needs stock markets and brokers, Electronic Communication Networks (ECNs) are already taking business from traditionally organised stock exchanges. Young and Theys (1999) forecast that stock exchanges will be the early victims of the internet revolution. Direct trading by individual investors on the internet is already undermining the role of traditional broking houses. It is thus possible that 'direct finance' could ultimately mean just that, lending direct to borrowers and cutting out both banks and brokers.

Banks and organised capital markets currently play an important role in enterprise financing because of lack of full information on the credit standings of borrowers.[2] In this context, banks fill the vacuum created by market failure. Borrowers normally have more information about their credit standing than lenders and consequently there is asymmetry of information (see, for example, Brock and Evans, 1997). In the real world of information asymmetry, which is most acute in developing countries, banks not only exist but dominate the financial system (Bhatt, 1994).[3] As banks gain expertise in lending, essentially appraising credit risks, they can develop well-diversified loan-asset portfolios to underwrite their commitment to repay deposits, on demand, and at their full nominal value. In the process of developing such portfolios in pursuit of profit they should allocate capital to its most efficient uses and ensure that it continues to be used efficiently. As more information accumulates and is disclosed by borrowing firms, capital markets naturally develop (Kumar and Tsetseko, 1992). The capital markets then take over the role of financing investment by the larger firms; for example, there seems to be a strong link between stock market activity, enterprise profitability and further investment (Blanchard et al., 1993).

However, it is rational for depositors to panic if they suspect a bank to be unsound (Diamond and Dybvig, 1983; Diamond, 1991). As emphasised by Mazumdar (1997), banking systems are inherently unstable and must be regulated to protect depositors and to underwrite economic stability. Implicit deposit insurance through 'lender of last resort' intervention to rescue and prevent bank failure leads to a moral hazard problem.[4] Managers, knowing that they no longer face the threat of mass deposit withdrawals, can take on excessive risk. This moral hazard problem, combined with the risk of fraudulent use of funds for direct personal gain or through patronage, necessitates the regulation of banking (Tsiddon, 1992). The first step in developing a policy for efficient enterprise financing is therefore to ensure that the banking system is well regulated and supervised.[5]

Apart from bank domination, another common feature of financial systems that is important in the context of enterprise financing in developing countries is government intervention. This raises the wider issue of whether governments should be in the business of directing credit or interfering with commercial interest rate setting at all. If, however, governments do not intervene, what form should social, regional and industrial policy take? Is industrial policy about proactively 'picking winners', in the form of individual firms or particularly sectors, and directing funds towards them, or is it about privatising enterprises (including banks) and allowing a well regulated and supervised financial system to allocate credit and capital efficiency. In developed countries there has been a move away from directed credit and interference in interest margin and spread setting towards market allocation of finance, but there remains a tendency to try to pick winners. The UK government, for example, has been trying to ensure that high-technology sectors receive adequate finance and support. The policy instruments are, however, subsidies and tax incentives, rather than directed lending and interest rate controls. The role for development banks then becomes one of addressing market failures through the provision of loan guarantees and finance, perhaps in partnership with the private sector for infrastructural projects (Murinde, 1997a).

Monetary policy may come into conflict with industrial and general development policy in setting the nominal interest rate and aligning it to the appropriate real rate?[6] Attempts to hold real rates at low or negative levels to ensure cheap industrial finance are not only likely to be inflationary, but also counterproductive. As the seminal model by McKinnon (1973) and Shaw (1973) demonstrates, financial repression will result because the development of intermediated finance and bank lending is inhibited.[7]

Credit rationing is another consequence of asymmetric information (Stiglitz and Weiss, 1981). It is most acute where information asymmetry is greatest. Small and medium-sized enterprises (SMEs) are thus likely to be most exposed to credit rationing because less is known about them and their credit standing; most often they have no collateral (Stiglitz and Weiss, 1987). Specifically, credit rationing occurs because banks have insufficient information to accurately rank SMEs according to their credit standing. The banks have a pretty good understanding of the average credit risks involved in lending to various sectors (e.g. farming and retailing), but are less able to accurately assign 'credit ratings' to individual borrowers. They thus tend to charge interest rates which are an average (for the sector) risk-related mark-up ('spread') over the base rate. This discourages better (low risk) borrowers, who regard the rate as too expensive, and encourages 'bad' (high risk) borrowers, who regard the rate to be cheap. An 'adverse selection' problem then arises as the average risk faced by the lender is pushed up by the preponderance of high-risk borrowers. To protect themselves against this tendency, banks reduce their exposure by rationing the supply of credit to markets where they are unable to adequately assess the individual credit ratings of borrowers. SMEs in developing countries are thus likely to face more severe rationing than SMEs in developed countries because the latter have stricter reporting and auditing requirements (imposed by the tax and company registration authorities). Also, smaller and newer firms tend to face more severe rationing than older and larger firms.[8] Rural enterprises (especially farms) in many developing countries are commonly small and thus rural credit is a particular problem given the added geographical problem of delivering of financial services to remote areas. These rural enterprises are the engine for higher consumption and saving in developing countries (see Durojaiye, 1991). Small urban enterprises, however, also suffer from this 'financial exclusion' problem. There are obvious implications for economic growth overall, although these arguments are presented in the context of an economic growth model (see Barro and Sala-i-Martin, 1995).

To deal with the rural 'financial exclusion' problem, special initiatives may well be required, such as mobile banking units and microfinance (Murinde and Kariisa-Kasa, 1997). Microfinance, along with other community finance institutions (CFIs) involving mutual financial institutions (credit unions, social investment funds, mutual guarantee schemes) may also help resolve urban 'financial exclusion' problems. More generally, the problem of credit rationing is commonly addressed by state-run loan guarantee schemes. These provide subsidised loan guarantees to reduce banks' exposure to credit risk, thereby encouraging greater bank lending to SMEs. It should be noted that loan guarantees can be targeted (with different levels of risk cover and subsidy) on different sectors of the economy in pursuit of industrial policy (e.g. encouraging lending to start-up and 'growth' enterprises in the high-technology sector) and social policy (e.g. rural credit, lending to women, lending in deprived urban areas).

Against this background, this chapter argues that strategies for stimulating business enterprise finance should revolve around: establishing effective bank regulation and supervision; setting an appropriate positive real interest rate; reducing unnecessary government intervention, guidance and direction (which also presents opportunity for corruption and 'cronyism'); and addressing the credit rationing and financial exclusion and other problems, such as inadequate competition in banking, that arise from market failures (Table 7.1).

Table 7.1 Financial sector problems and possible causes

Problems	Possible causes
1 Monetary instability: high (and variable) inflation	High level of monetary financing of the budget deficit; low real interest rates
2 Financial instability	Inadequate banking and wider financial sector regulation and Supervision (and monetary instability); capital outflows[a]
3 Shortage of Bank Loans (credit rationing/crunch). (NB The shortage may be concentrated e.g. on rural areas, SMEs, and low-income sectors)	Inadequate domestic saving; limited access to foreign capital[a]; asymmetric information; loan fixed cost problems; weak bankruptcy laws; post financial crisis effects
4 Weak Corporate Governance	Monetary instability (making monitoring difficult); government directed lending; low real interest rates; weak bankruptcy and antifraud laws
5 Shortage of Private/Venture Capital	Weak reporting and bankruptcy laws; inadequate tax incentives; no-exit – see footnote to Table 2
6 Underdeveloped Money and Capital Markets	Low levels of transparency; weak market regulation and supervision; lack of liquidity (few buyers and/or sellers of stocks and/or too few stocks); low levels of capital inflows[a]

Note

a Capital inflows and outflows are in turn related to wider economic and political instability (and related expectations about exchange rate fluctuations) and to the extent of capital account liberalisation.

The remainder of this chapter is structured into several sections. The section on 'The role of the financial sector' discusses the role of the financial sector in promoting the growth of enterprises. Financial system restructuring, as a means of correcting broad policy-induced distortions, is examined in the section on 'Financial sector restructuring'. Financial sector regulation and supervision is explored in the section on 'Financial sector regulation and supervision'. The section on 'SME financing' considers the financing of SMEs. The role of development banks is examined in the section on 'The role of development banks'. The section on 'Developing venture capitalism' focuses on the possibility of utilising venture capital. The prospects for developing capital markets is discussed in the section on 'Developing capital markets'. Concluding remarks are offered in the last section.

The role of the financial sector

Most enterprises finance their investment and further growth using a combination of internal finance (retained earnings) and external finance (e.g. new equity issues, bank loans or funds raised by issuing debt instruments including bonds).[9]

Thus, the capital structure of most enterprises includes retained earnings, debt and equity (see Bertero, 1994; Brealey and Myers, 2002).

As new enterprises develop, there comes a point at which external finance is required to accelerate their growth. Internal finance, however, remains the major source of funding for ongoing investment in all firms.[10] Much less commonly, because of fear of loss of control, private enterprises also accept investment from venture capitalists. This typically takes the form of equity investment by individuals ('business angels') or through a private equity fund which pools the personal wealth of investors. In general, however, as explained in Murinde et al. (1999), enterprises tend to have a 'pecking order of financing choices'; they prefer using internal finance before they resort to external finance.[11] Again, this may be because of the governance implications of expanding the share of external funding.

The importance of internal finance and external private capital means that careful consideration has to be given to the tax treatment of profits earned from direct private investment, especially when they are re-invested, or from indirect private investment (via venture and other private equity funds, or direct equity participations). 'Over taxation' is likely to result in under investment and/or tax evasion. 'Under taxation' can, however, lock capital into established enterprises, for example, the *Chaebol* in South Korea, starving potential new and more efficient users of capital. A careful balance, thus, has to be struck and it may well be that re-invested profits from start-up and early stage growth investments should be taxed at a lower rate than profits from ongoing internally generated capital investment by larger and older firms.

The overall goal should be to establish a financial system that allocates capital efficiently on a dynamic and ongoing basis. Hence, capital needs to be continuously allocated to its most efficient uses. This will involve re-allocation of capital, withdrawing it from inefficient uses and re-allocating it to the most efficient users or projects identified at any point in time. To achieve this, effective bankruptcy laws[12] are required and the use of capital allocated by the financial sector should be continuously and efficiently monitored (and be seen to be monitored). Effective, well designed, fair to creditors and debtors and well prosecuted bankruptcy laws are, thus, a key component of an efficient financial sector.

It is also clear from the above that the role of the financial sector in corporate governance needs to be carefully considered. As the allocator of debt (loan and bond) and equity finance, the financial sector is a major stakeholder in the economy. It also has fiduciary duties to other vicarious stakeholders, namely those that have deposited and invested their monetary and savings balances with financial sector. The infrastructure required for the efficient operation of a financial sector, thus, includes an effective corporate governance system, of which bankruptcy laws can be viewed as part. As the financial sector develops, the importance of institutional investors (pension and mutual funds and insurance companies) tends to increase relative to that of banks (Asikoglu et al., 1992).

As holders of the majority of shares (equity) and bonds (debt contracts), the way in which the institutional investors exercise their voting rights becomes increasingly important. Because of asymmetric information, however, banks will remain the major suppliers of debt (loans/credits/overdrafts) to SMEs and thus will also have a crucial role to play in corporate governance through their role in monitoring their exposure to credit and other risks.

For most developing and transition economies, banks are still the major source of external capital for large as well as small enterprises, and indeed for the private sector and the economy as a whole.[13] Debt finance thus tends to dominate equity finance, and bank debt (loan) finance dominates bond finance. This argument is consistent with the 'pecking order of financing choices'. Moreover, it takes time to develop capital (bond and equity) markets, a matter to which we return in the final section.

Given the widespread dominance of the banking sector in most financial systems, it is important to ensure that the banking sector operates efficiently. The potential for banking sector instability and its damaging effects have been illustrated by the 1997/1998 Asian and other financial crises in the last couple of decades. As explained in the introduction, it is clear that banks need to be regulated. As a result of the work of the Basle Committee on Bank Regulatory and Supervisory Practices, there is increasingly widespread acceptance of how banks should be regulated. What is required is the establishment of the appropriate legal structures and of a supervisory body to ensure that banks adhere to the regulations. Banks should monitor firms that have borrowed from them to ensure that they use capital efficiently and banking firms should themselves be monitored to ensure that they use their capital and the deposits they collect, and thus people's monetary and savings balances, efficiently and non-fraudulently.

There is clear evidence that banking instability and the subsequent recapitalisation of banks are expensive in terms of charges on the state budget and lost growth (see IMF, 1993; World Bank, 1994). Hence, prevention of instability should be given a high priority (Ciarrapico, 1992; Simon, 1992). There is also evidence that macroeconomic instability discourages investment and reduces economic growth (Fingleton and Schoenmaker, 1992). Clearly, a bout of financial instability, including bank failures, will generate macroeconomic instability, but macroeconomic instability can occur independently[14] of financial instability and result in increased financial fragility. Furthermore, fluctuations in growth and inflation clearly have an impact on the balance sheets of firms and banks. Bank supervisors must take this into account. The monetary authorities may or may not be responsible for both supervising banks and controlling inflation using monetary policy. Those responsible for inflation control must, however, recognise that the monetary brakes can only be applied, in the form of an aggressive rise in interest rates, if the financial sector is stable. This will only be the case if banks and the firms to which they lend are well governed. However, the symbiosis between financial and macroeconomics stability works in the other direction too. The more stable the macroeconomic environment, the easier it is to develop strategic investment plans and the more efficiently will capital be allocated. Faster economic development will result. Macroeconomic stability and the efficiency of the financial sector are, thus, highly interrelated and good macroeconomic (and microeconomic) policy formulation must take this into account (see Green and Murinde, 1993).

Finally, as capital markets develop, and indeed to facilitate their efficient development, attention needs increasingly to be paid to ensuring that the wider financial sector, not just banks, is effectively regulated and supervised.

Financial sector restructuring

The need for restructuring

Financial system restructuring derives from the need to correct government policy-induced distortions in the financial sector (Murinde, 1997a) and is inspired by the theory and evidence in favour of supply-leading finance rather than demand-following finance (see Murinde and Eng, 1994; Lyons and Murinde, 1994). There are two main forms of government intervention. The first form relates to intervention in the implementation of monetary policy (see Murinde and Ngah, 1995). The choice is between direct controls on levels of bank lending, which are likely to be distortionary, or a more market-oriented approach of manipulating the price of money, that is, interest rates.[15] Many countries have intervened in bank interest-rate setting in pursuit of social goals (e.g. cheap housing or rural finance) by setting ceilings on nominal lending or deposit levels or fixed margins over some base interest rate set by the monetary authorities (Hermes et al., 1998). Further, direct controls on bank lending are often imposed in pursuit of regional or industrial policy, and the directed credit instructions frequently conflict with monetary policy goals (Fry, 1995).

The second main form of intervention invalues final repression through holding real interest rates at low or negative levels (see McKinnon, 1973; Shaw, 1973). Econometric work on variants of the McKinnon–Shaw model has shown that many developing countries are either currently experiencing, or have experienced, 'financial repression' (see Fry, 1995; Murinde, 1997a; Hermes et al., 1998). However, the appropriate level of the real interest rate for a developing country is hard to gauge. It is important to note, however, that both overly high positive real rates and inappropriately negative rates are damaging (see Abebe, 1990). Low positive real rates are likely to stimulate more saving and higher aggregate bank deposits, leading to more lending than in the case of artificially low (negative) real rates; this is notwithstanding the criticism that 'cheap credit' undermines rural development in developing countries (Adams et al., 1984). Meanwhile, excessively high rates aggravate 'credit rationing'.

Restructuring is also required following a financial crisis. It should be considered as a means of increasing both financial stability and the efficiency of capital allocation; and thereby facilitating the achievement of a more rapid economic development.

In some cases, financial restructuring may take a form of financial liberalisation (see Murinde, 1997a,b). Key liberalisations are likely to include interest rate deregulation and removal of restrictions on branching, foreign bank entry and the range of financial activity (scope). The objectives are to increase the supply of loanable funds and achieve higher savings levels. However, the responsiveness of investment to financial liberalisation is debatable (see Hermes, 1994; Gupta and Lensink, 1997). In some countries, financial liberalisation has been associated with financial sector failures (Honohan, 1993). Rural credit markets may be completely wiped out by the introduction of financial liberalisation (Herath, 1994). In addition, financial liberalisation may not be able to stave off capital flight (Hermes et al., 1998). It is, however, not merely the levels of investment that matter, for in the larger sea it is the quality of the investment that assures competitiveness.

The efficient allocation of capital is only possible if there is efficient financial intermediation.[16] Broadly, this requires setting real (i.e. inflation adjusted) loan and deposit interest rates at levels that reflect the risk of lending and allow an adequate return to be made by lenders and other financiers on their capital whilst rewarding savers and depositors appropriately. Where there is widespread state ownership of banks and other financial institutions, there may be a temptation to hold lending rates down to stimulate growth. This is likely to be counterproductive and distortionary, with negative consequences for the pace of economic development in the medium term. It can lead to overborrowing, overinvestment (investment that earns an inadequate return on capital) and overindebtedness (too much debt relative to equity and income streams). The distortionary effects will be compounded if lending is directed, in accordance with some government plan, to preferred sectors; regardless of the risk to return ratios involved. Whilst much directed lending is well intentioned, most prominently in the formerly centrally planned economies, it is also subject to abuse and corruption (particularly *graft*) and political interference and can lead to the sort of problems faced by the *Chaebol* (particularly Daewoo) in South Korea and the state-owned enterprises in China in the late 1990s.[17] The result is an increase in financial instability since the lending banks and other creditors will face raising bad debt problems as a result of lending to inefficient enterprises.

Low interest rates also discourage the public from placing monetary and savings balances with banks, and may well lead to a lower level of savings (Fry, 1995). Hence, low (particularly negative real) interest rates reduce the volume of funds to be intermediated or lent by banks and other financial intermediaries.

Seemingly paradoxically, raising interest rates to positive real levels may well not only stimulate more saving and lending, but, in making borrowing more costly, discourage inefficient lending and overinvestment. Given the widespread dominance of the banking sector in most financial systems, it is important to ensure that the banking sector operates efficiently. The potential for banking sector instability and its damaging effects have been illustrated by the 1997/1998 Asian and other financial crises in the last couple of decades. As explained in the introduction, it is clear that banks need to be regulated. As a result of the work of the Basle Committee on Bank Regulatory and Supervisory Practices, there is increasingly widespread acceptance of how banks should be regulated. What is required is the establishment of the appropriate legal structures and of a supervisory body to ensure that banks adhere to the regulations. Banks should monitor firms that have borrowed from them to ensure that they use capital efficiently and banking firms should themselves be monitored to ensure that they use their capital and the deposits they collect, and thus people's monetary and savings balances, efficiently and non-fraudulently, making borrowing more costly and discouraging inefficient lending and overinvestment (Murinde and Mullineux, 1999). The net result is likely to be a more efficient allocation of an increased volume of capital, and a more rapid development of the financial sector (through reduced 'financial repression') and the economy as a whole.

This line of argument also suggests that it is best to reduce the amount of directed credit and free the financial sector to pursue profits, through the efficient allocation of their capital, whilst ensuring that regulatory infrastructure is in place and banks are adequately supervised.

In order to give banks the right incentives and help protect them from political interference, they should then be privatised as soon as possible. However, privatisation of banks is only possible once their bad debt problems have been addressed (Doukas et al., 1998). It should also be noted that financial sector restructuring is likely to take place within the context of wider policy reforms. If a financial crisis has occurred, or needs to be prevented, then financial sector restructuring is likely to take place whilst attempts are being made to assure macroeconomic stability, and to improve corporate and economic and political governance in general, that is, as part of widespread legal, political and economic reform.

Bad debt problems and bank privatisation

The recent examples of bad debts and bank privatisation in the transition economies of Eastern and Central Europe have prompted policy makers and analysts to rethink their strategies (Doukas et al., 1998). The dominant view is that privatisation of banks and other state-owned enterprises must be preceded by a work-out of outstanding debt relationships between them, especially when the banks are faced with bad or doubtful debts (Murinde and Mullineux, 1999). The resolution of the banks' bad debt problems paves the way for the privatisation of other state-owned enterprises. The banks' bad debt problems are usually resolved using some combination of removing bad loans from their asset portfolio and recapitalisation. There have been numerous recent examples of how this can be done, for example, the US savings and loans crises and the Nordic banking crisis. As banks' bad loan assets are debt problems for the borrowing enterprises, their resolution leads effectively to a financial restructuring of indebted firms. If the enterprises are state owned, as is often the case in formerly centrally planned economies, then the financial restructuring of the banks' asset portfolios leads to a financial restructuring of state-owned enterprise liability portfolios, paving the way for their privatisation too. If, as in Poland for example, the financial restructuring of banks and enterprises is done in tandem as part of the process of preparing them for privatisation, then more comprehensive enterprise restructuring, involving the establishment of new management structures, can be made a condition for government financial assistance; which is invariably required. Mass privatisation in the absence of prior restructuring under such conditionality, as executed in Russia and Czech Republic, has proved problematic. In the Czech case, because governance structures were weak, and in the Russian case, because the process facilitated the formation of oligarchies.

The financial and enterprise restructuring process may well require the breaking up of large conglomerate units; such as the *Combinats* in East Germany, and the *Chaebol* in South Korea. This will prove difficult to achieve unless adequate bankruptcy laws and mechanisms for welfare provision by the state are in place.

Restructuring in an international context

The process of globalisation and trade liberalisation is exposing more and more sectors in more and more economies to competition (see Banuri and Schor, 1992). As trade barriers are dismantled it becomes increasingly important that capital is allocated to the sectors in which the country or region has a comparative advantage. Trade liberalisation has gone hand in hand with capital account liberalisation in recent years, making it easier for countries to import capital to supplement domestic savings and thereby to accelerate development (see Anderson and Khambata, 1985). As with domestic sources of capital, it is important that capital inflows from abroad are efficiently invested and do not lead to overinvestment and/or inflationary pressures (and 'bubbles'). The problem of excessive capital inflows, leading to artificially low interest rates and an exposure to capital flight, has been recurrent in recent years and seems to have been a root cause of the Asian financial crisis (Dickinson and Mullineux, 2000).

The combination of capital account liberalisation, as strongly advocated by the IMF, at least prior to the Asian Crisis, and the General Agreement of Trade in Services (GATS), especially relating to financial services, in the Uruguay round of the General Agreement on Tariffs and Trade (GATT) concluded in the mid-1990s, have acted as drivers for increased financial sector competition. Entry by foreign banks and other financial institutions has increased in many countries. Their entry has often driven down interest rate margins and brokerage fees and facilitated the transfer of skills and 'know-how', leading to enhanced efficiency. Not surprisingly, entry is not welcomed by vested interest groups, including indigenous financial institutions, but the overall impact seems to be beneficial and some recent studies on the effects of GATS in developing countries have concluded that financial liberalisation *enhances* stability (Murinde and Ryan, 1999). It is well known that banking sectors subjected to increased competitive pressures can become more fragile if they are not carefully managed, hence bank supervisors should be especially vigilant during periods following liberalisation.

The possibility of excessive short-term capital inflows also needs to be carefully monitored and this has implications for exchange rate policy (Hermes *et al.*, 1998; Dickinson and Mullineux, 2001). A lesson of the Asian crisis is that exchange rate pegs (including currency boards) can outlive their usefulness as stabilisation devices. Once they do, then flexibility (floating or crawling pegs) or participation in a currency union become the only viable alternatives. It should be noted that, if directed lending is to be progressively abandoned, then the traditional role of development banks is called into question (see Bhatt, 1993). We will turn to this issue in the section on 'The role of development banks'.

Financial sector regulation and supervision

Banks typically hold liquid reserves that are a fraction of their demand deposit liabilities and generally the average term to maturity of their assets (usually predominantly loans) is greater than that of their liabilities (predominantly transactions and savings deposits). They are, thus, vulnerable to liquidity crises and unexpectedly large withdrawals of deposits can cause them difficulties unless the central bank is willing and able to supply liquidity in its role as 'lender of last resort' (Mullineux, 1987a,b).

The potential for widespread panic withdrawals leading to 'bank runs' is increased by the high level of uncertainty about the quality of banks' asset (especially loan) portfolios created by the asymmetry of information between the borrowers from banks and the banks' loan officers, and between the depositor's and the banks' management. Generally, the borrowers have more information about their circumstances than the banks and the banks' management has more information about the quality of the loan portfolio than the depositors and shareholders. Once depositors' confidence in the safety and soundness of a bank is undermined, then it is rational for uninsured depositors to panic and rush to withdraw their deposits. Banks' reserves are commonly in the region of 10 per cent of deposit liabilities, so it pays to be at the head of the queue to withdraw deposits. If the bank survives, perhaps after being bailed out by the central bank, then the money can always be redeposited in the bank the next day (and little interest has been lost). If the bank fails, or uncertainty continues, then the money can be deposited with a seemingly safer bank, in which deposits are deemed to be insured; perhaps because the bank is regarded as being too big to be allowed to fail. If there are no such banks, then bank runs are likely to become more widespread because all the other banks are also believed to have asset portfolios of dubious quality and a 'systemic banking crisis' involving a flight to cash (and possibly also capital flight) will have occurred.

Systemic banking crises are likely to undermine the wider financial sector and cause major interruptions in economic development. The cost of resolving the resulting bad debt problems and recapitalising the banking system is extremely high (Murinde and Mullineux, 1999) and takes a considerable amount of time. In the case of Japan in the 1990s, it has literally taken years. In the mean time, bank lending is impaired and the credit rationing which SMEs normally face (Stiglitz and Weiss, 1981), becomes acute; further inhibiting economic development by cutting off the supply of debt finance to the potentially most vibrant firms. In other words it may result in a 'credit crunch'.

Prevention of systemic banking crises is thus highly desirable. A well-designed bank regulatory and supervisory system can help achieve this goal. Benston and

Kaufman (1997) spell out the role of bank regulation.[18] The supervisors should be well trained and amply rewarded to assure their retention[19] and the supervisory agency should be well staffed and amply resourced. Further, the legal system must prosecute financial fraud and malpractice vigorously.

An important element of the regulatory mechanism for banks is implicit or explicit deposit insurance. Specifically, deposit insurance schemes are designed to provide a mechanism with which the bank regulatory authority can protect deposits in banking institutions. Under implicit deposit insurance, deposits are protected by the bank monitoring and regulatory authority such as a central bank acting as lender of last resort. An explicit deposit insurance scheme is a fund to which deposit-taking financial intermediaries (usually banks) make contributions. It provides uninformed and unsophisticated depositors with a financial safety net. This instills greater confidence among depositors and increases the likelihood of financial stability (Okeahalam, 2002). The lesson of the US savings and loans (S&L) crises of the 1980s is that deposit insurance premia must be risk related in order to offset the moral hazard problems and the US accordingly revised its scheme through the Federal Deposit Insurance Corporation Improvement Act (FDICIA), 1991.

Given the possibility that their lender of last resort role might conflict with their monetary control responsibilities, there is an ongoing debate about whether central banks should be involved in bank supervision. One view is that central banks should be given independent responsibility for monetary control (essentially setting interest rates) and should be required to concentrate on that task without the distraction of day-to-day supervisory responsibilities. An alternative view is that central banks, as lenders of lost resort and thus the ultimate insurers of deposits, need to be involved in supervision (and perhaps regulatory design) in order to protect the taxpayers, who will ultimately foot the bill for lender of last resort intervention, from abuse by recalcitrant banks. A wider issue to be faced, as financial sector liberalisation and development proceeds and banks diversify, is whether there should be a single supervisor for the whole financial sector, or sectoral supervisors for banking, insurance and capital markets, etc. In developed economies, the trend is towards the latter. Japan established a Financial Supervisory Agency in 1998. The UK established a Financial Services Authority in the same year and there is also a longer standing Financial Supervisory Authority in Sweden. These institutions have responsibility for the supervision of banks, insurance companies and securities firms.

Good supervision cannot be conducted cheaply and countries need to consider the extent to which reliance can be put on auditors appointed by the shareholders of banking and other financial firms. If heavy reliance is to be placed on auditors, then the quality of their training needs to be monitored and legislation will be required to clarify their role in the statutory supervisory procedures. This will involve checking the data which the regulators require to be published and probably also a requirement to enter into a dialogue with the supervisory authorities, regardless of confidentiality issues. There is, however, growing concern about potential conflicts of interest as the internationally reputed accountancy firms increase the share of profits they make from consultancy and other services, relative to auditing. Some of their best clients are, after all, banks. If the taxpayer is to be adequately protected from abuse, by banks and other financial firms, of the deposit and investment insurance that the state (usually via the central bank and/or the Treasury/Finance Ministry) commonly provides, then adequate budgetary provision for a state-run supervisory agency must be made. The financial firms under supervision, and thus ultimately the insured depositors and investors, can be required to make a contribution through fees charged by the supervisory agency. The extent of the state subsidy of the fees is an issue for debate given that financial stability is a public good.

A bank's exposure to the risk of insolvency clearly depends on the amount of capital it holds. The Basle Committee has laid down minimum standards for capital adequacy. These are supported by the IMF and are being adopted increasingly widely. These requirements are also risk-weighted and are currently under review. Initially the Basle capital adequacy ratios only related to credit risks, but the new arrangements will also encompass market and other risks as well. The aim is to ensure that banks hold sufficient capital to cover the risks to which they are exposed; this is consistent with the Basle capital adequacy ratio requirement.

Given the credit risks involved, it is natural for defaults to occur on some loans and there may well be cyclical swings in the volume of loan defaults. Banks therefore need to develop a policy for pricing loans (setting interest rates) according to their riskiness and providing for known and anticipated loans losses. Thus, banks should supplement capital with reserves to cover projected loan losses. The US has the most rigorous provisioning policy. Good provisioning policy should ideally be anticipatory (Mullineux, 1999), building up reserves in the boom, so that they can be drawn down in periods of slower or negative economic growth. The build up of reserves in boom times will attenuate tendencies towards overlending and overinvestment and reduce the need to build reserves *after* the bad loans have begun to appear, and when the banks can least afford to build the reserves. All this happens too often, leading to an excessive and protracted period of acute credit rationing, or a credit crunch. A good provisioning policy can thus act as an 'automatic stabiliser' for the economy, curbing lending in the boom and reducing the severity of the subsequent credit crunch. Unfortunately, during booms, banks find it all too easy to persuade themselves and their supervisors that recessions are an anachronism. As a result, inadequate provisions against bad and doubtful debts are made when banks can most afford to make them, storing up problems for the next recession.

Regulatory best practice normally also tries to guard against overconcentration of risks through overexposure (lending too large a proportion of capital to particular sectors of the economy or particular borrowers). Overexposure can tie banks into inefficient lending patterns since withdrawal of funds from underperforming sectors, or large individual borrowers, can cause loan loses on a scale that threatens the bank's own solvency. Recent examples are Japanese bank lending to the property sector, Korean bank lending to the *Chaebol*, and Chinese state-owned bank lending to SOEs. Regulation typically sets maximum exposure/concentration rules. Problems can also arise from bank ownership of non-financial firms and non-financial firm ownership of banks. Practice varies considerably around the world, but supervisors should carefully monitor interlocking shareholder relationships between banks and non-financial enterprises.

To facilitate the monitoring of banks by supervisors, shareholders, bondholders and depositors, disclosure and reporting rules can be imposed. As noted above, the auditor's job is to assure the quality of the reported data. The trend is towards more disclosure. As this occurs there will be a reduction in information asymmetry and an increase in the role of market, relative to state agency, supervision. There are, however, limits to this process since the work of auditors must also be monitored (given the aforementioned conflicts of interest within accountancy firms). It should, however, be recognised that disclosure of proprietary information undermines the banking franchise by creating a free rider problem – competitors can enter without incurring the costs of collecting and processing information. The information problem is most acute with regard to the credit risks involved in lending to SMEs, and it is to SME financing that we turn next.

SME financing

There is growing recognition of the role of the SME sector in employment creation. Even in the developed industrial economies it is the SME sector, rather than the multinationals, that is the largest employer of workers (Mullineux, 1997a). As globalisation progresses and international competition increases, the largest corporations have tended to shed labour, at least in their country of origin, in pursuit of cost efficiency. Job shedding by large enterprises has been particularly evident in the transition economies as former SOEs have sought to achieve international competitiveness. Invariably it is to SMEs that countries look for the job creation necessary to absorb the labour shed by the larger firms. During periods of industrial restructuring, including those involving privatisation programmes, it is particularly important that the SME sector thrives and creates new jobs.

As indicated in the introduction, SMEs normally face credit rationing as a result of an adverse selection problem resulting from information asymmetry (Stiglitz and Weiss, 1981). Credit rationing of this sort, which results from the inability of banks to accurately gauge and price risks and the inability or unwillingness of SMEs to provide adequate collateral or third party guarantees, becomes more acute in financial crises. It is also exacerbated by an aggressive tightening of monetary policy. In such acute cases, it is often dubbed a 'credit crunch'. Credit rationing can also result from 'financial exclusion', which occurs when potential borrowers wish to borrow amounts that are seemingly too small for banks to lend at a profit. Essentially, the 'fixed costs' of making the loan appear to be too high relative to the potential return given the uncertainty about the riskiness of the loan applicants.

To help resolve the risk exposure problem, developed countries commonly employ loan guarantees. Government funds, usually managed by an agency, are used to guarantee (cover) a proportion of the credit risks incurred by banks in extending loans to SMEs that qualify for loans under the scheme. Schemes vary in the extent of the guarantee and the degree of state subsidy. Subsidised insurance premia are commonly paid by firms in the form of a supplement to the interest rate charged by the lending banks, which normally manage the loans. The accumulated premia rarely cover the disbursements from the fund resulting from defaults on loans. By reducing banks' risk exposure, it is hoped that banks would lend more than they otherwise would to SMEs.

Schemes such as the UK's 'Small Firms Loan Guarantee Scheme' were introduced on an experimental basis with the objective of encouraging banks to engage in and learn the business of lending to small firms. Having succeeded in that aim, the UK's scheme was retained, but increasingly the guarantees have been focused in pursuit of industrial policy. Firms with employment growth potential and high-tech firms, have, for example, been increasingly targeted in the 1990s. This has left smaller firms still facing credit rationing. In the US, the targeting has also been used as part of social policy. The US Small Business Administration runs schemes targeting minority ethnic groups and women, for example. Table 7.2 outlines the recently created scheme in Barbados.

Loan guarantees in developing countries

The existing provision of loan guarantees by development banks to assist small and medium scale industries in most developing countries is patchy and short term in nature. Table 7.2 sets out the policies and practices of the major multilateral financial institutions involved and some bilateral institutions.

What have we learned from the above existing schemes?

Many developing countries have experimented with loan guarantee schemes of the variety presented above. The major drawbacks of the existing schemes are as follows:

- They are poorly publicised among industrialists, policy makers and corporates.
- They are difficult to access and use.
- The actual situations that would be most appropriate for their use are unclear to both the end users and the banks.

Overall, however, there is potential for guarantee schemes which underwrite loans to state-owned and parastatal companies, Treasury Bill and bond issues and international issues. Here, guarantee of loans to the economically weaker borrowers can reduce price and increase access to funds.

Experience with loan guarantees in developed countries

The amount of bank lending to SMEs under the UK's scheme appears to have been sensitive to the extent of the guarantee and the take-up by SMEs, not surprisingly,

Table 7.2 Loan guarantee schemes

Provider	Main features of loan guarantee	Pricing
International Bank for Reconstruction and Development (IBRD)	IBRD provides loan guarantees for privately financed projects. IBRD guarantees are flexible as they can cover loans in various currencies and guarantee credit obtained through bond offerings, private placements and commercial bank loans. The guarantees are protected by an indemnity or cross-guarantee from the country where the operation takes place. Recently introduced partial credit guarantee through the issue of a Put Option in lieu of a formal guarantee. Under this structure, IBRD provided a Put Option to holders of long-term project debt under which they have an option to sell their loans to IBRD after a number of years. The effect of this instrument is identical to that of a partial credit guarantee because it insulates the lenders from the credit risk. The IBRD programme is relatively modest. For example, only three guarantees were granted in 1997.	For this service, the IBRD charges the investor a standby fee and a guarantee fee. The standby fee is 25 basis points per annum on the present value of the Bank's guarantee exposure when the guarantee is not callable. The guarantee fee is in effect during the years when the guarantee is callable and the IBRD is at risk. The fee ranges from 40 to 100 basis points per annum on the outstanding debt and consists of a base fee of 15–75 basis points plus an amount dependent on the level of coverage and subsequently the value to the borrower.
International Development Agency (IDA)	The IDA provides partial risk guarantees to private investors on a pilot basis where other private and public agencies offer insufficient insurance coverage against country risk. Where projects in IDA countries are expected to generate large foreign exchange revenues and require substantial financing compared to the amount of IDA resources allocated for such countries, the IBRD will work with the IDA by providing 'enclave' loans and guarantees.	The guarantee fee ranges from 40 to 100 basis points per annum on the outstanding debt and consists of a base fee of 15–75 basis points plus an amount dependent on the level of coverage and subsequently the value to the borrower.

(Continued)

Table 7.2 (Continued)

Provider	Main features of loan guarantee	Pricing
International Finance Corporation (IFC)	The IFC guarantees are in the form of syndications (B-loans) which serve the same purpose of the other guarantees, that is of risk sharing. It recruits commercial banks, finance companies and other financial institutions to form a syndicate under the comfort of the IFC umbrella which is perceived as reducing the currency transfer risk through IFC's preferred creditor status.	The IFC charges 1% for its syndication role, plus a small annual administration fee. IFC also reduces commercial risk through swaps and other hedging instruments.
Multilateral Investment Guaranty Agency (MIGA)	MIGA specialises in the provision of political risk insurance for investments. It is the only global multilateral political insurer. After a slow start, it has been a relatively successful scheme. It provides long-term (up to 20 years) investment guarantees against the political risks of currency transfer, expropriation, war and civil disturbance, and breach of contract. MIGA can insure new cross-border investments originating in any MIGA member country destined for any developing member country. MIGA may insure equity investments, for up to 90% of the investment contribution, plus an additional 450% of the investment contribution to cover earnings attributable to the investment. MIGA can also insure up to 90% of the principal to cover interest that will accrue over the term of the loan. The project limit is currently US$ 50 million per project, and the country limit is US$ 225 million.	Rates vary with the sector insured and are applied to the amount of investment currently at risk ('current') and the amount expected to be at risk in the future ('standby'). Current risk premia amount to broadly between 2% and 3% for a full range of political risk cover. In addition to these rates MIGA charges an application fee of US$ 5,000 for guarantees up to US$ 25 million and US$ 10,000 for guarantees for more than US$ 25 million for manufacturing, natural resources, oil and gas, infrastructure and services. For exceptional underwriting costs incurred in the evaluation of projects a processing fee of US$ 25,000 is charged also.

The Inter-American Development Bank (IDB)	The IDB's PRI offer the same guarantee programmes as those provided by the IBRD's partial risk and the partial credit guarantees and the terms and conditions are similar. Guarantees can be issued in conjunction with a loan, or on a stand alone basis, and can be used to guarantee debt only and not equity. This programme has been relatively slow to start. It has, however, made at least two large guarantees in the past year.	A facility fee and a guarantee fee are charged. The facility fee is 0.25% per annum on the full amount of the guarantee while the guarantee fee is within a range of 15–75 basis points reflecting the risks involve in each transaction.
Overseas Private Investment Corporation (OPIC)	OPIC is a US government agency for assisting US businesses doing business in the developing nations and lowering their operational risks. OPIC offers political risk insurance, as well as project finance and investment funds. Its insurance programme covers risks associated with currency inconvertibility, expropriation and political violence. This insurance is available for investments in new ventures or expansions of existing enterprises. It guarantees the private sector solely, and wholly owned US companies investing overseas or joint ventures in which a US company owns at least 25% of the equity in the project. Its guarantees are typically used for larger projects with direct loans for smaller ones. OPIC does not require counter government guarantees, instead looking for repayment from the cash flows generated by the projects. OPIC can insure up to US$ 200 million per project. It will insure 100% of principal and interest for loans and leases from financial institutions to unrelated third parties. The coverage multiples that OPIC generally issues equals 270% of the initial investment, 90% representing the original investment and 180% to cover future earnings. OPIC can issue full as well as partial guarantees.	The rates vary with the sector insured and are risk specific. They amount to some 2–3% for the full range of political risk cover and may be increased or decreased by about one-third, depending on the risk profile of the project.

(Continued)

Table 7.2 (Continued)

Provider	Main features of loan guarantee	Pricing
The Commonwealth Development Corporation (CDC)	The CDC is the arm of the British Governments aid and development programme which assists the private sector in developing countries. The CDC can give guarantees of all sorts, although it is doing so only to a very limited extent now and a historic 60 million sterling ceiling on the provision of guarantees is not under pressure. Its reservations about guarantees stem from the possible effect they may have on its market rating in the run-up to privatisation. CDC accounts for guarantees on one to one basis as if they were a loan.	CDC looks for a net return from guarantee charges equivalent to that it could earn from a loan.
Bilateral guarantees	Several export credit schemes (e.g. the UK's Export Credit Guarantee Department) will provide some guarantees against political risk for their resident companies investing overseas.	Rates are variable, but generally range from 40 to 100 basis points per annum on the outstanding debt.
Private Insurer AIG	AIG provides political risk insurance in 130 countries. It limits its coverage to US$ 150 million per investment with a maximum term of ten years. Its rates are commercially confidential, but, as a commercial insurer, they can be assumed to be comparable with others in the market place.	Guarantee fees vary with the range of assurance provided. They range from 1% for an IBRD government backed guarantee (with the IDA charge even lower) upwards with the major grouping at some 2–3%.

seems to be related to the size of the premia and hence the degree of subsidy (Mullineux, 1994).

Failure rates amongst SMEs, especially start-ups, are relatively high. This is in part due to the inexperience and lack of training of the managers and the difficulty of accessing good advice. Loan guarantee schemes are thus frequently linked to the provision of pre and post finance training and business advice. This is true of the US Small Business Administration (SBA) and the UK's new Small Business Service.

The SBA has provided a model for the UK's new Small business financing, training and support scheme. An alternative model is the German system which revolves around the development bank (KfW) and regional guarantee banks, which are in turn supported by Chambers of Commerce (see Mullineux, 1994). The KfW provides implicitly subsidised medium to long-term loans to SMEs and helps underwrite loan guarantees extended by the guarantee banks. The guarantees are also partially covered by state (Länder) governments. The Chambers of Commerce provide expertise to help screen the loan applications submitted by the potential lenders (banks) to the guarantee banks. The originating banks are expected to screen the loans prior to applying for guarantee and to manage the loans once guarantees are agreed. The process of screening the loans thus encourages the development by banks of expertise in SME lending, and helps to facilitate the spread of best practice in screening loan applications by providing direct or indirect feedback by experts to banks on their screening processes. In so doing, the tendency for banks to rely too heavily on collateral to secure or underwrite loans is reduced.

As noted above, the 'fixed costs problem' creates an additional hurdle for smaller and new or young SMEs to jump before they can gain access to external finance. They can face 'financial exclusion' if the sums they wish to borrow are not large enough to cover the fixed costs of originating a loan. This is a problem commonly faced by farming communities in developing countries who operate on high fixed costs but do not have enough equity to cover the costs. Such problems can be exacerbated by geographical remoteness. Many urban communities face similar financial exclusion problems, however. It should be noted that members of poorer communities are less likely to be able to secure loans with suitable collateral.

A number of schemes are being developed in the US and Europe to tackle financial exclusion problems of this sort. These include microfinance schemes, which were pioneered in rural areas in developing countries. Microfinance schemes involve banks or agencies lending small amounts from specially created funds on an unsecured basis. They often target women, whose repayment record is generally good and as a result, default rates are normally remarkably low and they are much more profitable than might be expected. However, they rarely generate the return on capital that commercial banks have come to expect and they are usually run by agencies with social and early stage development objectives. The most famous example is the Grameen Bank.[20]

Mutual and cooperative banks have emerged in many countries as means of self-help providing finance and education on money management. There are numerous examples. Cooperative banks have been formed to serve agricultural communities in many countries and have been prevalent in France and Japan, for example. 'Raffeisen banks' emerged in a number of European countries to service the needs of urban craftsmen. Like their credit union cousins, which service the financial needs of individuals rather than enterprises, these banks tend to operate through the pooling of savings for lending to members. In this way information problems are reduced and peer pressure can help assure prompt and full repayment of borrowed funds. The Raffeisen banks were essential urban cooperative banks or business credit unions. The general principles of mutuality have also been applied to insurance and house building and purchase (UK building societies and US saving and loan associations).

A related idea is that of a mutual guarantee scheme (MGS). In this case, the participating enterprises pool savings which they use to provide a guarantee to leverage loan finance from banks to corporates (Barclay and Watts, 1995). In so doing they overcome the fixed costs problem by borrowing sufficiently large amounts and also the lack of collateral problem. As with the other examples of mutual or cooperative self-help, however, savings have to be built up first. These savings represent internal investment that has been foregone in the hope of accessing larger amounts of external funds in the future.[21]

In developed countries, many of the older mutual and cooperative banks have grown to the point where they are operating more like commercial banks. Their close relationship with the members has consequently been lost. There is, thus, a case for facilitating the creation of 'new mutuals' to tackle financial exclusion. The establishment of such CFIs could be encouraged by special tax and legal treatment and access to subsidised loan guarantees. The CFIs can also be used as a means of providing pre- and post-finance training. There are numerous schemes in the US and many of the ideas are likely to be transferable to developing countries. The provision of finance in combination with training appears to be the key to success (Mayo et al., 1998). One way of achieving this is to establish 'incubators' in 'business parks'. The incubators provide offices and clerical, administrative, accounting and marketing etc. support to new enterprises for a limited period until they are established and able to set up on their own (or not, as the case might be).

The main role of the incubators, which may be publicly and/or privately funded and may or may not seek a share of the profits, and the microfinance, mutual and cooperative schemes is to allow participating enterprises to grow to the point where it becomes profitable for mainstream banks to enter into a normal (but perhaps with the help of loan guarantees) lending relationship. Given that CFIs provide a supply of potential new clients to banks, the banks may be encouraged to provide capital to support their activities (see Mayo et al., 1998). CFIs could also be used by banks to supply basic banking services cheaply to people who would otherwise be financially excluded.

Thriving or 'growth' SMEs with good access to bank finance are soon likely to need access to equity finance in order to avoid 'overgearing' (an excessively high debt to equity ratio). It is to providers of private equity ('business angels' and venture capital funds) that they are likely to turn in the first instance. It should, however, be noted that if the founding entrepreneurs intend to continue to manage the expanded business then they will probably need further training and access to more sophisticated business advice. We discuss the role of private equity and the development of capital markets in the sections on 'The role of the financial sector' and 'SME financing', but we must first consider the future role of development banks in addressing market failures in enterprise finance.

The role of development banks

Developing countries which embarked on an industrialisation strategy in the early 1970s soon discovered that the private sector lacked access to medium and long-term funding for investment because of the rudimentary nature of domestic and international capital markets. The financial structure, in which commercial banks were supplying only short-term loans to industries while central banks were responsible for regulating the financial system, was clearly inadequate for the investment financing of the corporate sector and SMEs enterprises faced severe credit rationing. Development banks were therefore seen as institutions which could fill the medium to long-term finance gap or market failure in the financial systems of developing countries. The idea was that global development institutions, especially the World Bank, could channel funds through regional and national development banks, according to the location and size of the projects (World Bank, 1994). The banks were also expected to mobilise long-term investment funds by developing local capital markets. Moreover, it was thought that the banks would strengthen national development strategies by investing in strategic economic regions, for example rural areas (Baum and Stockes, 1985; Rudnick, 1993). In general, apart from providing long-term loans, development banks were expected to promote projects, enhance managerial skills, develop entrepreneurship and help develop technological capabilities of developing countries (Jequir and Hu, 1989; Fitzgerald, 1993). It is notable that a number of these roles are akin to those of the US SBA and Germany's loan guarantee system, which are focused on SMEs.

Subsequently, however, a number of criticisms have been levelled against development banks. These have been fuelled by the poor performance, in terms of very low returns and poor quality of portfolios of most development banks during the last three decades (Murinde, 1997a). These criticisms have opened debates on whether development banking was the right channel to use for financing the industrial development of developing countries (Yaron, 1994) and on their future role.

Industrial lending by development banks: lessons from history

Historically, industrial lending by development banks (also known as industrial banking), was used to provide medium-term and long-term finance for industrial projects. Industrial banking was therefore a subset of development banking in the sense that development banks generally provided medium-term and long-term finance for development projects, some of which were in the industrial sector.

Cameron (1972) notes that the industrial lending by development banks originated in the nineteenth century continental Europe and played an important role in consolidating the industrial revolution. The industrial banking model was widely implemented in Germany and was later adopted by Japan during the Meiji years (1878–1911), leading to the later formation of the Industrial Bank of Japan. Hence, Hu (1981) refers to the industrial banking model as the German–Japanese model. Case studies by economic historians lend support to the similarity between the German and Japanese practice of industrial banking. For example, the early work by Gerschenkron (1977, p. 13) and Clapham (1977, p. 390) shows that the role of industrial banks in the industrialisation of Japan, as well as the conducive bank–firm relationship, was very much like the German case of bank–firm relationships, extensive investment banking, and strong government intervention and assistance. Further, in a case study of Japanese industrialisation, Yamamura (1972, pp. 178–198) observes that, with government support, industrial banks provided industrial capital and entrepreneurial guidance to the budding industrial sector. The literature, thus, suggests that a special feature of industrial banking was the existence of a close relationship between the bank and the firm. The banks would provide working capital and medium and long-term loans; they would even take up equity to reduce the risk exposure of the companies they financed. In addition, it was not unusual for industrial banks to assume an active entrepreneurial role in industrial sector activities.

It may be argued that the success of industrial banks in fostering the industrialisation process in some countries in Europe, and later in Japan, may have acted as a spur in the setting up of development banks in developing countries. Generally, development banks now exist in almost all developing countries, where they operate in three tiers, namely the global development bank, regional development banks and national development banks (Murinde, 1997a: 200–214). At each tier, the development banks are actively involved in financing the industrial sector.

Given that the success story of industrial banks in the industrialisation process in Europe and Japan may have provided the impetus for the setting up of development banks in developing countries, it remains to be seen whether development banks, in the same measure as industrial banks, have played the catalytic role in industrial development. Hu (1981) observes that although the present development banks are expected to achieve similar results to those earlier achieved by industrial banks, the development banks are closer to the nineteenth century Anglo-Saxon model than they are to the historical German–Japanese model of banking. The nineteenth century Anglo-Saxon model, reflected by the English commercial banks of the time, was characterised by financial orthodoxy in the sense that the emphasis was on short-term financial performance; the banks did not involve themselves in the long-term managerial and financial aspects of the firm (e.g. new technology, higher targets for output and sales, and capital budgeting).

Jequier and Hu (1989) identify additional distinguishing features between industrial banking and development banking. It is noted that industrial banks had a clearly defined mission, namely industrial development, and served as instruments of national policies with respect to long-term investment in industry. In addition, the banks were large in size and had sufficient financial power to take risks normally encountered in industrial promotion; they were always ready to assume entrepreneurial and corporate governance roles as they possessed the necessary expertise. Modern development banks, on the other hand, do not get actively involved in the management affairs of firms because they do not have the same size, financial power and technical expertise to serve as industrial banks (Bhatt, 1993). The promotional role, the active involvement in equity ownership and the provision of managerial assistance by modern development banks are very limited. Indeed it has been observed that most of the development banks have taken a passive position of waiting for entrepreneurs to approach them instead of going out and cultivating an entrepreneurial talent (Hu, 1981; Bhatt, 1994). It is also noted that while modern development banks tend to emphasise healthy cash flows (irrespective of whether or not the cash flows are dominated by non-project operations), historically the industrial banks placed weight on the success of projects.

Modern development banks have to fulfil three main functions

For purposes of future policy strategies, it should be emphasised that modern development banks have to achieve three functions in a mutually reinforcing manner, namely: the financial function; the developmental function; and the technological development function.

The financial function derives from the argument that, because a development bank is basically a financial institution, its performance and efficient utilisation of resources should be determined on the basis of its financial statements (Jain, 1989; World Bank, 1994). The idea is that if the bank is performing profitably, its investments are doing well in terms of income generated. In addition, the profitability of the bank also depends on the efficiency of the bank itself. In this tradition, the analysis of the adequacy of profits and the overall rates of return is often based on the standard financial ratios. In other words development banks should be required to allocate capital efficiently; otherwise they will have a distortionary influence within the financial sector as a whole. That having been said, due account should be given to the external benefits derived by the infrastructure investment they finance.

The developmental function is based on the argument that this function represents the main objective of setting up development banks (Meeker, 1990). Given that the main business of development banks is project financing, it may follow that unless these projects are successful, the bank will not have succeeded in its main objective (Grzyminski, 1991). In judging success, or failure, the positive externality, as well as financial ratios, should be taken into account. However, the ability to repay the loan cannot be used as the sole criterion for the success of development banking; for example, financial statements could show a rosy picture if projects used non-project sources of income to repay the project loan. For the bank to have a positive developmental function it must get involved right from the beginning through identification, promotion and financing of business opportunities. The bank may also promote projects in accordance with the overall national development policies. This is particularly important because it has been observed that in developing countries there are few viable projects presented to development banks in a proper form (Sender, 1993). Thus, the bank needs to identify a number of opportunities and choose those which it thinks are the best in terms of profitability, economic feasibility and easy implementability (Hu, 1981; Bhatt, 1993).

The second step is to identify the personnel who will manage the enterprise; the entrepreneur's ability, integrity and commitment have to be identified. In a case study, Dugan (1990) cites Zraly, Senior Manager at Zivnosteuska Bank in Prague, who believes that their main objective is to look at the quality of management for a project proposal. If the sponsor is lacking in managerial qualities, the project is bound to be plagued by problems, however profitable it may appear. It can therefore be argued that development banks should have in place a mechanism and relevant criteria which will facilitate their judgement in this direction (Westlake, 1993). Development banks also need a system which will give early warning signals once the project commences; the system would enable the bank to intervene as soon as trouble is detected. Baum and Stockes (1985) observe that there is a need for close bank–client relationship between the borrowers and development banks to facilitate guidance and financial discipline.

The technological development function reinforces the developmental function. It is widely held that developing countries lack the required technology to achieve higher levels of economic development (Dowrick, 1992). Specifically, according to the 'two gap' theory, developing countries lack the necessary domestic savings as well as the foreign exchange to facilitate technological development (Todaro, 1997). Development banks are strategically placed to act as technological development institutions during the process of importing intermediate inputs and to facilitate development of intermediate technologies. Since the banks finance industrial projects and the projects need industrial equipment and technology to be successful, the banks can serve to fill this gap and shape the type of technology which should be imported or produced domestically. However, the bank must have the technical capability to assess not only the suitability of a suggested technology but also the trend of technological developments taking place in other economies. In circumstances where it does not have the required expertise, the development bank can enlist the services of research and consultancy firms.

A case study by Bhatt (1993) on the use of research firms to assess technological developments reports that the Korean Development Corporation has quite often relied on the expertise at the Korean Institute of Science and Technology. Technological policy also seems to be working in India and Brazil, where development banks have created engineering consulting subsidiaries to ensure that appropriate technological policies are promulgated (Jaquier and Hu, 1989). Even if all the stages of project identification, promotion, appraisal and implementation are properly done, unforeseeable events in the macroeconomic environment can affect the performance of development projects and, by implication, the development bank. In addition to external shocks like oil prices, many developing countries are characterised by high inflation, slow or stagnant growth and balance of payments difficulties (IMF, 1993; Murinde, 1993). These factors are not conducive to industrial development as they make it difficult for firms to plan ahead. Given that a development bank cannot take direct policy action to change the macroeconomic environment, it is necessary that the bank takes a direct interest in the implementation of projects to ensure that timely action is taken to insulate the project against macroeconomic shocks.

It is important that development banks aid, rather than impede, adjustments required in response to good liberalisation. Thus, financial sector liberalisation and privatisation if banks may necessitate the reorientation, restructuring and perhaps even corporatisation of domestic development banks, as in Sri Lanka.[22] As part of 'private finance initiatives' modern development banks can be expected increasingly to engage in co-financing arrangements with banks and other financial institutions. Their role should, thus, increasingly be to ensure that positive externalities are reaped in cases where private finance is unlikely to be sufficiently forthcoming, for example, in sponsoring infrastructural projects, or projects where social benefits are substantial, even though financial rewards are not sufficiently high to sustain private financiers. In other words, the case for intervention by development banks should rest largely on identification of market failures that result in an inadequate supply of private sector finance.

As access to international capital markets has increased and as, over time, domestic capital markets develop, we can expect to see the role of development banks as financiers of mainstream business activity decline. Increasingly they should engage in the co-financing, with the private sector, of infrastructural projects; perhaps drawing on external financial assistance. They (or some other agency) should also address the market failure in the supply of finance to SMEs and provide training and other support services to small businesses. In developing loan guarantee schemes, developing banks could make a major contribution to the successful development of the crucially important SME sector. In addition, development banks might take on the role of developing venture capitalism, the topic to which we turn next.

Developing venture capitalism

Venture capitalism involves the provision of equity capital for the start-up and development of enterprises (see Kitchen, 1992). The capital is usually raised from investors in the form of a fund which is used to make private equity investments in businesses. There is also a growth in private equity investment in SMEs by rich individuals, or 'business angels'. The private equity investors essentially provide capital to supplement that sunk by the initial entrepreneurs and in return normally require a say in how the business is run. Taking on private capital from outside is, thus, a big step for SMEs because it entails dilution of control over their enterprises. Banks generally interfere much less, so long as businesses are performing satisfactorily. Many potential growth SMEs, thus, fail to exploit their full potential because they do not wish to dilute control.

As mentioned in the introductory overview, it is important to ensure that the tax system is conducive to the provision of private risk capital, both by insiders (the start-up enterprises) and outsiders, who come on board at a later date. It is also important that firms and private investors wishing to re-invest profits, rather than take them out of their businesses, are given a fiscal incentive to do so. To encourage the participation of business angels, access to information on potential clients is also important. A government business support agency, possibly the development bank, could construct a database to be accessed via the internet, for example. Indeed, there seems to be an important role for business support agencies in fostering the supply of private equity or venture capital. Development banks could adopt a market development role here, perhaps through co-financing or through a dedicated fund – this is a role which the European Investment Bank has recently been assigned.[23] As the venture capital market becomes established, development banks would be expected to withdraw after spinning off or winding down active funds.

Increasingly, venture capital funds and business angels are looking outside their country of origin for opportunities, perhaps because they have developed sectoral expertise (Sagari and Guidotti, 1992). They are, thus, likely to become a growing source of foreign capital if the tax incentives and legal infrastructure are conducive[24] and could play a key role in enterprise restructuring programmes. There is a tendency for older and larger funds (e.g. 3i in the UK) to withdraw from start-up and early-stage investments and to concentrate on investing in growth enterprises and sectors (e.g. Communications and Information Technology). Their focus is increasingly on financing management buy-outs and buy-ins. Whilst they perform a useful role in the restructuring process and the financing of medium-sized enterprises, a new gap in small enterprise equity finance can emerge. For this reason a sympathetic tax regime should be retained to encourage 'business angels' and new entry into the venture capital business and the development bank might maintain an interest in providing start-up capital in the medium term.

Just as SMEs are likely to graduate, from CFIs or mutual financing, to bank financing, so 'growth SMEs' may soon outgrow venture and business angel financing. The external providers of the equity funds will anyway want to liquidate their investments for re-investment in the next batch of growth SMEs. At this stage it is common for the former SMEs to become a fully fledged public companies by investing in growth enterprises and sectors (e.g. CIT). Their focus is increasingly on financing management buy-outs and buy-ins. Whilst they perform a useful role in the restructuring process and the financing of medium-sized enterprises, a new gap in small enterprise equity finance can emerge. For this reason a sympathetic tax regime should be retained to encourage 'business angels' and new entry into the venture capital business and the development bank might maintain an interest in providing start-up capital in the medium term.

Just as SMEs are likely to graduate, from CFIs or mutual financing, to bank financing, so 'growth SMEs' may soon outgrow venture and business angel financing. The external providers of the equity funds will anyway want to liquidate their investments for re-investment in the next batch of growth SMEs. At this stage it is common for the former SMEs to become fully fledged public companies by abandoning their private status. This can be done through stock exchange floatations or initial public offerings (IPOs). Essentially shares are issued to the general public and the private capital providers (initial entrepreneurs, venturers and angels) can redeem their investments. They can, of course, reinvest some of the capital by purchasing shares in the public company. It is important to realise that private equity funding will be more plentiful in countries with well-developed capital markets that can provide the funders with an 'exit route' in the medium term. It is to capital market development that we turn next.

Developing capital markets

It should be emphasised that it takes time to develop well-functioning capital markets. A sound regulatory and supervisory framework needs to be in place and buyers and sellers in the market will not have confidence in it until it is tried and tested. As confidence grows, then more and more buyers (investors) and sellers (issuers) of securities will come to the market and its liquidity (and stability) will increase. This cannot be achieved by decree and it is also necessary that there are attractive stocks and shares which can be traded on the markets. Foreign portfolio investment can help deepen markets if the capital account has been liberalised, but recent experience suggests that capital can flow out as quickly as it flows in, causing large swings in stock prices. Transparency, sound regulation and vigilant supervision are important means of encouraging foreign investors to take the additional risk (including exchange rate risk) of investing overseas. It is not just financial risks and economic stability that matters, capital inflows and outflows are as much, if not more, affected by the level of political stability, it should be noted.

In developing and transition economies, the privatisation of state-owned enterprises can provide a supply of shares that are potentially attractive to investors. Indeed, there are some lessons which the emerging stock markets can learn from the experience of the developed markets, such as the London Stock Exchange, in terms of their potential for financing enterprises as well as the impact of government intervention in the market (see Green et al., 2000). The attractiveness of the shares will of course depend on how well the enterprises have been financially and organisationally restructured prior to privatisation. This is as true of banks as non-financial firms and, as mentioned in the section on 'The role of the financial sector' it is important that plans are in place to deal with outstanding bad debt problems, or perhaps preferably that such problem have already been resolved.

The growth of the insurance sector and the creation of funded pension schemes will help to 'deepen' the capital markets by creating a relatively stable demand for equities and bonds to satisfy the long-term portfolios needs of such institutions. Also, as noted in the introduction, such institutional investors should be encouraged to play a role in corporate governance.

Some impediments may constrain the developmental role of stock markets (Atje and Jovanovic, 1993). The high costs of complying with stock exchange registration requirements may, however, discourage IPOs by 'growth SMEs'. For this reason, many developed countries, particularly in Europe (e.g. France, Germany and the UK) have introduced special stock markets designed to attract IPOs by 'growth SMEs', who later hope to graduate to the main stock exchange. Generally, these exchanges have a lighter regulatory regime (with reduced reporting requirements etc.) that is less costly to comply with. It is these specially designed stock markets that increasingly commonly provide the necessary exit route (through IPOs) for private investors.

As financial systems evolve, new marketable financial instruments develop. These allow companies to issue a range of debt instruments across the maturity spectrum, from commercial paper and bills through to medium to long-term corporate bonds. In larger firms these securities are increasingly replacing bank intermediated debt finance. This process of disintermediation is a natural product of financial sector evolution (Arndt and Drake, 1985).

Corporate bill and bond markets require benchmark interest rates against which risk premia can be gauged. These are typically provided by treasury bill and government bond markets and, thus, the development of the latter can be regarded as part of the process of developing capital markets to facilitate corporate finance. Central banks commonly play a major role in helping develop these markets for government debt, but development banks might well have a role to play in helping to establish interbank and corporate bond markets.

It should be noted that banking sectors will be unable to allocate capital efficiently without access to an effective interbank market. Banks with surplus funds and a shortage of profitable lending opportunities should be able to lend to other banks in the opposite situation. Such markets typically use treasury bills and short dated bonds for benchmarking, making the development of a treasury bill market a key foundation of the wider development of interbank and other money markets. The development of money markets is in turn important for the sophisticated operation of monetary policy. It should also be noted that good monetary policy implementation is itself conducive to enterprise development and stable economic growth (Green and Murinde, 1992, 1993).

Initially, short-term tradable corporate securities are likely to take the form of bank accepted (underwritten or guaranteed) 'commercial bills'. As the reputation of the issuers develop, acceptance by banks, for which they charge fees, becomes an unnecessary expense and tradable commercial paper issued instead. Over time, other debt instruments have been developed in the Euromarkets and elsewhere, for example, floating rate notes (FRNs) and so too have hybrid debt/equity instruments (e.g. perpetual FRNs).[25] These too have displaced a great deal of bank variable interest rate lending. Larger corporations can issue corporate bonds, provided there is a suitable government bond to serve as a benchmark and if their credit ratings are high enough.[26] This development should be encouraged since it reduces the demand by large corporations for bank financing, focusing banks on the needs of smaller companies and SMEs. It also reduces the need for domestic bank borrowing from the international banking markets through which they raise foreign currency for onlending. This borrowing in turn increases short-term capital inflows, as in the case of Thailand; where banks borrowed heavily in dollars through the Bangkok International Banking Facilities (BIBFs) in the run up to the summer 1997 banking crisis (Mullineux, 1999).

As the capital markets develop towards the stage already reached in the US and being approached by the countries participating in the Economic and Monetary Union (EMU) in Europe (or 'Euroland'), 'junk bond' markets are likely to develop. These are markets in bonds issued by companies which the credit rating agencies are unable to give an investment grade rating. The bonds issued, thus, have relatively high and uncertain credit risks and earn high rates of return for investors in reward for their risk exposure. Over time smaller and smaller companies may gain access to such debt markets, but only to the point that issuance is cheaper than raising bank debt finance. Falling fixed costs of issuance will facilitate this process, but it is not limitless and is likely to be gradual.

Also in the later stages of financial sector development, markets in financial derivatives (swaps and options etc.) will develop and domestic banks will offer customised derivatives 'over the counter'. Financial derivatives facilitate the hedging of risks, as well as speculation. They are, thus, useful tools for risk management by banks and other enterprises, but it is by no means clear that countries should rush to develop domestic products and markets given that they are accessible through offshore and other major international financial centres.

It is important to ensure that regulatory development and supervisory competence keeps pace with the liberalisation of the financial sector and the capital account of the balance of payments; and the financial innovation and increased competition that results. The development of sound capital and banking markets will have the added benefit of attracting longer term foreign capital in the form of foreign portfolio and direct investment. The less restricted the outflows, the greater the inflows tend to be (see Greene and Villanueva, 1991; Fischer, 1993). Increased longer term inflows will in turn reduce reliance on less stable short-term capital inflows and thus help reduce private sector short-term debt exposures. It has been argued that if Thailand and the other worst affected South East Asian countries had had more developed capital markets, then the crisis would have been less severe. With the benefit of hindsight, this may be true, but it must be stressed that the development of sound capital markets has taken a long time in the US and cannot be achieved overnight in emerging market economics. Indeed, highly developed capital markets are a rarity and there is a case for carefully embarking on their development without delay, but there are few shortcuts. Reputation must be earned. As noted, good regulation and supervision is one of the foundations of stability, but there is less consensus about how best to regulate capital markets than there is for banks.

Concluding remarks

Government intervention, in the form of regulation and supervision of the wider financial sector and in the provision of loan guarantees and other small business support services, is necessary due to market failures (information asymmetry and fixed-cost problems in the main), but market-led financial reforms are the key to the evolution of the financial sector and enterprise development. The private sector should be brought into partnership with the government (e.g. in overcoming financial exclusion through bank financing of CFIs) and markets should be used to provide incentive compatible solutions (e.g. risk-related capital adequacy and deposit insurance) wherever possible.

Table 7.3 Financial sector problems and possible solutions (See also Table 7.1)

Problems	Possible solutions
1 Monetary instability	Reduce budget deficit financing (lower budget deficit, more bond finance); raise real interest rates (at least to positive rates)
2 Financial instability	Improve transparency (more reporting, better auditing) and bank regulation and supervision (and reduce monetary instability)
3 Shortage of bank loans	Raise savings levels (lower taxes/higher real interest rates), loan guarantee schemes and microfinance schemes (perhaps overseen by a development bank). More effective bankruptcy lows; foreign bank entry; reduce/eliminate capital controls
4 Weak corporate governance	Increased monetary and financial stability; eliminate government directed lending; privatise banks; positive real interest rates; effective bankruptcy laws and prosecution of fraud
5 Shortage of private/venture capital	Tighter reporting requirements; higher auditing standards; more effective bankruptcy laws; increased tax incentives; develop capital markets to provide an exit.[a] The development bank might initially manage a venture fund
6 Underdeveloped money and capital markets	Tighter reporting requirements; higher auditing standards; develop market regulation and supervision (increase the number of stocks traded (through privatisation); and institutional shareholding (mutual, pension and insurance fund development); reduce/eliminate capital controls

Note

a By exit we mean the ability to liquidate the initial private equity stake. This is commonly done via an Initial Public Offering (IPO) of shares on a stock market.

Banks will continue to play a leading role in promoting the growth of enterprises, especially the SMEs, although over time stock markets will become increasingly important in most developing and transition economies. Financial system restructuring, as a means of correcting broad policy-induced distortions, is an important ingredient of the symbiotic growth of the financial and corporate sectors. To sustain the momentum for growth, mechanisms for financial sector regulation and supervision have to be in place. Pending the development of domestic capital markets or access to international capital markets, development banking will remain important for funding long-term investment and infrastructural projects. Developing countries will continue to use development banks to tap into international capital flows by offering co-financing prior to the development of fully fledged capital markets. The development banks (or some other agency) should also develop loan guarantee schemes and provide training and other services to the SME sector. In other words, development banks should focus on addressing market failures. As the market failure in the provision of long-term capital to larger enterprises declines in importance (as their access to capital markets increases), the development (and commercial) banks should increasingly focus on meeting the needs of SMEs (Table 7.3). For it is SMEs that will be the engine of future development.

Notes

1. This is notwithstanding the well documented views on the role of banks in economic development (see Cameron, 1972; Edwards and Fischer, 1994; Mayer, 1994).
2. Information collection and processing is costly. Hence, full information may be achieved at such a high cost that in the long run it does not pay to attempt to achieve it because it requires, not only substantial and costly disclosure, but assurance, through auditing, of the quality of the disclosed information.
3. Traditionally, retail banks have predominantly held loans as assets and earned some fee income from providing payments, accounting (statements) and safe-keeping services. As banks diversify into securities (broking and dealing) or 'investment' banking businesses, then the sources of fee income multiply and fee income grows in importance relative to income derived from the 'spread' (or 'margin') between lending and borrowing interest rates.
4. As Miller (1996) has shown with repect to Argentina, explicit rather than implicit deposit insurance may also create instability (of a moral hazard nature) in developing countries.
5. Indeed, as the wider financial system develops, attention must increasingly be paid to its effective regulation and supervision.
6. The real interest rate is usually calculated by approximation as follows: $r = i - \pi$ where r = real interest rate, i = nominal interest rate, and π = inflation rate.
7. In general, however, financial repression is diagnosed by looking beyond the real interest rate symptoms. It is also characterised by high reserve requirements (which reduces the amount of loanable funds) and misaligned (especially overvalued) exchange rates; see, for example, Lensink et al. (1998) on the simultaneous use of these three measures of financial repression for a sample of African economies.
8. This is partly because the larger SMEs are commonly older and have a 'track record' and partly because of a 'fixed costs' of lending problem. Smaller firms commonly require smaller loans and the size of loan required may simply be too small to cover the (relatively fixed) costs of originating the loan.

9. This analysis focuses exclusively on the formal financial sector. However, there is evidence to show that the informal financial sector is in some cases paramount in fostering enterprise development (see Thomas, 1993; Montiel et al., 1993).
10. This is true of large firms as well as SMEs, where internally generated finance (undistributed profits) is the major source of capital investment funding, as well as for start ups, where access to private capital, in the form of the personal wealth of the entrepreneur and family and friends, is of crucial importance.
11. The pecking order theory of financing choices is underpinned using a simple asymmetric information model in Myers (1984) and Myers and Majluf (1984). The pecking order theory offers two important hypotheses. First, firms prefer internal finance because asymmetric information creates the possibility that they may choose not to issue new securities and may therefore miss a positive net present value (NPV) investment. Second, when firms resort to external finance, they prefer issuing safe securities (i.e. debt) before risky ones (i.e. equity). Thus, firms face a pecking order of financing choices, with internal finance at the top, and equity at the bottom, of the pecking order; as they climb up the pecking order, firms face increasing costs of financial distress inherent in the risk class of debt and equity securities (see Baskin, 1989).
12. An appropriate balance between the rights of creditors and debtors must be struck, so that entrepreneurship is not stifled and yet banks remain willing to lend, and the courts must deal with the cases efficiently (promptly and consistently etc.).
13. Brealey and Myers (2002, pp. 378–379) report that for all non-financial corporates (NFCs) in the US over the decade 1990–2000, internally generated cash was the dominant source of corporate financing and covered a high percentage of capital expenditures, including investment in inventory and other current assets; the bulk of required external financing came from borrowing; net new stock issues were very minimal. The observation is consistent with the findings by Rajan and Zingales (1995) in their international comparisons of capital structures in seven OECD countries, as well as the evidence by Corbett and Jenkinson (1997), Bertero (1994) and Edwards and Fischer (1994) in selected OECD countries. However, these studies also find some evidence of a shift from bank loans to direct financing from the capital (and particularly the bond) markets as part of the securitisation process associated with the financial liberalisation of the 1980s and 1990s.
14. For example, as a result of a supply shock such as an unexpected rise in oil prices.
15. Some of the liabilities of banks are normally regarded as money (e.g. demand deposits). Further, the proportion of bank liabilities regarded as money is normally higher in developing than developed countries. As bank lending increases and the share of bank liabilities in the money supply rises, it is increasingly important to curtail the growth of bank lending in order to curb the growth of the money supply.
16. Financial intermediation is strongly linked to economic growth; see Bencinvenga and Smith (1991).
17. The *Chaebol* and industrial conglomerates which borrowed heavily from banks in order to expand the scale and scope of their activity. By the mid-1990s they had very high debt to equity ('gearing') ratios. When interest rates were raised following the onset of the 1997 crisis they struggled to service their debts and were encouraged by the government to sell assets in order to reduce their debt levels. The Chinese SOEs are heavily indebted to state-owned banks and are not sufficiently 'profitable' to service the debts properly. The problem is likely to be aggravated following China's entry into the WTO, which will expose the inefficient SOEs to further competition.
18. See also Chick and Dow (1997) on different types of regulation for different types of financial institutions.
19. Training should be at least to the level of senior bank staff. A postgraduate University degree involving money, bank and finance followed by a professional training in banking and finance would be appropriate and salaries should not be too out of line with those senior bank staff.

20 The Grameen Bank extends microfinance in rural Bangladesh, targeting its lending on primarily women; who have been found to be more reliable debtors than men. This is believed to reflect their more highly developed sense of family responsibility and their susceptibility to peer group pressure (shame of non-repayment) – see Yaron et al. (1997). There were naturally some problems with the repayment of loans to the Grameen Bank following the floods in Bangladesh in 1998/1999. The objectives of the scheme frequently include education and training about the management of money and business.
21 The government of Mauritius commissioned a study to advise on establishing a Mutual Guarantee Fund to help foster the development of mutual guarantee schemes. The consultants were asked to advise on whether the Mauritius Development Bank or some other public or private sector institution or agency should manage the Fund.
22 See Wignaraja (1998).
23 The government of Mauritius commissioned a study to advise on the establishment of a State sponsored Venture Capital Fund and which public or private sector institution or agency should manage it. The Mauritius Development Bank was considered alongside alternatives.
24 It would seem sensible to offer similar tax and exemption packages to those aimed at attracting FDI, thereby differentiating the supply of private equity capital from portfolio investment. As regards the legal infrastructure, the amount of capital that flows in is likely to be greater the easier it is to get it out (liberal capital account) and the greater the legal redress (good bankruptcy and anti-fraud laws).
25 A floating rate note is a note whose interest payment varies with the short-term interest rate. The word 'note' is used here to refer to unsecured debt with a maturity of up to ten years. Perpetual floating rate notes are floating rate notes whose maturity is indefinite (i.e. is for ever and ever).
26 Credit rating agencies assign ratings to borrowers on the basis of the risk of default; some of the main credit rating agencies are Standard and Poor's and Moody's (see Murinde, 1997b).

References

Abebe, A. (1990), 'Financial Repression and its Impact on Financial Development and Economic Growth in the African Least Developed Countries', *Savings and Development*, 14(1), 55–85.

Adams, D.W., Graham, D.H. and Von Pischke, J.D. (eds) (1984), *Undermining Rural Development with Cheap Credit*, Boulder, CO: Westview Press.

Anderson, D. and Khambata, F. (1985), 'Financing Small-scale Industry and Agriculture in Developing Countries: The Merits and Limitations of Commercial Policies', *Economic Development and Cultural Change*, 33(2), 349–371.

Arndt, H.W. and Drake, P.J. (1985), 'Bank Loans or Bonds: Some Lessons of Historical Experience', *Banca Nazionale del Lavoro Quarterly Review*, 155: 373–392.

Asikoglu, Y., Kutay, S. and Ertuna, I.O. (1992), 'Mutual Funds in Developing Capital Markets', in K.P. Fischer and G.J. Papaioannou (eds), *Business Finance in Less Developed Capital Markets*, Westport and London: Greenwood Press.

Atje, R. and Jovanovic, B. (1993), 'Stock Markets and Development', *European Economic Review*, 37, 732–740.

Banuri, T. and Schor, J. (1992), *Financial Openness and National Autonomy*, Oxford: Oxford University Press.

Barclay, S. and Watts, G. (1995), 'The Determinants of Corporate Leverage and Dividend Policies', *Journal of Applied Corporate Finance*, 7, 4–19.

Barro, R.J. and Sala-i-Martin, X.X. (1995), *Economic Growth*, New York: McGraw Hill.
Baskin, J. (1989), 'An Empirical Investigation of the Pecking Order Hypothesis', *Financial Management*, 18, 27–35.
Baum, C. and Stockes, M.T. (1985), *Investing in Development – Lessons of World Bank Experience*, Oxford: Oxford University Press.
Bencinvenga, V.R. and Smith, B.D. (1991), 'Financial Intermediation and Endogenous Growth', *The Review of Economic Studies*, 58, 195–209.
Benston, G. and Kaufman, G.G. (1997), 'FDICIA after Five Years', Working Paper Series, Issues in Financial Regulation, WP-97-1, Federal Reserve Board of Chicago.
Bertero, E. (1994), 'The Banking System, Financial Markets and Capital Structure: Some Evidence from France', *Oxford Review of Economic Policy*, 1014, 68–78.
Bhatt, V.V. (1993), 'Development Banks as Catalysts for Industrial Development', *International Journal of Development Banking*, 11(1), 47–71.
Bhatt, V.V. (1994), 'Main and Lead Bank Systems', *International Journal of Development Banking*, 12(1), 53–79.
Blanchard, O., Rhee, C. and Summers, L. (1993), 'The Stock Market, Profit and Investment', *The Quarterly Journal of Economics*, 108(1), 115–137.
Brealey, R.A. and Myers, S.C. (2002), *Principles of Corporate Finance*, New York: McGraw-Hill (7th Edition), pp. 378–379.
Brock, W.A. and Evans, L.T. (1997), 'Principal–Agent Contracts in Continuous Time Asymmetric Information Models: The Importance of Large Continuing Information Flows', *Journal of Economic Behaviour and Organisation*, 29(3), 523–535.
Cameron, R. (1972), *Banking and Economic Development*, Oxford: Oxford University Press.
Chick, V. and Dow, S. (1997), 'Competition and Integration in European Banking', in A. Cohen, H. Hagemann and J. Smithin (eds), *Money, Financial Institutions and Macroeconomics*, London: Kluwer, pp. 253–270.
Ciarrapico, A.M. (1992), *Country Risk: A Theoretical Framework of Analysis*, Aldershot: Dartmouth.
Clapham, J.H. (1977), *The Economic Development of France and Germany*, Cambridge: Cambridge University Press.
Corbett, J. and Jenkinson, T. (1997), 'How is Investment Financed? A Study of Germany, Japan, the United Kingdom and the United States', *The Manchester School*, 65, 69–93.
Diamond, D. and Dybvig, P. (1983), 'Bank Runs, Deposit Insurance and Liquidity', *Journal of Political Economy*, 91, 401–419.
Diamond, D. (1991), 'Monitoring and Reputation: The Choice Between Bank Loans and Directly Placed Debt', *Journal of Political Economy*, 99, 789–721.
Dickinson, D.G. and Mullineux, A.W. (2001), 'Lessons for the East Asian Financial Crisis: A Financial Sector Perspective', *Geoforum*, 32, 133–142.
Doukas, J., Murinde, V. and Wihlborg, C. (1998), *Financial Sector Reform and Privatisation in Transition Economies*, Amsterdam: North-Holland.
Dowrick, S. (1992), 'Technological Catch up and Diverging Incomes: Patterns of Economic Growth 1970–1988', *The Economic Journal*, 102, 700–710.
Dugan, A. (1990), 'Banks in the Firing Line', *Euromoney*, September: 119–133.
Durojaiye, B.O. (1991), 'Rural Household Consumption–Savings Behaviour in Low-income Nations', *African Review of Money, Finance and Banking*, 1, 85–95.
Edwards, J. and Fischer, K. (1994), *Banks, Finance and Investment in Germany*, Cambridge: Cambridge University Press.
Fingleton, J. and Schoenmaker, D. (1992), *The Internationalisation of Capital Markets and the Regulatory Response*, London: Graham & Trotman.

Fischer, B. (1993), 'Success and Pitfalls with Financial Reforms in Developing Countries', *Savings and Development*, XVII(2), 111–135.

Fitzgerald, E.V.K. (1993), *The Macroeconomics of Development Finance*, Baltimore: The John Hopkins University Press.

Fry, M.J. (1995), *Money, Interest, and Banking in Economic Development*, London and Baltimore: The John Hopkins University Press.

Gerschenkron, A. (1977), *Economic Backwardness in Historical Perspective*, Cambridge, MA: Harvard University Press.

Green, C.J. and Murinde, V. (1992), 'The Potency of Budgetary and Financial Policy Instruments in Uganda', in C. Milner and A.J. Rayner (eds), *Policy Adjustment in Africa*, London: Macmillan (chapter 8).

Green, C.J. and Murinde, V. (1993), 'The Potency of Stabilization Policy in Developing Economies: Kenya, Tanzania and Uganda', *Journal of Policy Modelling*, 15(4), 427–472.

Greene, J. and Villanueva, D. (1991), 'Private Investment in a Developing Country', *IMF Staff Papers*, 38(1), 33–58.

Green, C.J., Maggioni, P. and Murinde, V. (2000), 'Regulatory Lessons for Emerging Stock Markets from a Century of Evidence on Transactions Costs and Share Price Volatility in the London Stock Exchange', *Journal of Banking and Finance*, 24, 577–601.

Grzyminski, R. (1991), 'The New Old-fashioned Banking', *Harvard Business School*, 69, 87–98.

Gupta, K.L. and Lensink, R. (1997), *Financial Liberalisation and Investment*, London: Routledge.

Herath, G. (1994), 'Rural Credit Markets and Institutional Reform in Developing Countries', *Savings and Development Quarterly Review*, XVIII(2), 179–191.

Hermes, N. (1994), 'Financial Development and Economic Growth: A Survey of the Literature', *International Journal of Development Banking*, 12(1), 3–22.

Hermes, N., Lensink, R. and Murinde, V. (1998), 'Does Financial Liberalisation Reduce Capital Flight?', *World Development*, 27(7), 1349–1378.

Honohan, P. (1993), 'Financial Sector Failures in Western Africa', *Journal of Modern African Studies*, 31(1), 49–75.

Hu, Y.S. (1981), 'The World Bank and Development Finance Companies', *Journal of General Management*, 7, 47–57.

International Monetary Fund (1993), *World Economic Outlook*, Washington, DC: International Monetary Fund.

Jain, P.K. (1989), 'Assessing the Performance of a Development Bank', *Long Range Planning*, 22, 100–107.

Jequier, N. and Hu, Y.S. (1989), *Banking and the Promotion of Technological Developments*, New York: International Labour Office and St Martin's Press.

Kitchen, R. (1992), 'Venture Capital: Is It Appropriate for Developing Countries?', in K.P. Fischer and G.J. Papaioannou (eds), *Business Finance in Less Developed Capital Markets*, Westport and London: Greenwood Press.

Kumar, P.C. and Tsetseko, G. (1992), 'Securities Market Development and Economic Growth', in K.P. Fischer, and G.J. Papaioannou (eds), *Business Finance in Less Developed Capital Markets*, Westport and London: Greenwood Press.

Lyons, S.E. and Murinde, V. (1994), 'Cointegration and Granger-causality Testing of Hypotheses on Supply-leading and Demand-following Finance', *Economic Notes*, 23(2), 308–317.

McKinnon, Ronald I. (1973) *'Money and Capital in Economic Development'*, Washington, DC: Brookings Institution.

Mayer, C. (1994), 'The Assessment: Money and Banking: Theory and Evidence', *Oxford Review of Economic Policy*, 10(4), 1–13.

Mayo, E., Fisher, T., Conaty, P., Doling, J. and Mullineux, A.W. (1998), *'Small is Bankable Community Reinvestment in the UK'*, York: Joseph Rowntree Foundation.

Mazamdar, S.C. (1997), 'Bank Regulation, Capital Strucuture and Risk', *Journal of Financial Services Research*, 19(2), 209–228.

Meeker, L.G. (1990), 'Development Finance: Doing the Undoable Deals', *Journal of Commercial Bank Lending*, 72, 13–27.

Miller, G.P. (1996), 'Politics of Deposit Insurance Reform: The Case of Argentina', *Federal Reserve Board of Chicago Proceedings*, 19(May), 473–489.

Montiel, P.J., Agenor, P. and Ul-Haque, N. (1993), *Informal Financial Markets in Developing Countries*, Oxford: Blackwell.

Mullineux, A.W. (1987a), *UK Banking after Deregulation*, London: Croom-Helm.

Mullineux, A.W. (1987b), *International Banking and Financial Systems: A Comparison*, London: Graham and Trotman.

Mullineux, A.W. (1994), 'Small and Medium-sized Enterprise Financing in the UK: Lessons from Germany', London: Anglo-German Foundation Report.

Mullineux, A.W. (1997a), 'The Funding of Non-Financial Corporations (NFCs) in the EU (1971–1993): Evidence of Convergence?', *Mimeo*, Department of Economics, University of Birmingham.

Mullineux, A.W. (1997b), 'Banking Sector Restructuring, Debt Consolidation and Small and Medium-Sized Enterprise Financing in Formerly Centrally Planned Economies', in S. Gangopadhyay (ed.), *Institutions Governing Financial Markets*, New Delhi: Allied Publishers Ltd, pp. 74–87.

Mullineux, A.W. (1999), 'Lessons from the Financial Crisis in Pacific and East Asia', *Working Paper*, Department of Economics, The University of Birmingham.

Murinde, V. (1993), *Macroeconomic Policy Modelling for Developing Countries*, Aldershot: Avebury.

Murinde, V. (1997a), *Development Banking and Finance*, Aldershot: Avebury.

Murinde, V. (1997b), 'Financial Markets and Endogenous Growth: An Econometric Analysis of South East Asia', in N. Hermes and R. Lensink (eds), *Financial Development and Economic Growth*, London and New York: Routledge.

Murinde, V. and Eng, F.S.H. (1994), 'Financial Development and Economic Growth in Singapore: Demand-following or Supply-leading', *Applied Financial Economics*, 4, 391–404.

Murinde, V. and Kariisa-Kasa, J. (1997), 'The Financial Performance of the East African Development Bank: A Retrospective Analysis', *Accounting Business and Financial History*, 7(1), 81–104.

Murinde, V. and Mullineux, A.W. (1999), 'Introductory Overview: Issues Surrounding Bank and Enterprise Restructuring in Central and Eastern Europe', in C.J. Green and A.W. Mullineux (eds), *Financial Sector Reform in Central and Eastern Europe: Capital Flows, Bank and Enterprise Restructuring*, Cheltenham: Edward Elgar, pp. 1–20.

Murinde, V. and Ngah, J.S. (1995), 'Central Bank Financing of the Budget Deficits: The CFA Franc Zone Versus Eastern Africa', *International Journal of Development Banking*, 13(2), 3–12.

Murinde, V. and Ryan, C. (1999), 'The Implications of The General Agreement on Trade in Services (GATS) for The Banking Sector in The United Arab Emirates', Geneva: UNCTAD.

Murinde, V., Agung, J. and Mullineux, A.W. (1999), 'Patterns of Corporate Financing and Financial Sector Convergence in Europe', *Mimeo*, University of Birmingham.

Myers, S.C. (1984), 'The Capital Structure Puzzle', *The Journal of Finance*, XXXIX(3) 575–592.

Myers, S.C. and Majluf, N.S. (1984), 'Corporate Financing and Investment Decisions when Firms have Information Investors do not have', *Journal of Financial Economics*, 13, 187–222.

Okeahalam, C.C. (2002) 'Deposit Insurance and International Banking Regulation', in A.W. Mullineux and V. Murinde (eds), *Handbook of International Banking*, Cheltenham: Edward Elgar.

Olomola, A.S. (1991), 'Credit Control Policies for Improved Agricultural Financing in Nigeria', *African Review of Money, Finance and Banking*, 1, 23–37.

Oshikoya, T.W. (1992), 'Interest Rate Liberalisation, Saving, Investment and Growth', *Savings and Development*, XVI(3), 305–320.

Rajan, R.G. and Zingales, L. (1995), 'What Do We Know About Capital Structure? Some Evidence from International Data', *Journal of Finance*, 50, 1421–1460.

Rudnick, D. (1993), 'ADB Boosts Local Markets – Borrowing or Raising Funds on Capital Markets', *Euromoney*, April, 17–18.

Sagari, S.B. and Guidotti, G. (1992), 'Venture Capital: The Lessons from the Developed World for the Developing Markets', *Financial Markets, Institutions and Instruments*, 1(2), 1–59.

Sender, H. (1993), 'Banking More for Less', *Far Eastern Economic Review*, 57, 52–53.

Shaw, Edward S. (1973), *'Financial Deepening in Economic Development'*, New York: Oxford University Press.

Simon, J.D. (1992), 'Political-risk Analysis for International Banks and Multinational Enterprises', in R.L. Solberg (ed.), *Country-risk Analysis*, London and New York: Routledge (chapter 7).

Stiglitz, J.E. and Weiss, A. (1981), 'Credit Rationing in Markets with Imperfect Information', *American Economic Review*, 71(3), 393–410.

Stiglitz, J.E. and Weiss, A. (1987), 'Credit Rationing and Collateral', in J. Edwards, J. Franks, C. Mayer and S. Schaefer (eds), *Recent Developments in Corporate Finance*, Cambridge: Cambridge University Press, pp. 101–135.

Thomas, J.J. (1993), *The Informal Sector in Developing Countries*, London: Macmillan.

Todaro, M.P. (1997), *Economic Development* (7th Edition), London: Addison Wesley Longman.

Tsiddon, D. (1992), 'A Moral Hazard Trap to Growth', *International Economic Review*, 33, 299–321.

Westlake, M. (1993), 'Only the Best Will Do', *Banker*, 143, 23–29.

Wignaraja, G. (1998), *Trade Liberalisation in Sri Lanka: Exports, Technology and Industrial Policy*, Basingstoke: Macmillan Press.

World Bank (1994), *Assessing Development Effectiveness: Evaluation in the World Bank and International Finance Corporation*, Washington, DC: World Bank.

Yamamura, K. (1972), 'Japan 1878–1930: A Revised View', in R. Cameron (ed.), *Banking and Economic Development*, Oxford: Oxford University Press.

Yaron, J. (1994), 'Assessing Development Finance Institutions: A Public Interest Analysis', *World Bank Discussion Papers*, 174, Washington, DC: The World Bank.

Yaron, J., Benjamin, M.P. and Piprek, G.L. (1997), 'Rural France: Issues, Design, and Best Practice', World Bank Series.

Young, P. and Theys, T. (1999), *Capital Market Revolution*, London: Financial Times.

Part III

Incentive policies for competitiveness

8 Privatisation, regulation and domestic competition policy

Christos N. Pitelis

Introduction

The aim of this chapter is to provide a literature survey for policy makers in developing countries for three interrelated topics: privatisation, regulation and domestic competition policies (CPs). Its main focus is to critically assess the background theory, review practice and provide guidance on implementation.

The following section deals with domestic CPs in theory and practice, while the next two sections deal with privatisation and regulation and provide guidance for implementation on privatisation–regulation, and on domestic CPs. The final section has concluding remarks.

Domestic CPs: theories and international experience

Definitional issues

Domestic CPs refer to the stance governments adopt towards the role of competition between firms in economic development and the measures they take to implement their objectives.

Competition policies usually attempt to influence the degree of competition in industries, such as, the food or textile industry. In this context they are part and parcel of a more general category, that of supply-side and industrial policies (IPs).

The term 'industrial policy' refers to a set of measures taken by a government aiming at influencing a country's industrial performance towards a desired objective.[1] As all government measures and policies affect industry one way or the other, boundaries between competition-IP and other policies, such as even technology policy, regional policy, structural policy, competitiveness policy and macroeconomic policy are not always clear. The nearer we can get to a demarcation line is arguably by referring to government's own perception of what they aim CP and IP to be plus an underlying body of theoretical knowledge hopefully informing such perceptions. The government's objective is assumed to be the improvement of the welfare of its citizens. This happens when resources are allocated efficiently and wealth creation is taking place at a pace preferably faster than in other countries (improved international competitiveness). Industry is believed to be an important contributor to the wealth creation process for numerous reasons, for example, the tradability of its products, its positive link with technology, innovation and productivity growth, and even its close links with services. It follows that a government wishing to improve welfare will be well advised to design measures that lead to efficient allocation of resources and the creation of wealth, that is, to a strong, productive, competitive and therefore wealthy economy. There is universal agreement among economists that competition between firms in industries can be a potent means of facilitating the desired objective. However, views differ as to the role, the type, the degree, and even the nature (including the definition) of competition.

It is not possible or even useful to discuss all these issues in detail in this chapter, but a bird's eye view of alternative perspectives can facilitate understanding, including the link between competition, privatisation and regulation.

The neoclassical perspective

Concentration and market power

The dominant perspective on CP today is the mainstream (neoclassical) theory of competition, monopoly and industry organisation (IO). In the context of this approach, competition is seen as a type of industry structure. This can be perfect or imperfect. Perfect competition is characterised by the existence of numerous firms, which produce very similar (homogenous) products, full and symmetrically distributed knowledge about firm and industry condition (demand and cost curves, in particular), and free mobility of resources, for example, no barriers to entry and exit of firms in the industry. Under such conditions firms are price-takers; they cannot influence the prices which are determined by the interplay of supply and demand in the industry. In addition, such prices only cover average costs, and there are only normal profits.

The opposite to perfect competition is monopoly. Here we have only one firm in the industry, with blockaded entry. When this is the case, a monopolist that maximises profits will charge a price that is higher than the price of the perfectly competitive firm, by restricting output. As a result, consumers will end up with lower quantities of goods for which they pay higher prices. In addition to this being bad for consumers, it also leads to misallocation of resources, because output is being restricted compared to that of perfect competition.

The neoclassical view is that the leading concern of an economy should be to allocate efficiently its scarce resources and in so doing to maximise the welfare of its consumers. In this context monopoly leads to market failure due to the 'wrong' type of market structure (thus it is called structural market failure). When such failures exist, it is believed that the government can step in to correct these. However, there are problems. First, in reality, the two polar opposites, monopoly and perfect competition, are recognised to be unrealistic, the most prevalent form of industry structure being some sort of 'imperfect competition', such as 'monopolistic competition' or more likely 'oligopolistic competition'. An industry structure is characterised as 'oligopoly' when there is interdependence between (usually a relatively small number of) firms. When one firm acts, the other is affected and needs to react, sometimes pre-emptively. How is oligopoly linked to resource allocation? A second problem is that the comparison so far between monopoly and perfect competition assumes that these face the same cost and demand conditions, that they have the same information-knowledge, technology, resources and competences and therefore the same impact on economic performance. Is this true? If not, one has to take into account any differences.

Starting from oligopoly, the theory of IO has built on the above discussion and developed models of oligopoly. Most well known are the 'limit pricing' model, the contestable markets one and that of generalised oligopoly. Various contributors, such as Modigliani (1958) developed the first one in the 1950s. The model recognises that in an industry barriers to entry exist, such as economies of large-scale production (scale economies), product differentiation, initial capital requirements, advantages due to differences in the absolute costs that different firm face, etc. In addition, it is recognised that such barriers cannot fully protect established firms (incumbents) from entry, except in rare cases. In normal cases, firms that wish to charge the maximum price, which does not induce entry, will have to find a lower, 'limit' price. This is the price that is low enough not to attract entry. In the limiting case where the only barrier to entry is the scale of output, the limit price is found by simply taking away the minimum efficient scale (MES) of output from the perfectly competitive output. If we assume that entrants will enter at MES in order not to suffer cost disadvantages, the limit price ensures no entry. This is because even if one firm enters with MES, the resulting output and, therefore, price becomes the perfectly competitive one. However, entrants are attracted by above normal profits. Realising that their very action of entering the industry reduces profits to normal, they will choose not to enter.

If oligopolistic firms charge limit prices, consumers will pay less than in the case of monopoly, but still more than in the case of perfect competition. However, assume that there are no barriers to entry and exit is costless (there are no sunk, that is, non-recoverable costs). In this case, it can be argued that any price departures from average costs will invite entrants, who can then exit costlessly, if prices are driven down. This 'hit and run' behaviour by potential entrants will tend to ensure prices that are perfectly competitive, even in the case of oligopoly (Baumol, 1982). If, however, we assume instead that incumbents can somehow blockade entry, it can be shown that if they also maximise profits and collude to increase prices, they can achieve again monopoly prices and outputs (Cowling and Waterson, 1976).

Clearly, whether prices are competitive (or contestable), limit prices or monopoly prices will depend on the existence of barriers to entry and exit (mobility barriers). These need not be only structural (MES), as assumed by the limit-pricing model. They can also be strategic, namely the result of conscious action by firms to restrict entry. Indeed, limit pricing itself can be seen as such a policy. Instead of reducing prices, however, firms can do other things; they can advertise, innovate, invest in excess capacity and/or produce many apparently competing products (product proliferation) with the express purpose of reducing entry. Depending on the extent and degree of success of such actions, the resulting industry price–output outcome can be anywhere between competition and monopoly! This long detour has actually not helped very much, other than in pointing out that every case is a different one and that what we need is empirical evidence from real-life cases, as well as some measures of the losses in allocative efficiency due to the prevailing degree of monopoly.

To find out the degree of monopoly, economists usually use measures of concentration (see Box 8.1). If an industry is highly concentrated it is presumed that there is prima facie evidence for reduced competition, possibility for collusion, strategic barriers and high prices. However, the link between concentration and market power in the form of higher profits is questionable; it can be that more efficient firms grow larger (which increases concentration), and are also more profitable. The crucial issue in this context is collusion over prices, and barriers to entry. The first is normally illegal in most countries, thus notoriously difficult to identify. Studies on barriers to entry have confirmed their existence and importance. Even they, however, could be seen as an inducement to innovation, see below.

To obtain an indication of the importance of monopoly welfare losses at the economy-wide level, one has to try and measure such losses. There are various ways to do this and a large empirical literature. The results vary widely, but it is generally believed that some static losses do exist, therefore monopoly is a potentially serious problem.

As already suggested, such findings fail to account for any difference between perfectly competitive firms and oligopolistic firms with a degree of monopoly power, in terms of efficiency. There are various dimensions of this issue. One refers to differences in costs. Williamson, for example, has argued that monopolies may have lower cost curves, which implies an efficiency gain vis-à-vis the perfectly competitive industry. This gain should be traded-off any static losses (Williamson's trade-off). Perhaps most importantly, oligopolistic market structures may be more prone to invent and innovate. This can result in dynamic productivity benefits, which must also be taken into account. The literature on the link between market structure and innovation has failed to detect any significant differences in relative innovative records. In this context, one might be justified in looking at the static losses and concluding that monopoly is a problem and needs to be dealt with by the government (see Scherer and Ross, 1990).

In principle, the government could step in to ensure perfectly competitive markets by, for example, discouraging mergers and acquisitions (M&As), encouraging mobility and even breaking up large firms, etc. If this took place simultaneously in all industries, consumer welfare would be maximised. However, if this is not the case, perfectly competitive structures in one industry but not in others need not improve overall welfare. This is the problem of 'second best'. In a second best world, what can be the scope of CP? Many economists believe that a degree of 'workable competition' is still desirable. This could take the form of guarding against the acquisition and abuse of monopoly power. In practice, this can be done by, for example, disallowing a firm from achieving a certain market share

Box 8.1 Measures of concentration

> Concentration is the number and size distribution of firms in an industry, or more simply the extent of which a relatively small number of firms accounts for a measure of industry size (e.g. output, sales, employment, assets, value added). Given the potential (but controversial) link between concentration, market power and restrictive practices, there is a strong interest in measuring concentration. There are numerous measures, far too many to discuss here, see for example, Scherer and Ross (1990) for more detailed exposition. Measures include inequality measures, such as the well-known Lorenz curve and Gini coefficients, which are used in measuring income inequality. Better known and specific for concentration in industries are the concentration ratio (CR) and the Herfindahl index (H). The former refers to a measure of size, for example, output, accounted by the largest X firms in an industry, where X is usually 3, 4 or 8. Thus, CR_4 refers to, let's say the output of the top four firms divided by the output of the industry as a whole.
>
> CR is easy to measure but cannot detect changes within or without the chosen X firms. In addition, it can provide conflicting information when comparing industries. An industry A, for example, can be more concentrated than B, for CR_4 but less for, let us say, CR_8. The Herfindahl index does not have this problem. It is defined as the sum of squared market shares ($\sum_{i=1}^{n} s_i^2$). H has the additional advantage that, by squaring the market share, it gives more weight to larger firms, as one would expect and hope for. A problem is that squaring market shares may be inappropriate in some cases, where raising it to a power less or more than two could be better. Hannah and Kay (1988) have proposed doing just that. Choosing, however, the elasticity parameter for S_i requires detailed and agreed upon information, thus it is subject to the criticism of being arbitrary. For this and its links with theoretical models of price–cost margins, for example, Cowling and Waterson's (1986) the H index is widely regarded as the best static measure of concentration. It is static because it focuses on a point in time, that is, it fails to account for change over time. There are no dynamic measures of concentration that command popularity and use in empirical studies as the CR and the H-index. Importantly, however, it is worth reminding us that concentration is not the problem *per se*, as it is often the result of differential firm efficiency, for example, in terms of differential innovativeness. The problem instead is collusion combined with entry deterrence. It is there that the focus on anti-trust authorities should be.

and/or pursuing certain restrictive practices, namely practices that restrict competition (see Box 8.2). This view is adopted by many countries. For example, it is explicitly acknowledged in the Treaty of Rome (original Articles 85 and 86) and the anti-trust policy of the United States, see Pitelis (1994).

Box 8.2 Mergers, acquisitions and restrictive business practices

Firms grow internally or externally through M&As. M&As represent a most significant firm strategy and contributor to firm growth. There are three major categories of perspectives on M&As and many particular proposed reasons. There is a market power perspective, an efficiency perspective and hybrid views (that allow for both effects). Market power views derive in the main from early IO of the Bain (1956) and Modigliani (1958) tradition. In brief, the idea is that *horizontal mergers* increase concentration. This, in turn, increases market power, as it allows larger firms in concentrated oligopolistic industries to raise barriers to entry and thus prices. Models like Modigliani's 'limit pricing', assume barriers to entry in the form of economies of scale, collusion by incumbents and thus a positive link between concentration and prices. Similar results are obtained from generated oligopoly models like Cowling and Waterson's (1976). This result is questioned by Baumol's (1982) 'contestable markets' model, which assumes easy entry and costless exit, and finds no link between concentration and prices.

Even in the absence of contestability, one should look for any efficiency aspects of M&As. Such may be internal to the firms of more systemic. Authors in the neoclassical tradition, point to 'synergy' effects which emerge from rationalisation of resources, or even the '1+1 = 3 effect'. A more dynamic 'synergy effect' comes from Penrose's (1959) theory, which points to acquisition of management talent through M&As and knowledge-creation through 'teamwork', facilitating firm growth. Penrose, however, goes on to point out the beneficial and harmful effects of 'big business competition', namely the innovation-inducing aspects of big business competition, but also the possibility of big firms exploiting their market power to raise prices.

There is a huge literature on M&As and restrictive business practices, such as, in particular, collusive and predatory behaviour and strategic entry deterrence in the IO literature, surveyed, for example, in Scherer and Ross (1990). Price collusion allows oligopolists to increase prices. It is assumed by many models, for example, limit pricing and generalised oligopoly, and it is generally considered to be a policy that firms would like to pursue. However, it can be unstable, as firms have an incentive to 'cheat', by reducing prices. Open collusion is illegal in most countries, but 'tact' collusion can be widespread and it is hard to identify. Strategic entry deterrence involves firms raising barriers to entry to deter potential competitors from entering an industry. Well known in the literature are investments in excess capacity and product proliferation strategies. Limit pricing can also be used as a form of strategic entry deterrence. In general, here the name of the game is to make pre-entry credible commitments, that is, commitments you will need to honour post-entry because, in doing so, you are better-off in terms of post-entry profit, see Dixit (1982). Credible commitments can deter entry, thus allow incumbents to keep prices high.

There is a host of other ways that firms can limit competition. Vertical restraints, predatory behaviour against existing rivals, and even product differentiation, vertical integration and other firm strategies can serve such a purpose. Since it is in the narrow self-interest of industry participants to use such strategies or practices, and given that these move away from the 'perfectly competitive' ideal, governments are called in to correct the emergent market failures. The usual way to guard against the acquisition and abuse of monopoly power, through, for example, a Mergers and Monopolies Commission (MMC), such as this of the United Kingdom. The role of such commissions is to assess the potential impact of M&As on the degree of competition within industries and intervene, for example, by blocking proposed mergers, when it is expected that these will lead to dominant positions.

As already noted (with the exception of the Penrosean view), much of the above devices from a 'neoclassical perspective', and it fails to consider the particularities of developing countries. If we focus on wealth creation, the important question becomes the role of M&As on innovation and productivity. The process of growing, as Penrose (1959) had pointed out, is almost always efficient. However, the outcome is often inefficient, given incentives to firms to reduce competition. This should be taken into account by the MMC in order not to throw out the baby (innovation) with the bathwater (concentration).

Transaction costs

A way through which firms may increase their size and, in this way, industry concentration, is if they internalise activities previously taking place at the market place. For example, firms may take over their suppliers or distributors rather than dealing with them at arm's length. There are numerous reasons why firms could thus 'integrate', including the pursuit of market power, reduction in the forces of competition (Porter, 1980), etc. However, one possibility is that firms integrate because market exchange is costly; finding and dealing with other firms can lead to high exchange or transaction costs. Coase (1937), and many other economists believe that reducing market transaction costs is an important reason for the existence, and boundaries (therefore the size, too), of firms, see e.g. Pitelis (1991).

If increasing firm size is the result of transaction cost reduction, such efficiency gains should also be taken into account by regulatory bodies. A vertical acquisition for example, could be motivated by efficiency, not market power motives. If so, the relevant authorities should take this into account. Overall, such considerations suggest a more lenient attitude towards large oligopolistic firms. An ideal scenario would be transaction costs-motivated integration in contestable markets. The observed industry structure would be the result of efficiency, and if oligopolistic, it would still charge competitive prices.

In reality, it is unlikely that transaction costs will be the exclusive determinant of firm size and/or that markets will be contestable. Nevertheless, both views seem to have influenced CP experts, and they should certainly not be ignored.

Differential competences and related perspectives

It is arguable that the dominance of the neoclassical view on competition, monopoly and CP is currently under serious threat. This is the result of the emergence and current popularity of an alternative perspective, which can broadly be defined as the differential competences perspective. Within this broad category, there is a diverse group of contributors. However, they share between them the view that competition is not just a type of market structure, and that not only the efficient allocation of scarce resources is important, but also the creation of value and wealth. There is a wide belief that firms are very important contributors to value/wealth creation, and also that each firm is an individual entity, which differs from other firms primarily in terms of its distinct resources-capabilities or competences.

The lineage of this perspective is impressive indeed. It can be claimed to include founding fathers, such as Smith (1977) and Marx (1959), and more recently influential economists such as Schumpeter (1942), Penrose (1959) and Richardson (1972).

In brief, classical economists, such as Smith and Marx focused on wealth creation, not just allocation. They both saw competition as a process, regulating prices and profit rates, not a type of market structure. Smith described the amazing productivity gains through specialisation, the division of labour, the generation of skills and inventions within the pin factory. Marx suggested there is a dialectical relation between monopoly and competition (whereby the competition leads to monopoly and monopoly can only maintain itself through the competitive struggle) on technological change, the rate of profit and the 'laws of motion' of capitalism at large. Marx focused in addition to competition within the factory, and at the society at large, between employers and employees.

Building critically on Marx, Schumpeter described competition as a process of creative destruction through innovations. He saw monopoly as a necessary and just, yet only temporary, reward for innovations. He attributed firm differential performance on differential innovativeness and saw concentration to be the result of such innovativeness.

Penrose's now classic 1959 book on *The Theory of the Growth of the Firm*, is arguably a glue that binds all together. In it, firms are seen as bundles of resources in which interaction generates knowledge, which releases resources, which are thus an incentive for endogenous growth and innovation. Differential innovations and growth lead to concentration which, however, can also be maintained through monopolistic practices. The world is seen as one of big business competition, where competition is god and the devil at the same time. It drives innovativeness, yet it is through its restrictions that monopoly profit can be maintained. Building on Penrose, Richardson (1972) observed that firms compete but also cooperate extensively. Such cooperation is not just price collusion as the neoclassical theory assumes. It lies between market and hierarchy and occurs when firm activities are complementary but dissimilar (require different capabilities).

There are many more contributions in this perspective, but they have not been taken into account, yet, in any systematic form on the issue of CP. Yet, they have obvious implications. First, the focus on value and wealth creation through differential competitiveness suggests a broader welfare criterion, than just the consumer surplus. Second, differential competences provide another efficiency-based reason for concentration. Third, competition as a dynamic process of creative destruction through innovation implies a need to account for the determinants to innovate, when considering the effects of 'monopoly'. Fourth, competition with cooperation (co-opetition) as in Richardson, implies the need to account for the potential productivity benefits of co-opetition, in devising CPs.

An extra dimension on competition relates to its strength, and the role of proximity and location. This links to the work of Richardson, but has been developed by Porter (1990). Porter claims that local competition is more potent than foreign competition. This may have important implications in devising domestic CP, see below.

International practice and lessons from experience

Despite its limitations, the neoclassical perspective has arguably dominated CP thinking in the western world for many decades. For example, the various anti-trust legislation in the United States, as well as the original Articles 85 and 86 of the Treaty of Rome in Europe, seem to be directly informed and influenced by the aforementioned perspective. At least, this seems to be the case in theory. The practice has varied from theory, and also between countries and in time. As we have argued elsewhere (Pitelis, 1994), the European policy, for example, can be described as *ad hoc*, discontinuous and even inconsistent. *Ad hoc* because the theoretical basis of various policies was not clear. A notable example is the 'national champions' or 'picking winners' policy, which various European countries pursued in the 1960s and 1980s. This involved identifying potentially successful firms or industries and using a number of measures like subsidies, tax breaks, etc. to promote them. It also involved a lenient and even encouraging attitude towards mergers and in cases (often in pursuit of considerations of fairness and distribution) nationalisation of utilities but also other 'strategic' industries, such as the car industry. Underlying was the hope that such firms could compete successfully with foreign rivals, thus raise export surpluses and the country's competitiveness. Evidently, this tended to exacerbate structural market failures, and was also inconsistent with the theoretical pursuit of 'perfect competition'. The policy was also pursued at a pan-European level, in search for a pan-European company, which could outcompete large American multinationals.

It is widely accepted that such policies blunted incentives for protected firms to compete, and gave rise to 'problematic enterprises', or 'lame ducks'. After rescuing them for a number of years, European governments led by Mrs Thatcher's Britain eventually resorted to deregulation and privatisation as well as a switch of focus to small firms. This also resulted in discontinuity of policies, from large firms and the government, to small firms and the market.

The approach of Japan and the so-called 'tigers' of the Far East was different. The policy of Japan, for example, led by the Ministry of International Trade and Industry (MITI) was not informed by neoclassical economics. It involved a strongly interventionist approach by the government aiming at creating dynamic advantages in sectors of the future. Such sectors were chosen on the basis of being high value-added, high income elasticity of demand and gradually knowledge-intensive. In such sectors, MITI provided financial and other support and guidance. It regulated the degree of competition (neither too little, nor too fierce) by aiming at an 'optimum' number of firms in it, and protected these sectors at the same time from foreign competition. It also paid attention to the benefits of cooperation (Best, 1990). Overall, the approach to competition could be described as domestically focused balanced competition, with cooperation (co-opetition). The approach of the East Asian 'tigers' was similar, albeit some of them, especially Singapore, extended the art of 'technology transfer' (practised by the Japanese), through an aggressive inward investment policy. The performance of the Japanese economy and that of the tigers' has been very impressive until recently. It is not surprising that commentators attributed this success, in part, to its approach to competition and IP (as well as to other characteristics of the Far Eastern economies, like education, equitable distribution of incomes, high saving ratios, etc.).

To attribute the success of the Far East just to its approach to competition and its interventionist IP, given especially similarly interventionist policies by Western governments in the past, implies either misconceived policies by the latter, or a differential degree of (in)competence. This may well be the case, but there is also a second potential argument. In contrast to the West, the Japanese rejected (were thankfully unaware of?) the neoclassical perspective and favoured an approach that focused on resource creation (not just allocation) through dynamic competition for innovations, thus growth, further innovation, productivity and competitiveness. This approach that seems to combine Schumpeterian and Penrosean ideas with its accompanied focus on production and organisation (Best, 1990) may well be the *differentia specifica* of the Far Eastern approach. It has been associated with major innovations, such as total quality, 'just-in-time' life-time employment, the coexistence of competition with cooperation (co-opetition) and others.

There have been numerous developments in economics and management in recent years (the knowledge-based perspective, the new international trade theory, endogenous growth theory, new location economics, 'new competition', etc.), some described above. These arguably offer support to the Japanese perspective and policies. In part due to both these, recent approaches to competition and IPs in the Western world have tended to move away from the neoclassical perspective, to an approach and policies aiming at improving competitiveness at the firm- and macro-levels. There are various versions of this new approach. The 'new industrial policy' approach, for example, retains its neoclassical flavour but emphasises input, linkages and technology policies as incentive-compatible means of improving firm and industry competitiveness, see Audretch (1998). More general competitiveness models, such as Porter's, focus on the role of firm clusters and other determinants of competitiveness (Porter, 1990). Cluster policy is seen as the new IP (Porter, 1998), based on co-opetition. Recent focus by the European Union on education, (soft) infrastructure, technology and innovation and (clusters of) small firms moves in this direction.

These approaches, however, still lack a systematic effort to link competition and cooperation (co-opetition) to firm and national competitiveness, and thus to economic development.

An integrative framework

The aforementioned discussion suggests the need for a framework on domestic CP, which takes into account the broader considerations presented in this section, notably the role of competition (and cooperation) as a process in value/wealth creation.

Building on the works presented so far, it can be argued that productivity and (thus) value/wealth creation within firms is determined by firms' organisation and structures (infrastructure), human resources (labour, management, entrepreneurs), innovativeness, and unit cost economies (of scale, scope, learning growth, transactions, external and economies of pluralism). At the national level these should be supported by a supply-side friendly macro-policy and facilitatory institutional context, see Pitelis (1998).

Domestic CPs can be derived within this context. Competition and cooperation (co-opetition) influence all determinants of productivity and value creation. All the same productivity enhancements may lead to differential advantages that can be used to restrict competition. The need for a domestic competition and cooperation (co-opetition) policy thus arises from the need not to thwart the beneficial effects of co-opetition on productivity and value creation. Firm co-opetition strategies (e.g. firm clusters) that enhance productivity should not be stopped, but facilitated in this context.

Nonproductivity-enhancing forms of cooperation (like collusion) instead, should be forcefully discouraged. The same is true for other restrictive business practices. M&As should be examined on a case by case basis, as they may have productivity enhancing attributes, but may also lead to market power, which can eventually stifle incentives to innovation and productivity (see Box 8.2). Pluralism and diversity should be encouraged, as it provides benchmarks for comparison and thus information. Institutional changes facilitating a productivity enhancing culture and ideology and productivity-compatible legal frameworks should be aimed at.

A last observation here is the need for CPs to be compatible with macroeconomic policies, but also importantly to be supported by a facilitatory institutional context. North (1981) has shown the importance of institutions and institutional change in reducing transaction and transformation costs and increasing productivity and growth. Institutions, but also culture, attitudes and ideology can be hugely important factors in economic organisation. Governments can be a potent catalyst in institutional change, as they possess a monopoly of force and the ability to legislate and regulate. The devising of a facilitatory framework is part and parcel of domestic CP. The neoclassical 'market failure' theory of the state assumes that the institutional context is given. The possibility to vary it implies a more proactive role for the state. In this context, the state should not just intervene when markets fail. Rather, it should legislate and regulate proactively, so that markets, firms and itself should fail less, see below.

In sum, current developments in economics and management point to the need for a broader conceptual framework for domestic CP, to account for the role of cooperation, institutions and knowledge-creation through diversity and pluralism. The need for a strong CP measures that discourage the emergence and exploitation of market dominance, is maintained and even strengthened in this framework; but the focus is not just on consumer welfare, but overall productivity, value and wealth-creation. Despite similarities with the neoclassical perspective, the one developed here is claimed to also provide some new and different insights on domestic CP, which, in addition, are most relevant for developing countries, see below.

Developing countries

A problem with much of our discussion concerning the role of competition is that it presupposes the existence of well-functioning markets, an entrepreneurial class and the existence of a structural and institutional framework which facilitate the implementation of chosen policies. This is evidently not the case in developing countries for which the approach of Japan and the four tigers can be instructive. There is a need to both create firms, markets, entrepreneurship, competitive advantages and, in addition, to ensure the existence of competitive forces. In the early stages of economic development it is common for countries to rely on some sort of support for domestic industry, to include varying degrees of protectionism. This is in line with theoretical arguments of the infant industry type, and can be of benefit to a country, provided they are not permanent. In the latter case, the usual problem of disincentives takes stock. Firms in protected industries will lack incentives to innovate. Relatedly, they will tend to focus on the domestic market, which will tend to be more profitable. This will not expose firms to international competition, reducing further these incentives to innovate so as to become internationally competitive. Domestic and international competition, and also export rivalry can be useful complementary means of providing incentives to firms to improve efficiency and innovate, see Kikeri *et al.* (1992). However, it is widely agreed that international competition cannot serve as an adequate substitute for domestic competition, see Porter (1990), World Bank (1998). In addition, to expose domestic firms to (foreign) competition, in the absence of the prerequisites discussed above, could have detrimental effects on domestic industry. This is a fundamental issue for developing countries.

Generally, developing and transition economies are typically characterised by highly concentrated industries, large state-owned sectors and firms and firms (public or private) operating in industries protected by various barriers, thus insulated from the forces of competition. The issues of privatisation, regulation and domestic CPs should therefore be seen within this context, as follows: develop capabilities which allow you to liberalise, privatise and regulate inefficient state-owned enterprises (SOEs); adopt CPs which expose such firms to competitive forces, while at the same time facilitate the creation of new small to medium-sized enterprises (SMEs), which, alongside existing ones, co-opete between themselves (e.g. in the form of clusters) and with the larger firms to increase productivity and competitiveness.

Privatisation and regulation

Introduction

Privatisation and regulation can be seen as part and parcel of competition and IP: both influence the degree of competition in a particular industry and industry as a whole and can lead to changing industry structures, dynamics and (thus) performance.

Privatisation of state-owned assets became a major economic policy of many countries in the 1980s. This was part of a reconsideration of the nature and role of the state sector in developed and less developed economies, which was resolved in favour of the view that government involvement had been excessive. At the end of the 1980s, the demise of central planning in Eastern Europe led to an attempt to create markets there and, in a sense, to privatise whole economies. The drive to privatisation led to the need to regulate privatised firms with potentially high market power. It is interesting to examine why this has happened. Here we focus on the theoretical reasons offered for or against privatisation, and regulation.

Markets and states: general theoretical issues

The state is widely acknowledged to be one of the most important institutional devices for the organisation of economic activity, along with the market (price mechanism), and the firm. The role of the state has in fact been increasing steadily since the Second World War. In most OECD countries, government receipts and outlays as a proportion of GDP are very high, often in excess of 50 per cent. Many theories have tried to explain the growth of the public sector in market economies, the so-called Wagner's Law, originating from a number of ideological perspectives. In brief, 'neoclassical' theories tend to consider such growth as a result of increasing demand for state services by sovereign consumers, while 'public choice' theorists regard it as a result of state officials, politicians and bureaucrats' utility-maximising policies, which tend to favour enhanced state activity. In the marxist tradition, the growth of the state sector was linked to the laws of motion of capitalism – increasing concentration and centralisation of capital, and declining profit rates – which generate simultaneous demands by capital and labour on the state to enhance their relative distributional shares, for example, through infrastructure provisions and increased welfare services, respectively. There are variation on these views within each school (see Pitelis, 1991).

Besides explaining why states increase their economic involvement over time, many economists in the 1980s focused their attention on why public ownership could fail to allocate resources efficiently and, more particularly, on the relative efficiency properties of market vs non-market resource allocation. Particularly well known here are the views of the Chicago School, in particular Friedman (1962) and Stigler (1988). In a number of papers, Friedman has emphasised the possibility of states becoming captive to social interests of powerful organised groups, notably rich business people and trade unions. Stigler, among others, on the other hand, has pointed to often unintentional inefficiencies involved in cases of state intervention. Examples are redistributional programmes by the state that dissipate more resources (e.g. in administrative costs) then they redistribute. For these reasons – and the tendency generated by utility-maximising bureaucrats and politicians towards excessive growth – rising and redundant costs tend to lead to government failure. Wolf (1989) has a classification of such failures in terms of derived externalities (the Stigler argument): rising and redundant costs because of officials' 'more is better' attitude, and distributional inequities, for powerful pressure groups (as in Friedman).

On a more general theoretical level, the case for private ownership and market allocation has been based on three well-known theories. This is because of 'free riding' problems (i.e. everybody believing that someone else can take care of communal ownership) and overall blunting of incentives, given that no privately appropriable benefit is expected. Second and relatedly, Alchian and Demsetz's (1982) residual claimant's theory, building on the tragedy of commons view, suggests that private ownership of firms is predicated on the need for a residual claimant of income-generating assets, in the absence of which members of a coalition (e.g. a firm), would tend to free-ride, thus leading to inefficient utilisation of resources.

There is now huge literature on the merits and limitations of these theories (see for example, Eggertson (1990) for an extensive coverage). Some significant weaknesses have been exposed in each defence of private ownership and market allocation. Concerning the 'tragedy of the commons', it has been observed that, historically, communal ownership has often had efficiency-enhancing effects. Hayek's critique of pure planning loses much of its force when one considers choices of degree in 'mixed economies', which is virtually always the case, at least in market economies. In addition, Hayek fails to consider knowledge generation within firms, as in Penrose (1959). Lastly, the residual claimant theory becomes weaker when applied to modern joint-stock companies run by professional management groups with little share ownership. In addition, it downplays the potentially productivity enhancing attributes of cooperation, see Eggertson (1990).

Other well-known arguments relating to the problem of government failure are Bacon and Eltis' (1986) claim that services, including state services, tend to be unproductive; and Feldstein's (1984) view that pay-as-you-go social security schemes tend to reduce aggregate capital accumulation. The alleged reason for this is the view that rational individuals consider their contributions to such schemes as their savings, and they thus reduce their personal savings accordingly to remain at their optimal consumption–savings plans. Given, however, that the schemes are pay-as-you-go (i.e. contributions are used by the government to finance current benefits), no actual fund is available, so that individuals' reduction of personal savings represents an equivalent reduction of aggregate saving, equated by Feldstein to capital accumulation.

Some of the above reasoning is reminiscent of (and is supported by) some marxist criticism of the role of the state, for example, the views that state services involve unproductive (i.e. no surplus value generating) labour (Gouph, 1989). This is often linked to the falling tendency of the rate of profits, and the tendency for government spending under advance capitalism to exceed government receipts, for reasons related to demands by both capital and labour on state funds and resistance on both sides to taxation, which are particularly intensified under condition of monopoly capitalism.

The near universality of the attack on the state is informative of the general theoretical case underlying the drive to privatise (for one of the few attempts to defend the public sector, see Heald, 1983).

Private vs SOEs

Concerning specifically the relative efficiency properties of the private sector vs public sector SOEs, the focus of attention has been in the main on issues of managerial incentives, competitive forces and differing objectives. It can be claimed that public sector enterprises achieve inferior performance in terms of profits or of the efficient use of resources. While private sector managers are subject to various constraints leading them to profit-maximising policies, this need not be the case with public sector managers. Such constraints arise from the market for corporate control (i.e. the possibility of take-over of inefficiently managed firms by ones which are run more efficiently), the market for managers (that bad managers will be penalised in their quest for jobs) and the product market, including the idea that consumers will choose products of efficiently run firms for their better price for given quality.

Among other factors which tend to ensure that private sector agents (managers) behave in conformity with the wishes of the principals (share-holders) – by maximising profits in private firms – are, for example: the concentration of shares in the hands of financial institutions; the emergence of the multi-divisional (M-form) organisation, which is based on separate divisions under central control, which operate as profit centres; the possibility of contestable markets, that is, markets where competitive forces operate through potential entry by new competitors, as a result of free entry and costless exit conditions. SOEs may not be subject to such forces, at least not to the same degree, which implies the possibility that managerial incentives for efficient use of resources and profit maximisation may be less pressing in public sector firms. Profit maximisation may not be pursued also if public sector enterprises simply do not aim at such policies, for example, because they are used as redistribution vehicles by the government; and/or for non-economic reasons, such as the need for electoral support; and/or because they aim at correcting structural market failures, such as the high prices of private sector monopolies. All these factors tend to establish the economic-theoretical rationale for the efficiency of private firms, and therefore for privatisation. Vickers and Yarrow (1988) and Kay et al. (1986) offer extended discussion.

Various limitations have also been identified in the case for the relative efficiency of private firms. One limitation arises from the possibility that the various constraints on private sector firms' managers are not as strong as they are often suggested to be. For example, large size may protect inefficient firms from the threat of takeover; it may be difficult to tell when a manager has performed well, given the often long-term nature of managerial decisions; and bounded rational consumers may often fail to tell differences in the quality of similarly priced products. Concerning competition, a private sector monopoly is as insulated from it as a public sector monopoly, *ceteris paribus* (assuming no difference in the forces of potential competition). Furthermore, the absence of competition is not *per se* a reason for privatisation: it could well be a reason for opening up the public sector to such forces, for example, through competitive tendering and franchising (Yarrow, 1986). SOEs can be M-form, and when their stocks are traded, financial institutions may exert pressures. Such considerations have led many commentators to the conclusion that the issue is not so much of the change in ownership structures as the nature of competitive forces and of regulatory policies themselves (Kay and Silberston, 1984; Yarrow, 1986; Vickers and Yarrow, 1988; Clarke and Pitelis, 1993).

Rationale of public ownership

The reason for public sector enterprises has often been market, not government, failure (see, for example Rees, 1986). In mainstream economic theory, the first fundamental theorem of welfare economics shows that market allocation can be efficient, if market failures do not exist. Such failures, however, are widely observed, famous instances being the existence of externalities (interdependencies not conveyed through prices); public goods (goods which are jointly consumed and non-excludable); and monopolies, which tend to increase prices above the competitive norm. The observation, among others, that efficient government itself is a public good, has led to the idea of pervasive market failure (Dasgupta, 1986), which is viewed as the very *raison d'être* of state intervention (Stiglitz, 1986). The very reason why public sector enterprises are run by the state is that they have been seen as natural monopolies (firms in which the minimum efficient size is equal to the size of the market as a result of economies of scale, leading to declining costs). If private, it is assumed that these firms would introduce structural market failure in terms of monopoly pricing. The undertaking of the activities of such natural monopolies (often known as public utilities) by the state could solve the problem through, for example, the introduction of marginal cost-pricing policies. Although such policies need not necessarily re-establish a first best Pareto-optimal solution (given imperfections elsewhere in the economy), they could at the very least point to the limited value of any claims that public utilities do not maximise profits, given that this was not their objective to start with.

Whatever the case may be, the consensus of the 1980s has been that government failure is more of a problem than market failure. The disappointing performance of SOEs in many cases has strengthened this perception. Alongside the theoretical focus on government failures, and in part because of its perceived significance, privatisation of SOEs was one of the key elements of structural adjustment programmes, advocated by the IMF and the World Bank. All these can, in part, explain the privatisation drive of the 1980s and 1990s.

Especially in the case of developing countries, SOEs have been the result of more complex and often more political reasons. These include the replacement of weak private sectors, the transfer of technology to 'strategic' sectors, to generate employment, to provide goods of lower costs (Kikeri *et al.*, 1992). As evidence provided by the last mentioned shows, the performance of SOEs has not been satisfactory, leading, among others, to serious burdens of already hard-pressed public budgets. Attempts by various countries to improve performance without change of ownership have also met with limited success. This has contributed to the privatisation drive and suggests for, Kikeri *et al.* (1992), that ownership itself matters. This conclusion, however, is challenged by, among others Adam *et al.* (1992). In their view, while there is plenty of evidence that governments can pursue non-commercial aims, there is no intrinsic reason why this should always be the case. In their view

> The fact that there are many examples of SOEs (such as those operated in Singapore) which are not used in pursuit of non-commercial goals is adequate refutation. The presence of non-commercial objectives is more indicative of poor management than an intrinsic feature of public ownership.
> (Adam *et al.*, 1992, p. 15)

Objectives of privatisation

Much like in the case of nationalisation and SOEs before, and in addition to the general theoretical issues concerning market vs state ownership, and the relative efficiency properties of SOEs vs private enterprises, and the support for privatisation by the IMF and the World Bank, privatisation policies have been motivated by a host of reasons, sometimes overtly political. Drawing on, among others, Megginson and Netter (2000) and Pollitt (2000), we can identify the following objectives of privatisation.

1. Promote economic efficiency.
2. Introduce competition and expose SOEs to market discipline.
3. Reduce government involvement in industry.
4. Raise revenue for the state.
5. Curb the power of public sector unions and gain political advantage.
6. Promote wider share ownership and develop the national capital market.

Arguably, only the first three of these follow from the belief for differential efficiency at the private sector; the others relate to more political concerns. In particular, objectives 1, 2 and 3 are interrelated and follow directly from our discussion on government failure and the benefits of competition. The fourth objective relates to immediate needs by governments to reduce their private sector borrowing requirement (PSRB), while 5 and 6 are more political and aim at reducing state capture and/or redistribute power (objective 5), and legitimising changes and the prevalence of market capitalism as a whole (objective 6). Evidently, these last three objectives, too, have efficiency implications. Yet, they are motivated by extra-economic reasons, too.

Methods and speed of privatisation

There are various taxonomies concerning privatisation methods. Four principal ones are those proposed by Josef (1996), but see also Megginson and Netter (2001).

1. *Privatisation through restitution.* This refers mainly to land or other easily identifiable property expropriated in the past, that can be returned to the original owners or their heirs. This form has been applied exclusively in the ex-centrally planned economies, for obvious reasons.
2. *Privatisation through sale of state property.* Either direct sale (asset sales) of SOEs to economic agents (individuals or firm/firms) or share issue privatisation, that is, SOE sale to investors through public share offering.
3. *Mass or voucher participation.* Here eligible citizens are provided with vouchers for free or at a nominal cost, which they can use to bid for stakes in SOEs and/or other assets being privatised.
4. *Privatisation from below.* That is, through the start-up of new private businesses in ex-centrally planned economies.

Privatisation from below arguably belongs to a category of its own. While in the first three cases we are dealing with existing enterprises, in the last we refer to new firm creation. This is extremely important; it links with our discussion in the previous section on wealth-creation; and the problems of developing countries. New firm-creation is by definition of essence in developing countries, so 'privatisation from below' should arguably be the top priority. Other types of privatisation can evidently be pursued, alongside new firm/wealth creation.

Speed is a wexed issue in theory and in practice. It very much depends on the particularities of the country. In theory, speed helps deal with required adjustments, avoid oppositions and show commitment and credibility. Its disadvantage can be that it can result in unexpected problems, and require adjustments such as in the case of Poland. Gradualism has been adopted with success by countries such as China while speedy reformers have also met with differential degree of success in other countries, see Megginson and Netter (2001). Best practice here seems to be country specific. If the country is 'ready' as defined by the World Bank (i.e. it is in its interest, it is feasible and it is credible), then go fast. If not, one should better prepare first (see section below on implementation).

Regulation

Regulation refers to steps by government to ensure that firms that already enjoy market power are not able to abuse their dominant positions. The theoretical case for regulation follows directly from our discussion earlier in this section, notably the derivation and abuse of monopoly power and its impact on economic performance.

In the context of our discussion about 'government failure', it is arguable that the best regulation is no (state) regulation at all (which is market regulation) or self-regulation. Any government intervention may be subject to failures, and at the very least to some costs. As a result, if competitive forces can do the job, of regulating firm conduct, this is best. Failing this, CPs, already discussed, should do the trick. Failing this, too, self-regulation should be enough. An argument going back to Coase (1937) suggests that large firms behaving badly are easy to spot, and can/will be punished by alert consumers – that is, the market. This explains why firms are likely to be less opportunistic than employees. Partly in order to pre-empt this (protect capitalism from itself, as Keynes had put it), large firms (or their representing associations) can develop rules of conduct, which, when adhered to, would remove any need for government regulation. In all, market regulation and self-regulation would suggest that the best firm of regulation is no regulation at all, or regulation through CP.

In the above context, it is arguably surprising that in an era of government disengagement, deregulation and privatisation, re-regulation has become an important issue in the 1980s and 1990s. In main part, this has resulted from the extensive privatisation drive of that period. This has led to large privatised firms, with possible dominant positions, in markets with insufficiently competitive forces. This and the need of such firms to satisfy shareholders, has arguably rendered market and self-regulation inadequate, thus leaving CP and regulation as the more realistic means of protecting consumers. However, this has not been costless. It has led to the creation of new regulatory bodies, as well as serious related problems, such as the need to address the aims and scope of such bodies, to ensure their own regulation, to avoid the problem of 'jurisdification' (i.e. over-regulation or legal pollution), and also to devise incentive- and productivity-compatible regulatory systems. All these have been very demanding for government resources, indeed to these very governments that have vowed to reduce government! In discussing these issues Michie (1996) for example, observes that

> Faced with jurisdification, the need would be to reverse the spread of such regulation rather than develop it further. It may be, however, that the only feasible way of escaping this new generation of regulatory activity is by avoiding – or reversing – the very privatisation that spawned it.... However, and despite early hopes that regulation would only be temporary, this seems to be far from being the case.
>
> (Michie, 1996)

There are various issues pertaining to successful regulation, which we can touch upon here. First, the setting-up of regulatory bodies, such as, for example, Oftel (for telecommunications), Ofgas (for the gas industry), Ofel (for electricity) and Ofwat (for water) in the United Kingdom. Their objective is to do no less than regulate in a way that promotes sustainable competitiveness and keeps everybody happy (industry, consumers, etc.) in the short and long runs. This is a tall order. It involves, among others, a regulatory contract, between the regulator and the regulated, which should be able to ensure sufficient profits for happy shareholders and for longer term investments and fair prices at the same time. These are relatively new, relatively little researched and understood issues. Moreover, they require extensive knowledge of the industry's conditions.

Assuming all the above are resolved, there are additional issues. First, who regulates the regulators? Following a 'public choice' perspective one may suggest that regulators may tend to over-regulate. This is because rational regulators would be expected to maximise their own utility, seen in standard managerial terms (i.e. enhanced status, bureaux, etc.). Government regulations suffer from the very problem discussed by Alchian and Demsetz, of the absence of a residual claimant. Last, but not least, the existence of new regulatory bodies raises the need for coordination, domestically and internationally; domestically between regulatory and CP bodies, internationally between regulatory and CP bodies of other countries. All these are arguably a tall order.[2]

To conclude, regulation seems paradoxical at the theory level and could be difficult to implement. Regulators need to be chosen very carefully and satisfy stringent criteria in terms of knowledge, ability, independence, etc. All these easier said than done. The World Bank for example, advises governments of developing countries to set up regulatory bodies before privatisation in certain cases, see next section. However, in a very real sense, to be able to do so presupposes not being a developing country! If implementation is difficult, one may well need to re-consider the whole exercise of privatisation to start with, at least in cases involving post-privatisation dominant positions.

Evidence and lessons of experience

Given the theoretical debates, it is important to consider the 'lessons of experience'. Evidence summarised in Kikeri et al. (1992), which draws on the experience of the World Bank, suggests that, on balance, there have been benefits from privatisation in terms of productivity improvements and other performance measures in most cases. Moreover, these have not been at the expense of other groups, such as labour and consumers. Kikeri et al. (1992) focused on twelve cases in Chile, Malaysia, Mexico and the United Kingdom. They found that privatisation led to increased domestic welfare in eleven out of twelve cases, and productivity in nine. Many firms expanded their investments and growth; for example, the Chilean telephone company doubled its capacity in the four years following sale. In addition, labour as a whole was not worse off, while other stakeholders too, such as consumers gained. On the other hand, more recent work by Frydman et al. (1999) focuses on corporate performance in Central Europe. It provides evidence in favour of the idea that privatisation improves corporate performance only if ownership is transferred to outsiders but not to insiders.

> In the context of Central Europe, privatisation has no beneficial effect on any performance measure, in the case of firms controlled by inside owners (managers and employees), and that it has very pronounced effect on firms with outside owners.
>
> (Frydman et al., 1999, p. 1154)

Two recent studies critically survey the literature on the evidence so far. Pollitt (2000) focuses on the case of the United Kingdom, OECD's most significant privatisation programme. He points to the objectives of the UK government, and tries to evaluate the policies in terms of the original aims. After a review of multi-firm case studies, Pollitt offers two sets of conclusions.

> *The overall impact.* Privatisation itself does not seem to be associated with an acceleration of productivity growth or profitability. It seems that management changes within the public sector prior to privatisation did however lead to improvements in performance prior to privatisation. Privatisation does have a positive impact on financial performance rather than productivity. There is evidence that firms in regulated industries exhibit improvements in performance only when regulation is tightened or competition increased.
>
> *The performance of individual firms.* Some privatisations were a clear success: British Airways, Cable and Wireless, Amersham International. Some reorganisations prior to privatisation were a clear success: British Steel and British Coal. In the regulated industries BT and British Gas perform well in absolute terms but not relative to prior to 1980. The privatisation's of Jaguar and BAA seem to have yielded little benefit.
>
> (Pollitt, 2000, p. 134)

> ... that privatisation has generally improved consumer welfare via a combination of higher quality and quantity of output and lower prices. The improving technology of regulation has undoubtedly facilitated this. Shareholders have benefited via windfall gains. Workers do not seem to have received lower salaries as a result of privatisation if they remained with the company while those who left were re-employed elsewhere in the economy (unemployment fell from 1986) or went to early retirement. The government gained large asset sales and increased profit taxes. Competitor firms gained almost by definition in all but a few industries as entry barriers were removed.
>
> (Pollitt, 2000, pp. 138–139)

Megginson and Netter (2001) survey a wide range of empirical studies concerning the privatisation of SOEs worldwide. They ask a broad range of questions, ranging from firm financial performance to capital markets and corporate governance. Given the comprehensiveness of this work, covering developing and transition economies, we quote their findings extensively, (as summarised by the authors in p. 2 of the pre-publication version of their article).

> Privatisation programs have reduced the average world-wide level of state ownership by roughly one-half (to less than six per cent) over the past two decades, with the SOE share of national output falling especially rapidly in developing countries during the 1990s. ... First, the evidence is now conclusive that privately-owned firms outperform SOEs and empirical studies clearly show that privatisation significantly (often dramatically) improves the operating and financial performance of divested firms in both transition and non-transition economies. Second, governments have raised significant revenues through the sale of SOEs, with the cumulative value of such sales reaching $ 1 trillion during 1999. Third, privatisation is a major component in developing both capital and product markets within a country. The choice between privatisation via public share offering versus through asset sales is significantly related to factors such as firms size, government fiscal condition, the degree of shareholder protection, and the degree of income inequality. Further, those countries which have chosen the mass (voucher) privatisation route have done so largely out of perceived necessity – and face ongoing efficiency problems as a result. Governments have great discretion in pricing the SOEs they sell, especially those being sold via public share offering, and they use this discretion to pursue political and economic ends. While maximising revenues by setting high offering prices for SOEs is important to governments, many trade this objective off in favour of targeting sales to preferred buyers in direct sales and allocating shares to domestic investors (particularly SOE employees) in share offerings. On average, investors who purchase shares of firms being privatised earn significantly positive excess returns both in the short-run (due to deliberate underpricing of share issues by the government) and over one, three, and five-year holding periods. Finally, privatisations have contributed significantly to the development of national stock markets and corporate governance systems.
>
> (Megginson and Netter)

While all these sound very positive, and as already noted, others are less sanguine, about the benefits of privatisation. In addition, findings concerning positive effects from privatisation do not address the fundamental question whether privatisation can substitute for the role of the emergence of new private business. Kikeri et al. (1992) for example, suggest that 'in many instances, privatisation will be less important for the growth of the private sector than the emergence of new private business' (Kikeri et al., 1992, p. 1). This is in line with our discussion in the section on 'Domestic CPs: theories and international experience'.

The experience on regulation is arguably not too satisfactory. Even in best practice developed countries, with comparatively extensive experience, commentators find there is a lack of knowledge, and a lack of requisite 'regulatory innovation' to cope with radical technological change. Arguably the record of the developing countries is unlikely to prove better.

To conclude, when rightly motivated and implemented, privatisation and regulatory reform can be important means of affecting competition and economic performance. A fundamental dilemma, however, is that, in one sense, privatisation is worth its salt when government failure is more of a problem than market failure. However, it is the failing government that is called to implement it. Remove well functioning-markets, entrepreneurship and competences in developing countries, and the problem gets compounded. In such a context, the issue is clearly not just privatisation, but notably to create what is missing, and privatise and regulate inefficient SOEs. Privatisation alone could well create more problems than it solves, that is, private unregulated inefficient monopolies without competences in international markets, relying on captive domestic consumers for high profits for the few.

Privatisation, regulation and domestic CPs: issues of implementation in developing countries

Much of the preceding discussions have been developed in the context of, and are therefore more suited for, developed economies. For example, an emphasis on guarding against the abuse of monopoly power by private sector firms presupposes (in the absence of government created monopolies) that these firms had had time to acquire competitive advantages and dominant positions, in a more or less level playing field. The issue of government failure vis-à-vis market failure is linked to, first, a process of increasing state involvement in developed mixed economies, but also to the possibility that hitherto 'natural monopolies' (or parts of their operations thereof) are no longer 'natural monopolies' as a result of developments in technology which facilitate the outsourcing of such activities to the private sector.

As already noted, most developing and transition economies do not face similar conditions. In these economies the crucial issue seems to be the creation of new firms, the privatisation of inefficient SOEs and their regulation, so that they operate in a level-playing field with existing and newly created SMEs, which co-opete between themselves and with privatised–regulated ex-SOEs, plus of course those SOEs left in the hands of the state sectors, in order to create value.

The aim of this section is to address issues of implementation for developing countries in the above context. We start with implementation issues for SOEs assumed to be inefficient and in need for privatisation. Then we proceed to issues of regulation of such companies and complete the section with a discussion of domestic CP measures of relevance to developing countries, which aim at enhancing productivity and competitiveness, not just achieve efficient resources allocation.

Implementing privatisation

Implementing privatisation involves, first, deciding whether to go ahead, and then, adopting some best practice. The experience of the World Bank is useful in this context. Kikeri *et al.* (1992) provide a framework for decision making. According to this, there are two major conditions for decision making. First, the nature of the market. Second, country conditions. Putting the two together gives rise to the following proposed framework (see Table 8.1).

When it comes to implementing privatisation itself, the following steps are proposed, see also Donaldson and Wagle (1995). First, clearly define the objectives. Second, 'start small, learn by doing and move on to larger, more complicated transaction' (Kikeri *et al.*, p. 8). Third, privatise management. Fourth, prepare for sale. Fifth, deal carefully with the issue of valuation and sale price. For the World Bank, it is essential for the market to decide the price through competitive bidding. A problem here is that sometimes the market can be thin or imperfect. However, trying to value and price the assets of non-competitive concern is unlikely to lead to sale. Sixth, address the issue of financing (e.g. avoid tempting potential investors with attractive alternative investments, for example, tax free high interest rate government bonds, if at all possible).

Table 8.1 A decision-making framework for privatisation

Country conditions	Enterprise conditions	
	Competitive	Non-competitive
High capacity to regulate; market friendly	Decision • Sell	Decision • Ensure or install appropriate regulatory environment
Low capacity to regulate; market unfriendly	Decision • Sell, with attention to competitive conditions	Decision • Consider privatisation of management arrangements • Install market-friendly policy framework • Install appropriate regulatory environment • Then consider sale

Source: Kikeri et al., 1992, *Privatisation: The Lessons of Experience*, The World Bank.

Lastly, be transparent. Based on his own experience with government advising, this author would also add stability, fairness, predictability and reliability. In addition, in certain cases, when there is popular discontent and/or cultural hostility to privatisation, the gradual sale of parts of an SOE through the stock market might help with requisite institutional and cultural changes. Crucial, in this case, however, is to introduce competent management, independent from political pressures.

A most important condition for successful privatisation is the removal of anti-incentives or obstacles. Such include an unclear and undeveloped, and often contradictory, legal system; illiquidity of the local population and the absence of developed capital markets and the hostility that often exists towards privatisation. All these point to the need for institutional and structural change alongside privatisation, in line with it. In fact, privatisation is a form of institutional and structural change, supported by and supportive of, other such changes.

The framework proposed by the Bank, with the additional observation added here, is helpful, yet suffers from some limitations. These relate to the argument that few developing or transition economies are likely to belong to the top row (Table 8.1). On the other hand, low capacity to regulate may well be linked with low capacity to do what is proposed in the bottom row (Table 8.1). This suggests once again that building institutional and organisational competence, new firms, markets, entrepreneurship and competences should be seen as at least equally important to privatisation. The latter should be a means, not an end.

Reforming SOEs

Privatisation is not the only way to improve firm performance. Another way is to reform SOEs. This is particularly important for various reasons. First, because of the less than conclusive theoretical debate on market vs state. Second, because despite a long period of privatisation, there still exist many SOEs, especially in developing countries, see World Bank (1995). Third, because the co-existence of SOEs and private firms may well be a good means of information concerning relative efficiency properties. Competition between SOEs and private firms, moreover, could lead to systemic efficiency effects.

The World Bank (1995) points to three preconditions for SOE reform: political desirability, political feasibility and political credibility. Provided these conditions are met, evidence collected from twelve developing countries and on the basis of indicators such as SOEs' financial returns, productivity and saving–investment deficit, suggests the following components of reform to be determinants of success: divestiture, competition, hard budgets, financial sector reform and changes in institutional relationships between SOEs and governments. The study suggested that 'more successful reformers made the most of all five components. Indeed they used them not as separate options but as mutually supportive components of an overall strategy' (The World Bank, 1995, p. 5). Successful reformers divested more, introduced more competition, hardened SOE budgets, reformed the financial sector by strengthening supervision and regulation, relaxing control over interest rates and reducing directed credits. While all countries tried to improve the incentive structure by changing the relationship between SOE management and the government, only those who used 'successful contracts' succeeded.

Three types of contract were identified by the World Bank:

- Performance contracts (between government and a government employee-manager managing an SOE). These link managerial performance to rewards – pay.
- Management contracts (between government and a private firm-manager contracted to manage an SOE). These specify the government requirements from the private firm, as well as the latter's compensation.
- Regulatory contracts (between government and a private regulated monopoly), as discussed above.

Evidence from firm case studies in terms of information, rewards and penalties and commitment, suggested that regulatory contracts worked well for enterprises in monopoly markets, when properly designed and implemented. Performance contracts instead rarely improved incentives, and could do more harm than good. Finally, management contracts worked well but were not widely used.

Countries that did not meet the three conditions for reforms should not proceed. In such a case, macroeconomic reforms can prepare the ground. Proposed specific steps designed to increase desirability include the following:

- reductions of fiscal deficits;
- easing of trade restrictions;
- removing of barriers to entry;
- initiating financial sector reform.

Steps aiming to further feasibility include:

- eliminating obstacles to private job creation;
- uncoupling SOEs' jobs and social services.

Finally, to enhance credibility, governments can aim to

- improve their reputation;
- establish domestic and international constraints.

For countries ready to reform, the World Bank suggests policy makers to classify SOEs into two types:

- those operating in (potentially) competitive markets (manufacturing and most services);
- those operating in natural monopoly markets where regulation is required (some utilities and most infrastructure).

Enterprises in the first category can be divested so as to enhance competition and arrangements for the sale are transparent or at least open to the possibility of competitive bidding. In the second category, divestiture can also take place subject to transparency, competitive bidding and the introduction of a credible regulatory structure.

Important questions for policy makers include the following:

- begin with small or big enterprises?
- restructure financially before selling SOEs?
- lay off workers before selling?

Concerning monopolies the following findings are of interest:

- regulatory contracts work better when the government introduces competition, which reduces the incumbents' information advantage;
- price regulation is more effective when it allows firms to retain some of the benefits of improved performance, while at the same time reducing prices for consumers;
- credible regulatory contracts lower costs to the consumers.

For firms that cannot be divested and are unsuitable for management contracts, the World Bank suggests measures which introduce competition, cut government subsidies, eliminate soft credits and hold managers responsible for results, while providing the freedom to make required changes. While results are possible in these cases, implementation can be hard, in part due to difficulties with getting the details right.

The following 'decision tree' for SOE reform, proposed by the World Bank, summarises the above (see Figure 8.1).

Much like the framework provided for implementing privatisation, the decision tree provided by the Bank is informative and helpful. A problem once again is that most developing and transition economies may be willing to adopt SOE reform, yet lack the overall competence. From the three factors discussed in the tree (implementing other reforms, reduce worker opposition and improve reputation, boost credibility), implementing other reforms is the important one. These should aim in the main to creating competences, markets, skills, entrepreneurship, advantages, and institutional changes. Reducing worker opposition could be a mixed blessing, as consensus is important for productivity. Reputation and credibility should also be acquired in a way that carries people with it, not against it.

In the formerly centrally planned economies, innovative methods are adopted to address these problems, which include giving enterprises away, providing 'sweeteners' to labour, establishing mutual funds or holding companies, setting-up reviewing agencies to guard against the nomenclature making off with the assets in very low prices, decentralisation of implementation and the development of social nets and unemployment insurance (Kikeri et al., 1992). Methods of mass privatisation involve turning over ownership to current management or

266   Christos N. Pitelis

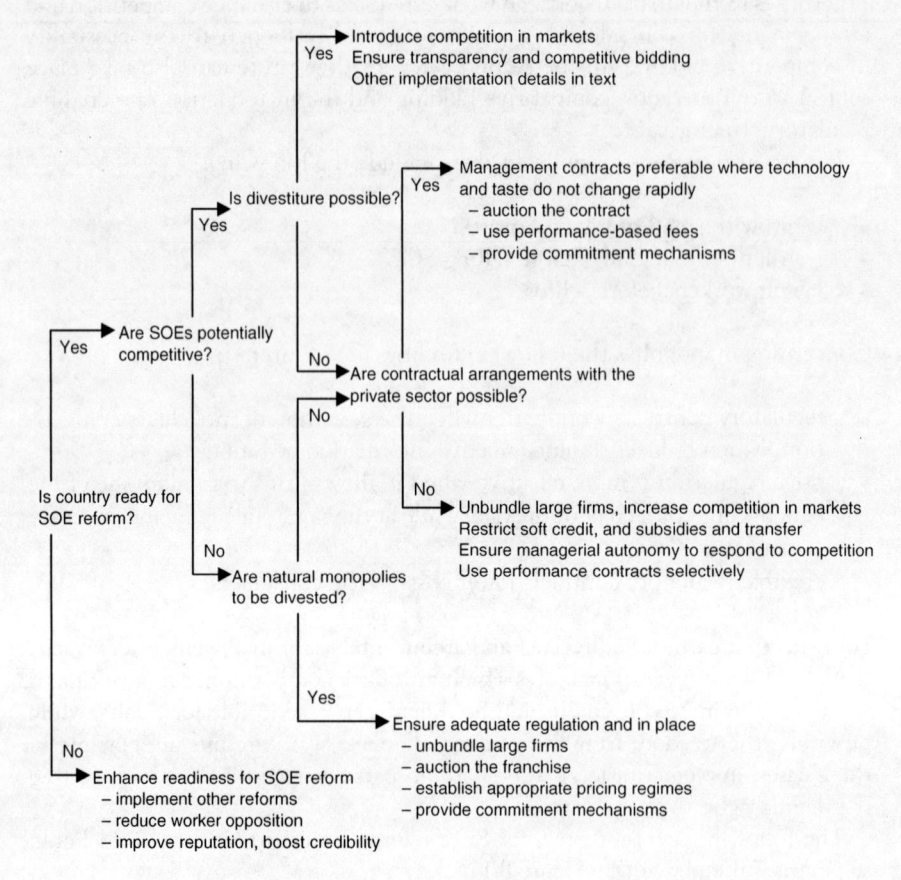

Figure 8.1 A decision tree for SOE reform.

Source: The World Bank (1995) *Bureaucrats in Business*, p. 16.

labour; making enterprises joint stock and distributing a percentage of share to management or labour; creating mutual funds-cum-holding companies and distributing shares to the public; distributing vouchers to the public or companies that entitle them to bid directly on share of individual firms; variations of the above, etc. (ibid., p. 88).

Implementing regulation

Two important issues pertaining to successful regulation is, first, the setting-up of regulatory bodies and, second, the devising of regulatory systems. The former is likely to be resource-heavy. It requires real commitments to resource such bodies with people with knowledge, competence and independence. This can be problematic in developed countries, and more so in developing ones. Governments often find it difficult for their own appointees to ignore their wishes. Yet, if such conditions are not satisfied, it is most unlikely that regulation will lead to results.

This questions the very decision to privatise. The crux of the matter is that privatisation without regulation in cases of firms with dominant positions can be harmful, so implementing privatisation without regulations can defeat the very reason for privatisation.

Assuming that the issue of the regulatory body is addressed, an important issue is to devise regulatory system(s) which satisfy the objectives of the regulators, that is, sustainable competitiveness with profitable, innovative firms and non-exploited consumers. Best practice here is provided by Britain. When faced with the privatisation of British Telecom (BT) and other utilities, the British government was suggested the following choices by Stephen Littlechild.

1. *Rate of return regulation.* This involves the regulation of the profits of utilities, on the basis of an allowed rate of return on assets. This is easy to administer and enforce and provides transparency, but it provides perverse incentives to companies, both to increase their rate base and not to seek cost reductions (the Averch–Johnson effect), see Pollitt, 2000.
2. *The output related profit levy.* This involves having a profit tax system for utilities with declining tax rates for higher output levels. It is aimed to encourage regulated firms to produce more output than under a constant tax rate, so as to maximise its profits.
3. *Retail price index (RPI-X).* Proposed by Littlechild, it involves setting a maximum price, which can be increased by the rate of inflation (as measured by the RPI minus some factor X, which captures future productivity growth expectations. It is simple and provides cost minimisation and productivity enhancement incentives.

The British government accepted RPI-X and set up a regulatory agency (the Office of Telecommunications, OFTEL), to oversee the regulation of prices and services provided by BT. Initially applied for BT, the system was later applied to other utilities. It 'has been evolving, but has provided transparency in regulation and the freedom for each regulator to focus on the regulatory issues particular to his or her industry' (Pollitt, 2000, p. 10).

Pollitt discusses in detail the various issues emerging from the experience of regulation in Britain. As already noted, such experiences may be hard to replicate, however, and the setting up of impartial regulatory bodies and the implementation of regulatory systems, often presupposes the very competences the absence of which defines the term developing and transition economy. Crucially, regulation, just like privatisation, should not be seen as an end in itself, and should be combined with other measures. These are discussed below.

Implementing domestic CPs

Much like the case of regulation, implementing CPs requires setting up competition or anti-trust authorities. Here, too crucial is the selection of competent, knowledgeable and independently minded individuals. They should aim at ensuring sustainable competitiveness, an important determinant of which is competition. They should not be 'captive' to business or other interests, and should coordinate with regulatory bodies and other authorities, domestically and internationally. They should be able to devise clearly articulated and transparent rules, concerning the acquisition and abuse of dominant positions and vehicles for their attainance, such as M&As, as well as restrictive practices. At the same time they should recognise that 'competition policy is not a panacea for competitiveness; competitiveness depends significantly on other factors such as investments in human capital and infrastructure' (The World Bank, 1998).

To the above we would add the other determinants of productivity discussed in the section on 'Introduction'. All these issues are closely interlinked. This implies the need for a systemic approach to CP that tries to simultaneously address the issues of doing, while also addressing the prerequisites, such as capability building.

Building on our earlier discussion, domestic CPs should not be linked to the degree of competition in industries, but should aim at improving productivity and efficient resource allocation. A prerequisite to achieving this is to encourage inter- and intra-firm competition so as to nurture conditions favourable to the creation of new ideas, techniques, products, processes, organisational and institutional forms, and, moreover to best exploit for this purpose the information-providing (and enhancing) attributes of economic organisations, notably markets, firms, states and people at large. CP should provide incentives, support, mechanisms and institutions for achieving productivity and competitiveness, for example, through linkages, joint inputs, and resource mobility (Audretch, 1998). They should address 'state' capture by sectional interests – in part through striving for conditions of contestability in private and (up to a point) political markets[3] – and a plurality of institutional and organisational forms, including, for example, support for (clusters of) SMEs, see below. Pluralism can also enhance the generation and use of new knowledge.

The exact measures, which need to be taken to achieve the above, can vary according to the conditions prevailing in every country. For example, the recognition of the benefits from cooperation, and therefore the need to ensure competition and cooperation, suggest the need for measures facilitating the 'clustering' of SMEs, see Best (1990), Best and Forrant (1996). 'Clusters' of SMEs can also be a potent source of indigenous development for LFRs, countering a dependence on TNCs and can themselves be a determinant of inward investment (see Box 8.3).

For the successful implementation of CP, a crucial factor is the institutional framework. For North, 'the central issue of economic history and of economic development is to account for the evolution of political and economic institutions that create an economic environment that induces increasing productivity' (North, 1991, p. 98). Also, the analysis of institutions and institutional change 'offer the promise of dramatic new understanding of economic performance and economic change' (North, 1991, p. 111). This is particularly important for developing countries.

Examples of required institutional measures include the delineation and enforcement of property rights and a pluralism of organisational forms and ownership structures, which exploit existing and generate new knowledge through economies of pluralism. Important is also an attempt to promote attitudes, values and generally culture conducive to dynamic competitiveness through innovativeness, thus to productivity, growth and convergence. All these are easier said

Box 8.3 Scale, clusters and competition

> In developing countries, some scale is essential for efficiency through economies of scale and learning. M&As is often an easy way to do this and to rationalise, especially in declining sectors. A minimum efficient scale is also essential to compete in export markets, often a *sine-qua-non* for small developing countries. An alternative to acquiring the benefits of scale can be through clustering (see Chapter 5). Theory and our own experience with consulting suggest the following best practice.
>
> 1. Explore the possibilities of clustering. When these are positive, develop clusters.
> 2. Allow M&As of SMEs where clustering is not possible, and benefits from minimum scale are required for competitiveness.
> 3. Promote competition through removing barriers to domestic competition, import competition and export rivalry.
> 4. Regulate any natural monopolies.

than done. A way through which these can be achieved is with the government assuming the role of a catalyst, by identifying and implementing in close cooperation with the private sector, changes proposed by those nearer at the action, for example, the private sector itself. Such *bottom-up* policies exploit dispersed knowledge and also promote subsidiarity and democracy. Exact actions, however, should be based on an analysis of each particular case. This is beyond the scope of this chapter, but the following methodology can be proposed.[4] First, a consensually agreed upon theoretical framework. Second, an audit of the external (international) environment. Third, and audit of the internal (national) environment. Fourth, deciding the *direction* of the strategy. Fifth, its dimensions. Sixth, the required actions. Seventh, addressing the issues of prerequisites, resources and mechanisms for implementation. Eighth, *feasible* actions. Ninth, control-evaluation. Tenth, new actions for implementation.

To conclude, domestic CP should focus on the nurturing of institutions, mechanisms and organisations which foster dynamic efficiency, productivity and growth, competition and cooperation (co-opetition) other than price collusion.[5]

Concerning the degree of competition *per se*, measures taken to promote domestic competition should work alongside import competition and export rivalry. When steps are taken to support domestic industry, these should not be allowed to consolidate and become anti-incentives for improved competitiveness. A clear phasing out strategy, should be in place. Measures to remove barriers to mobility are essential in this respect, as there seems to be the need to coordinate entry and exit policies (Frydman *et al.*, 1999). In this context, mergers should be discouraged when there is risk of monopoly power and (strategic) barriers to entry, and only encouraged between SMEs, which, moreover cannot be clustered.

Co-opetition should be of the type described in the case of clusters. Clustering can and should be seen as a new form of competition and IP (Porter, 1998). They can provide locally based development and also be an attraction to inward investment. They are of the utmost importance in that they could allow for the simultaneous removal of the constraints outlined at the beginning of this section, in that they facilitate entrepreneurship, decentralisation and locationally specific advantages, simultaneously.

Overall in their complex interrelationships, the exploitation of knowledge through the existence of a plurality of institutional and organisational forms, the benefits of co-opetition also arising from these and appropriate CPs the related amelioration of the problem from state capture, and the parallel exploitation of the benefits of clustering, can enhance productivity and development.

To summarise, domestic CPs should aim at maximising the benefits from competition and cooperation (co-opetition) for innovation and productivity. As noted, such benefits can fail to materialise when firms use market power to restrain competition. In this context, mergers that can lead to the acquisition of monopoly power should be discouraged, as well as restrictive and collusive practices. When support of infant industry is adopted, for infant industries and related reasons, there should be a clear and well understood phasing out clause. Protectionism can be a most potent disincentive for firms to become internationally competitive (Kikeri et al., 1992).

Summary and conclusions

We discussed privatisation, regulation and domestic CPs in theory and in practice, with an eye to drawing lessons from theory and experience on conceptual and implementation issues pertaining to these topics, of use for developing country policy makers.

In brief, we have found out that CPs are often motivated by neoclassical ideas, which are currently challenged by alternative views. In practice, CPs varied between and within countries and were often inconsistent with their alleged objectives. Privatisation has been an important policy issue, trying to address the problem of government failure. Interestingly, it has led to resource-intensive and institutionally demanding needs for regulation. There is an ongoing controversy on the evidence of privatisation, yet on balance support for the idea that it has led to improved efficiency. The evidence on regulation is less sanguine.

Developing countries often lack the competences required for successful implementation of privatisation, regulation and CP. In the absence of such competence proceeding with privatisation may fail to lead to benefits. In addition, for these countries it is arguably more crucial to develop markets, entrepreneurship and competences, including institutional and social capital, thus to privatise.

In this context, domestic CP, privatisation and regulation overall and in developing countries should all be seen within the broader context of enhancing productivity and competitiveness. At the theoretical level, domestic CP should aim at maximising the net benefits from co-opetition. In this context, privatisation can be a potent means of addressing problems of government failures, introducing competition and enhancing incentives for productivity. Establishing a regulatory framework can be important for the benefits of privatisation to realise; as it is to learn from experience. It is important to observe that the road to productivity and competitiveness is not one-way. Developing countries should exploit the informational benefits from the existence of a plurality of institutional and organisational forms, to include clusters and (even!) SOEs (see Pitelis, 1998). Theory and history suggest there are no panaceas. Domestic CPs, privatisation and regulation should be combined in a way that aims at enhancing innovation and productivity, rather than finding an optimal state. Privatisation should be a means to an end. If not properly conceived and implemented, and in the absence of a suitable regulatory framework, its benefits could be severely compromised. Mistakes are bound to occur.

Notes

1 Industry refers in the main to manufacturing. This, however, tends to recede, given an emerging fuzziness of the boundaries between manufacturing and services.
2 Michie (1997) observes that even 'quangos' (quasi-autonomous non-governmental organisations) and 'quasi-markets', such as those created in the British National Health Service (by splitting hospitals as service providers from local authorities as purchasers of services) have contributed to additional government regulation and intervention.
3 There can be too much contestability in public sector markets, in that it can increase the dependence of politicians, bureaucrats, etc. to pressures by organised interest groups, leading to regulatory capture.
4 This is based on the author's own experience with policy making in Greece, where he has coordinated the 'Future of Greek Industry Project', a consensus-based, bottom-up industrial strategy, orchestrated by the government and supported by the major social partners, see the section 'Summary and conclusions'.
5 This need not exclude (threats to) protectionism, either, both in support of such players and as a means of ensuring fair and open trade.

References

Adam, C., Cavendish, W. and Mistry, P.S. (1992), *Adjusting Privatisation: Case Studies from Developing Countries*, James Currey, Islington, UK, Ian Randle Publishers, Kingston, Jamaica and Heinemann Educational Books, Portsmouth, NH.
Alchian, A. and Demsetz, H. (1982), 'Production, Information Costs and Economic Organization', *American Economic Review*, 62(5): 888–895.
Audretch, D.B. (ed.) (1998), *Industrial Policy and Competitive Advantage. Volume 1: The Mandate for Industrial Policy*, Edward Elgar, Cheltenham.
Bacon, R. and Eltis, W. (1986), *Britain's Economic Problem: Too Few Producers*, Macmillan, London.
Bain, J.S. (1956), *Barriers to New Competition*, Harvard University Press, Cambridge, MA.
Baumol, W. (1982), 'Contestable Markets: An Uprising in the Theory of Industry Structure', *American Economic Review*, 82: 1–15.
Baumol, W. (1993), *Entrepreneurship, Management and the Structure of Pay-offs*, MIT Press, Cambridge, MA.
Best, M. (1990), *The New Competition: Institutions for Industrial Restructuring*, Polity Press, Oxford.
Best, M. and Forrant, R. (1996), 'Creating Industrial Capacity: Pentagon-Led versus Production-Led Industrial Policies', in J. Michie and John Grieve Smith (eds), *Creating Industrial Capacity, Towards Full Employment*, Oxford University Press, Oxford.

Buigues, P.-A. and Jacquemin, A. (1998), 'Structural Interdependence Between the European Union and the United States: Technological Positions', Paper presented at the International Conference on Industrial Policy for Europe, Royal Institute of International Affairs, London, 26–28 June.

Chandler, A.D. (1986), 'Technological and Organisational Underpinnings of Modern Industrial Multinational Enterprise: The Dynamics of Competitive Advantage', in A. Teichova, M. Levy-Leboyer and H. Nussmaum (eds), *Multinational Enterprise in Historical Perspective*, Cambridge University Press, Cambridge, chapter 2, pp. 30–54.

Clarke, T. and Pitelis, C.N. (eds) (1993), *The Political Economy of Privatization*, Routledge, London.

Coase, R.H. (1937), 'The Nature of the Firm', *Economica*, 4: 386–405.

Cowling, K. and Waterson, M. (1976), 'Price Cost Margins and Market Structure', *Economica*, 43: 267–274.

Dasgupta, P. (1986), 'Positive Freedoms, Markets and the Welfare State', *Oxford Review of Economic Policy*, 2(2): 25–36.

Dixit, A. (1982), 'Recent Developments in Oligopoly Theory', *American Economic Review*, 72(2): 12–17.

Donaldson, D.J. and Wagie, D.M. (1995), *Privatisation: Principles and Practice*, International Finance Corporation, The World Bank, Washington, DC.

Eggertson, Y. (1990), *Economic Behaviour and Institutions*, Cambridge University Press, Cambridge.

European Commission (1994), *Return to Growth, Full Employment and Convergence*.

European Commission (1996), 'The Competitiveness of European Industry', Working Document of Commission Services, November.

Feldstein, M. (1984), 'Social Security, Induced Retirement and Aggregate Capital Accumulation in the United States', *Journal of Political Economy*, 82: 905–926.

Friedman, M. (1962), *Capitalism and Freedom*, University of Chicago Press, Chicago.

Friedman, M. and Friedman, R. (1980), *Free to Choose*, Secker and Warburg, London.

Frydman, R., Gray, C., Hessel, M. and Papaczynski, A. (1999), 'When does Privatisation Work? The Impact of Private Ownership on Corporate Performance in the Transition Economies', *The Quarterly Journal of Economics*, November, 1153–1191.

Gouph, I. (1989), *The Political Economy of the Welfare State*, Macmillan Educational, London.

Hannah, L. and Kay, J.A. (1988), *Concentration in Modern Industry*, Macmillan, London.

Hayek, F.A. (1945), 'The Use of Knowledge in Society', *American Economic Review*, 35: 519–530.

Heald, D. (1983), *Public Expenditure: Its Defence and Reform*, Martin Robertson, Oxford.

Ioannides, S. (1992), *The Market, Competition and Democracy*, Edward Elgar, Cheltenham.

Josef, C.J. (1996), 'Privatisation is Transition. Or is it?', *Journal of Economic Perspectives*, 10(2): 67–86.

Kay, J. (1993), *Foundations of Corporate Success*, Oxford University Press, Oxford.

Kay, J.A. and Silberston, Z.A. (1984), 'The New Industrial Policy – Privatization and Competition', *Midland Bank Review*, Spring, 8–16.

Kay, J. Mayer, C. and Thompson, D. (1986), *Privatization and Regulation: The UK Experience*, Clarendon Press, Oxford.

Kikeri, S., Nellis, J. and Shirley, M. (1992), *Privatisation: The Lessons of Experience*. The World Bank, Washington, DC.

Kitson, M. and Michie, J. (1996), 'Does Manufacturing Matter?', Paper presented at the Eighth Annual Conference of the European Association for Evolutionary Political Economy (EAEPE), Antwerp, Belgium.

Marx, K. (1959), *Capital*, Lawrence and Wishart, London.

Megginson, W. and Netter, J.M. (2001), 'From State to Market: A Survey of Empirical Studies on Privatization', *Journal of Economic Literature*, 39: 321–389.

Michie, J. (1996), 'Privatisation and Regulation', *International Encyclopaedia of Business and Management*, Routledge, London, pp. 4097–4106.

Michie, J. and Pitelis, C. (1996), 'Demand and Supply-Side Approaches to Economic Policy', Paper presented at the Eighth Annual Conference of the European Association for Evolutionary Political Economy (EAEPE), Antwerp, Belgium.

Modigliani, F. (1958), 'New Developments on the Oligopoly Front', *Journal of Political Economy*, 66: 215–232.

Mueller, D.C. (1989), *Public Choice II. A Revised Edition of Public Choice*, Cambridge University Press, Cambridge.

Nolan, P. and O'Donnell, K. (1995), 'Industrial Relations and Productivity', mimeo, University of Leeds.

North, D.C. (1981), *Structure and Change in Economic History*, Norton, London & New York.

North, D.C. (1991), 'Institutions', *Journal of Economic Perspectives*, 5(1): 98–112.

Ostry, S. (1998), 'Technology, Productivity and the Multinational Enterprise', *Journal of International Business Studies*, 29(1): 85–99.

Penrose, E. (1959), *The Theory of the Growth of the Firm*, Basil Blackwell, Oxford.

Pitelis, C.N. (1991), *Market and Non-Market Hierarchies*, Basil Blackwell, Oxford.

Pitelis, C.N. (1994), 'Industrial Strategy: For Britain in Europe in the World', *Journal of Economic Studies*, 21(5): 2–92.

Pitelis, C.N. (1998), 'Productivity, Competitiveness and Convergence in the European Economy', *Contributions to Political Economy*, 18: 1–20.

Pollitt, M. (2000), 'A Survey of the Liberalization of Public Enterprises in the UK since 1979', in M. Kugumi and M. Tsuji (eds), *Privatisation Deregulation of Economic Efficiency*', Edward Elgar, Cheltenham, UK.

Porter, M.E. (1980), *Competitive Strategy*, Free Press, New York, NY.

Porter, M.E. (1990), *The Competitive Advantage of Nations*, Macmillan, Basingstoke.

Porter, M.E. (1998), 'Clusters and the New Economics of Competition', *Harvard Business Review*, November/December.

Rees, R. (1986), *Public Enterprise Economics*, Philip Allan, Oxford.

Richardson, G. (1972), 'The Organisation of Industry', *Economic Journal*, 82: 883–896.

Scherer, F.M. and Ross, D. (1990), *Industrial Market Structure and Economic Performance* (3rd edition), Houghton Mifflin, Boston.

Schumpeter, J. (1942), *Capitalism, Socialism and Democracy*, (5th edition), 1988, Unwin Hyman, London.

Smith, A. (1977), *An Enquiry into the Nature and Causes of the Wealth of Nations*, London, Dent.

Stigler, G. (1988), 'The Effect of Government on Economic Efficiency', *Business Economics*, 23: 8–13.

Stiglitz, J.E. (1986), *Economics of the Public Sector*, Norton, New York.

The World Bank (1995), *Bureaucrats in Business: The Economics and Politics of Government Ownership*, World Bank policy research report, Washington, DC.

The World Bank (1998), *Competition Policy in a Global Economy: An Interpretive Survey*, The World Bank, Washington, DC.

Vickers, J. and Yarrow, G. (1988), *Privatization: An Economic Analysis*, The MIT Press, London.

Wolf, C. (1989), 'A Theory of Non-market Behaviour: Framework for Implementation Analysis', *Journal of Law and Economics*, 22(1): 108–140.

Yarrow, G. (1986), 'Privatization in Theory and Practice', *Economic Policy*, 2: 324–377.

9 International trade policies

Sheila Page

Introduction

The policy environment for a country's international trade or for a trader depends only partly on direct interventions on trade by its own government. Many policies towards aspects of the economy: macroeconomic, infrastructure, sectoral, can be as significant as direct policy towards trade or towards firms involved in trade. And the trade and other policies by trading partners will have the same types of effects as policies by the home government. In an interrelated economy, at country and world levels, the policies need not even be directed at a partner; a policy towards a third party may change the relative attractiveness or costs of the traded sector in any country. International institutions may impose or forbid certain types of policy, both within and between countries. Trade is for most countries, and for many firms, a marginal activity: the result of a balance between domestic and foreign production, and therefore of all the forces acting on each of these. This position at the intersection of a range of policies and production conditions means that assessing the 'success' or 'effectiveness' of a policy may also require multiple criteria. A policy must promote, or at least not hinder, a variety of objectives. The objectives, the weighting of each, and the effects to be expected from them will change as a country develops, as the international environment changes, and as firms evolve.

This chapter is directed at identifying the role and effects of national policies specifically directed at trade. It will first examine briefly the evolution of our understanding of the relation between trade and development. It must then look at the policy and economic constraints and opportunities for country policy which are created by the international environment. After this background, it will examine the trade policies available within a country. A section on 'The role of national policies' will examine the role of international policy and the possibilities available to a country to influence it, before concluding with an assessment of the policies available.

The role of trade in development

Background

Trade was first considered a central element in a country's development path in the 1950s and 1960s; at the same time, planning was becoming the 'normal' way of development. Both these owed their new significance to the experience of the developed countries in the 1930s and then in the Second World War: the breakdown of the trading system in the 1930s and the further disruption by war had had a serious effect on many of them, and the need to mobilise all national resources in the war had underlined the importance and demonstrated the feasibility of planning. All the countries at war had used active government intervention to direct production sectors to maximise performance during the war. And there was also the tradition of public works from the depression of the 1930s. The developing countries, many becoming independent in the late 1950s and 1960s, appeared to face a single possibility if they wanted rapid economic growth. There were no examples of countries that were still clearly 'developing', but competing against 'industrial countries' in some industries. Empirical observation, and the history of primary product consumption within countries, suggested that demand for their primary exports would grow less rapidly than average demand in the developed countries, and much less rapidly than their objectives for their own growth. Therefore, it seemed that the only path open to them was to continue to specialise in primary products for export, but concentrate on increasing production of other goods for home consumption. Due to the constraint from the expected limited growth in demand for their exports, they would have to substitute an increasing proportion of their imports with home production to avoid having foreign exchange as a constraint on their growth. In terms of trade policy, as export promotion was (by assumption) not likely to be successful, this meant a concentration on policies to control imports, not only their quantity but their composition, to concentrate limited resources on the goods least replaceable by local production. Following the policy precedents of the 1940s, they would do this by active intervention on trade and production.

The importance of the experience of the newly industrialised countries (NICs) (the first generation of successful Asian economies, notably Hong Kong, Singapore, South Korea and Taiwan) was that they showed that it was not necessary for a country to develop an integrated national industry before competing with developed countries in manufactures. It was possible to specialise in exports of one or a few manufactured goods, and therefore secure a better export market prospect than from primary goods; there seemed to be an alternative strategy. But later analysis has emphasised that in most cases the successful exporters had first had a period of import substitution. There remains disagreement about whether this is because they were mistaken, and then found the better solution, or because the import substituting period was necessary as a preparation. Conventional trade theory and the macroeconomic effects it predicts from trade are not a guide on this, and the view that exporting itself is an engine of development lacks good empirical or theoretical backing.

Conventional theory. In traditional terms, opening trade should raise a country's income (welfare) by permitting it to change the composition of its output to a more efficient structure, that is, permitting it to specialise according to comparative advantage. This assumes either that prices are already operating as correct signals, or that they are altered to remove distortions as part of the opening of the economy. The increase is a simple shift to a higher point, not an increase in growth. This traditional efficiency gain is problematic faced with increasing evidence that external openness is not necessarily associated with reduced domestic price distortion. The effects on macroeconomic stability are ambiguous: liberalising increases the number of potential shocks that could affect the economy, but a large number may itself ensure some offsetting effects, and existence of trade flows provides a potential way for offsetting domestic shocks.

Export-led growth. But much of the literature on the role of the external sector moved beyond these efficiency effects.[1] The apparent association between high and rapidly growing exports and rapid growth of manufacturing and of total output suggested that a policy of opening an economy to external influences (liberalisation) or even a policy of deliberately biasing growth towards exports (export promotion) could improve rates of investment and growth, and raise efficiency not only through the conventional efficiency (and multiplier) effects, from efficient allocation of resources and increases in aggregate demand, but by increasing the 'dynamic efficiency of the economy'. What are the steps in this argument?

One possible case is if there is unemployed capacity and raising exports is the only way of stimulating growth. This could be because it is impossible to raise other domestic sources of demand directly or because a balance of payments constraint is binding on them (effectively the same argument used earlier for import substitution). But the argument usually used is a more radical extension: that external demand contributes to development by stimulating structural or behavioural changes, not merely a one-off gain from reallocation (as assumed by the traditional theory). Clearly the transition to efficiency could be prolonged (because of the great number of new possible markets and types of production), thus giving a period of growth, as a sequence of one-off changes. But the argument goes further, to assert that exposure to competition, from imports and in export markets, increases the X efficiency of firms, not simply by providing improved information or access to technology (these can be done without trade), but because the threat of losing markets (and profits) is more of an incentive to change than the potential to increase them.

The variant view, that import substitution and export promotion are steps in a sequence also implicitly takes a view that what is needed is to develop the efficiency of firms. The argument is that an individual industry or firm needs to develop skills, both training specific to the sector and managerial or marketing skills; perhaps it needs to reach a minimum level of output, if there are economies of scale. Therefore, it is necessary to start in a market which is 'easier', both because of less competition and because it is more familiar. But import substitution cannot continue indefinitely because structural change becomes more complex; the new production required to substitute for the inputs and the capital requirements of the first import substitutes may be increasingly difficult to produce in the country; thus a new import and foreign exchange constraint emerges, but by this time a country (or some firms) may be sufficiently mature to export.

If these are the ways in which trade contributes to development, we must think in terms of actions that have a direct impact on firms and entrepreneurs, not merely about action on the flows of goods or investment. But the role of incentives and responses also raises the question of how to make the transition from import substitution to export promotion. Are the incentives and responses appropriate to succeed in one adaptable to the other?

The possibility that there is a sequence of correct policies raises the question of whether there are general rules about the role of trade in development, or whether the particular characteristics of the country or the nature of the external environment need to be considered. The economic size of a country will at least influence the length of time an import substitution strategy may be viable; for some small countries this may be a negligible period, while large ones can have a long period. A small country may need to move to exporting before it is 'ready', and therefore may be more likely to need special measures to help exports. A large country may find it easier to find alternative growth paths to exports. The level of development will influence how effectively either policy can be pursued. Choosing which sector should be promoted in an export strategy requires a competent planning agency (protected from undue influence by private interests), able to assess either the economic viability of a proposal or the likelihood of success of the proposer. Administering a system of training or special tax or subsidy incentives or setting up systems to encourage coordination among firms requires as a basis a well-established fiscal system. Assisting with access to credit or foreign exchange requires good monetary institutions and instruments. The rate of growth of external demand will alter the incentives for either an import substituting or an export promoting strategy. Slow growth increases the pressure on income from traditional exports, making finding an alternative more pressing, but it potentially reduces the return also to new exports. High indebtedness and therefore high interest payments can be seen as a particularly inelastic and unsubstitutable import payment, requiring a transition to an export strategy.

Trade theory argues that increasing the openness of an economy (if non-distorting) improves the return to factors which are less scarce in the country than in the world as a whole (it moves their price, and therefore their return, nearer to the levels in the rest of the world). For countries with abundant labour, this is likely to mean an improvement in income distribution towards wages, but for those where natural resources, whether agricultural, mineral, or scenic (in the case of tourism), are the principal advantage, it may instead shift the distribution towards returns to holders of these, that is towards profits and rents.

The international regime as a constraint on strategy

Trade has been regulated at international level since the establishment of GATT in 1947, converted in 1995 to the WTO. The most well-discussed effect is from restrictions on countries' own trade policies. The restrictions on certain types of trade policy and the pressure to reduce even 'WTO-compatible' barriers to goods, and now services, remove certain policy tools. Indirectly, they also affect fiscal policy (tariff revenue losses). As regulations extend to domestic measures with an effect on trade (notably to subsidies or too-relaxed intellectual property rules which could assist exports or substitute for imports), other tools are removed. While the least developed countries (LLDCs) are exempt from most of these restrictions, and developing countries much less regulated than developed, the pressure to conform is a *de facto* restriction even on those technically exempt.[2]

How important is this? In the past, it would have conflicted with the use of interventionist policies on trade. But the NICs have already passed the point of heavy dependence on trade controls, and the 'new generation' of emerging countries has turned against these policies for the same reasons, whether of intellectual conviction or fashion, as the developed. There is, however, potential conflict if a country should ever want to reverse a liberalised strategy or promote a sector, so the new strictness and increased coverage of the international regulatory regime do limit the policy choice available to today's developing countries.

In its moves on standards and intellectual property, the last Trade Round (the Uruguay Round, ending in 1994) moved into new forms of international regulation. The increasing complexity of goods traded, and the increase in the share of manufactures, and also in the sophistication within manufactures, have been important forces for the imposition of minimum quality or other standards, reinforced by rising incomes, and therefore rising standards for health and safety. As national rules, for example, on product standards or intellectual property, have evolved and the share of trade has risen for most countries, international regulations have been needed to avoid conflicts. The WTO rules on areas like sanitary and phytosanitary standards for agricultural products specified general international standards which have to be used in most circumstances, except where a country can make a scientific case for its own rules. They also forbid discrimination between imports and domestic production. This goes beyond the traditional GATT rule of Most Favoured Nation (no discrimination by importers among suppliers) to National Treatment (no discrimination between imports and home production) and even further, with minimum standards, to international limits on national governments' behaviour. This was, therefore, a significant extension of international limits on national policy. Of course, the regulation of tariffs that is the oldest part of GATT could be considered an international standard, but extending rules to specifying national regulations greatly strengthened the WTO regime.

The requirements on standards may push countries into putting more resources into setting, administering and enforcing them, at an earlier stage, than would be strictly efficient on the grounds of their own needs (assuming that developed countries, which did not adopt either the standards or international conformity until recently, made rational decisions). They also offer savings: of being able to move directly to an international standard, rather than having to adopt a national one and then adapt. This is an area where assistance is available, and there are efficiency gains. It could be a permanent, rather than adjustment, issue if a country wants or needs national discretion on the nature of the standard, but this is less common than for tariff policy.

The requirements on intellectual property do impose real costs, not only to users of the technology, but to national income, as many countries have adopted cheap or free transfer of technology as a tool for accelerating technical innovation. There is no direct compensation (although there may be technical assistance in implementing the rules), and only temporary exemption for least developed countries. This is an area where even countries with strict standards have had very different rules, in the length of period of protection, in the nature of that protection and in provisions for new producers, so that the advantages of international standardisation for efficiency are not clear-cut. Countries have lost not only a general tool of development, but the ability to vary its application to suit their circumstances and policies.

Despite the change in fashion from such approaches in developed countries, planning and intervention remain the approach of the international agencies which influence the developing countries (although with stabilisation rather than development targets), with both the World Bank and the IMF committed to strong fiscal and monetary measures to achieve clear objectives. Similarly the donors support stabilisation and macroeconomic balance targets, and in some cases also encourage distributional or poverty objectives. But while some macroeconomic policies are thus consistent with the international constraints, there is a bias now against sectoral targeting, and thus neither the emphasis on leading industries of the import substituting model nor the paths followed by the NICs of concentrating on particular export sectors would be supported. In their actions on their own trade, however, the donor countries remain very sector-specific, with strongly differentiated tariffs, in particular peaks for clothing, some agriculture, and metal products, and non-tariff barriers and quota systems remain for clothing and agriculture. The preference schemes for developing countries, notably the Generalised System of Preferences (GSP), are also very differentiated, with the European GSP, for example, having four classes, according to degree of sensitivity, defined differently for industrial and agricultural goods, three classes of countries, by level of income, and further special treatment for countries with particularly specialised exports or with 'good behaviour' in certain environment or labour conventions.[3] In addition, there are special arrangements for various groups, including the ACP, the Andean drug countries, South Africa. By 1997, it offered free access to least developed countries for most manufactured exports (including Multi-Fibre Arrangement [MFA] goods, but with exclusions in leather, an important export for the least developed) and most agricultural goods (with slightly more exclusions than under Lomé). Other countries have graduated countries which became competitive with developed countries (wholly or for particular products), or added new groups of countries, notably South Africa and the former centrally planned, and also introduced criteria based on environment, labour or human rights standards and greater preferences for the least developed countries. The result is that developing countries face a trading environment which is highly policy-driven and differentiated, in contrast to the implicit assumption in industrial countries and in international institutions that it is largely market-driven, with efficient price incentives.

The international environment means that on the one hand, developing countries face constraints on the policy they can use, but on the other, they cannot assume that they face a non-distorted market in which the benefits of following market-based policies are achievable.

The role of national policies

Policies directed at trade

The traditional tools for influencing trade are direct controls on volume, through quotas, or changes in its price, through import tariffs or export subsidies.

There are other measures which serve the same purpose, indirectly, or avoid unintended interactions. Public procurement restrictions, requirements to purchase local goods for public contracts, restrict the quantity of imports. Local content requirements, requiring foreign investors, contractors to the government, or others which the government can regulate, to use some proportion of local goods, again restrict the quantity of imports. Tax incentives or marketing assistance to exports are the equivalent of subsidies. Duty-drawback schemes which exempt imports used in exports from some or all of normal import duties are intended to restrict imports from imposing costs in exports. Conventionally it is assumed that measures influencing prices are less distorting than those controlling quantities directly, and that general measures are less distorting than measures directed at particular goods or activities. As the discussion in the section on 'The role of trade in development' indicates, however, there may be advantages (and disadvantages) from trade other than the simple cost saving, efficiency ones which this distinction implies. More practically, if we assume that governments are behaving rationally, then we must assume that they are choosing their objective (a particular level or composition of imports or exports) and then choosing the instrument, rather than choosing the instrument, and then analysing the result.

Quotas have been extensively used by developing countries. Although most are illegal under GATT rules (Article XI restricts trade measures to 'duties, taxes or other charges'), developed countries would have been in a weak position in challenging them because they also used them extensively in agricultural trade (under a general exemption of agriculture from GATT controls) and under the MFA (a recognised derogation from GATT). The Uruguay Round brought both agriculture and clothing and textiles under GATT control, permitting increased pressure on developing countries to comply with the rules. In addition, the growth in membership of developing countries during the Round had an effect: those applying for membership were treated more severely than existing members, and then this could be used as justification for requiring greater compliance by existing members, to be equitable. The Round before the Uruguay Round, the Tokyo Round, had made the first effort to regulate non-tariff barriers.

The Uruguay Round came after a decade in which the developed countries had increasingly used non-traditional non-tariff measures (not only the MFA, but 'voluntary' export restraints, temporary import controls, etc.). The introduction of a 'standstill' on these during the Round, followed by the increased strength of WTO disciplines (discussed below) effectively ended new measures, and there has been a slow reduction in the existing measures. The MFA was to be phased out over ten years. The Round also restricted developing countries' use of direct controls. They had had greater latitude, under Part 4, to use these as balance-of-payments measures; this could no longer be used as a permanent exemption, and they were now expected to use tariffs instead. There had already been pressure on the more advanced to avoid using them; this was now extended to all developing countries (reinforced by the pressure towards trade liberalisation by the international financial institutions and by the growing belief by developing countries that these were not an efficient tool of development).

Tariffs have been under GATT discipline from the beginning, of course, but the control was less rigid than might appear. While countries were forbidden to increase any tariff that had been notified and 'bound' to the WTO, this left two widely used loopholes. Countries could notify tariffs without binding them, and they could bind them at a level substantially above the actually applied level (subject, in both cases, to other countries accepting this in the negotiations). Until the Uruguay Round developing countries did not normally bind their tariffs. But then, there was pressure to increase binding. Developed countries completed the binding of their manufactured tariffs (the share of products covered rose from 94 to 99 per cent) and, after 'tariffication' of their agricultural measures, bound these as well (increasing the share from 81 to 100 per cent). Developing countries were also encouraged to bind theirs: the proportion bound rose from 13 to 61 per cent; this was mainly accounted for by Latin America (100 per cent) and Asia (70 per cent). Africa, in general, did not increase bindings (or bound only at very high levels), so this mainly affects its exports, not its domestic trade policy. While for many, the bound rate was above the currently applied rate, giving some flexibility to raise rates, the margin was often of the order of 20 points, so that it is still an effective limit on major reversals of tariff policy. (Countries can break their binding, but only by offering equivalent compensation in other reductions.) There remains no restriction except what is acceptable to other countries in the bargaining on the level of tariffs or on their dispersion, although there has been pressure by the international financial institutions to reduce dispersion in favour of a small number of steps. It is therefore still possible to have a finely targeted policy, but may be difficult to alter it.

The Agreement on *Subsidies* and Countervailing Measures defined which subsidies were not allowed on traded goods. It exempted least developed countries (and countries with an income under $1,000) from the provisions on export subsidies, and gave other developing countries eight years to conform. For import replacing subsidies, least developed had eight years and other developing five. Developing countries are also allowed a stricter standard of proof in any complaint: for developed countries, there is a presumption that any subsidy equivalent to more than 5 per cent of the value of the product is damaging, but for developing countries it is necessary to prove damage. While some phasing out was to occur during these periods, the countries have no obligations, except on import-related subsides by 2003; they can apply for an extension (giving at least a year's notice) beyond this. Until then, their obligation is to notify the WTO of any subsidies. They need not notify permitted subsidies, for example, within the *de minimis* provisions (3 per cent for manufactures, 10 per cent for agriculture), or those specifically permitted, the green measures. But if a country does not notify a subsidy that does come under the regulations, then it loses its exemption period, leaving a dilemma if there are subsidies about which it was doubtful (although this becomes less important as 2000 and 2003 grow nearer). If countries have not notified subsidies, then if any are now challenged and found to be unacceptable subsidies, they will not be able to secure exemption. New subsidies are not allowed. This means that in the future any subsidies must not be directly related to exports. Other subsidies, based on product or region, for example, are allowed. In particular, income-support policies not linked to output were permitted, as are environmental programmes and domestic food aid. Certain assistance measures to promote agricultural and rural development were allowed in developing countries. Direct payments under production-limiting programmes, are also broadly exempt. The problem is that both these are more available to developed countries than to developing.

Under the Uruguay Round agreement, all *agricultural quantitative border measures* were to be replaced by tariffs, although these could be (and were) designed to provide substantially the same level of protection (and thus range up to 1,000 per cent in some developed countries). Then, tariffs and export subsidies were to be reduced by 36 per cent over six years by developed countries. Developing countries were allowed ten years, with a minimum reduction of 24 per cent. Least developed countries were not required to make any reduction. Domestic support was to be reduced by 20 per cent (developed countries), or 13.3 per cent (developing).

There is a plurilateral agreement (i.e. one which members of the WTO can choose whether to sign) on *government procurement*. Therefore it is still open to countries to use preferences for domestic suppliers (except in so far as developing countries are constrained by tied-aid provisions). This can be used as a way of supporting national services, and in particular to offset the distorting disadvantage of tied aid, but its potential cost is particularly high, as it directly increases the cost of providing those services considered sufficiently important to require government provision.

The general prohibition of controls on trade has been held to forbid restrictions tied to foreign investment (requirements for *local content or minimum exports*, for example), and again there has been increasing pressure during and since the Uruguay Round to reduce such measures. It would be controversial to introduce them now.

One form of export subsidy remains legal (presumably because it is extensively used by developed countries), *duty-drawback*. Countries are allowed to exempt imports used in exports from payment of import duties (or may refund the duty). This is even extended to allow countries to estimate the quantity of imports used, where imports and similar locally produced goods are used together. Countries which operate VAT schemes apply another subsidy, through the *exemption of exports from VAT*.

Monetary and fiscal policies

Some of these can be directly targeted at export or import flows, but general policies will also have an effect. Making *credit to exporters* more available or cheaper is clearly effectively a subsidy, but it has not been controlled by the WTO (it is a major instrument used by developed countries, and has been subject to a separate OECD limit on the degree of subsidy). If it is a more general measure, offering low cost credit to all producers or to the economy as a whole, it will still act as a subsidy, but this would certainly be permissible under trade rules because it would be non-discriminatory. The problem would be its cost to the economy, and as with any subsidy, the question is why it should be offered, if it is a general rather than a targeted intervention, and whether the distortions to price signals which it introduces vitiate any efficiency advantages of promoting trade policy.

As an alternative to subsidies, exporters (or all firms) can be given *exemption from taxes*, for a period or under other special conditions. As long as the conditions are not directly related to exports, these are allowable under WTO rules. The general tax regime in its impact on the economy can be used to promote home production or trade by altering the pressure of demand in the economy. The *exchange rate*, if the government is able to influence it, can be used to alter the relative price of imports and exports, and thus in the same way as a uniform tariff or subsidy. A lower exchange rate makes imports cost more relative to home goods (equivalent to a constant-level tariff) while making exports less expensive relative to competing goods (a subsidy) (Box 9.1). A too high exchange rate has the opposite effect. For all these macroeconomic measures trade is only one of

Box 9.1 Countertrade

> From time to time other types of government intervention are proposed: an example from the 1980s is 'countertrade': government organisation of product for product swap arrangements with another country. There can be no economic argument for this except an odd form of lack of market information: the selling and buying countries, for some reason, have a (correct) higher estimate of the value of their own products, and therefore gain advantages from trading with each other (which are not offset by the fact that each faces a higher valuation in the partner country). The non-market advantages may, however, be important in opening new markets or exporting new products: it attracts attention; it looks simpler than normal trade (this is in practice not the case, because of the difficulties of timing, valuation, security, etc.); and simply that it was (briefly) new and exciting. There is a potential case for governments trying to identify other 'gimmicks': there is probably an inherent reluctance to take the first step of improving and exporting, and if there is this inefficiency in the market, an action to encourage the first step may be efficient, even if in normal circumstances it would not be.

the objectives for which they are used. Is it the most important? And for some countries, is the economy sufficiently integrated and the administration sufficiently competent for measures to have the intended effect?[4]

Sectoral allocation of government spending

As well as measures directed at trade flows and indirect influences through the general stance of policy, government policy can affect trade semi-directly by promoting or discouraging sectors with particularly close connections to trade. The most obvious is infrastructure policies: trade depends more than the average on transport and communications, including fixed facilities, like roads and harbours, and services, like road, sea or air transport. High costs for these have the same effect in raising prices of either imports or domestic output as a tariff or an export tax. For large or land-locked countries, the costs are likely to be high if road or other internal infrastructure are poor (or if there are administrative or bureaucratic delays at the border of a land-locked country with its outlet to the sea). For small countries with ports, including islands, there may be costs if ports are inefficient, through poor infrastructure or government requirements. Both may be damaged by poor airport facilities or poor airline service: either could again be an infrastructure or a government regulation problem. For air services there is the additional difficulty in some countries that the national airline itself may be the subject of government protection (or neglect). Costs may also be high if the country's trade is too low to justify special stops by international transport and it is not on a recognised trade route, but this is more difficult to tackle by government action.

For all of these, there are potential conflicts between trade policy and other government objectives: to reduce spending, to avoid direct intervention in infrastructure (or to diminish this), and to protect other sectors, notably transport providers. If trade is the only objective, the government should ensure that investment is appropriate to the return (in additional income) compared to other types of investment and the cost of investment.

The interactions between trade and social infrastructure are similar, although it is of course much more difficult to identify the returns to investment. The Asian NICs placed a heavy weight on increasing education and training as part of their development programmes; other areas have had less emphasis. It is difficult, however, without political, historical, and social analysis to determine how much this difference was because of different assessments of the return and how much because of different national attitudes towards education. The connection in the opposite direction is that increased income because of growth allows higher 'consumption' of educational services. A third connection is the argument that growth, by increasing the demand for skilled labour, increases the demand for education, and this serves the national objective of improving the level of education in the population.

A more fundamental question is whether in some sense it is necessary to develop new types of behaviour, going beyond questions of education or training.

Under this view, there is some inadequacy, an underdeveloped ability to perceive or respond to existing opportunities, which holds a country to a point below its physically determined production frontier or budget-determined balance of payments constraint. The lack of analysis of what this is (it is embodied in old concepts like 'take-off' or newer ones like openness or readiness, not in operable models) means that it is difficult to know whether it is best tackled by appropriate education or training or by providing particularly strong stimuli to respond, and if so whether these stimuli should be through technology (industrial strategy), demand (macroeconomic policy), or competition (trade instruments).

Obstacles from government inefficiency have been mentioned in several cases, and it is clear from studies of a variety of developing countries that administrative, tax, and infrastructure provision can suffer as much from inexperience, lack of training, and lack of familiarity with best practice elsewhere as does private production. Some calculations have been done for the cost of such barriers to traders. Milner *et al.* (1999) found for Uganda that transport costs (including all the administrative costs of clearing borders) could provide effective protection as high as or higher than formal tariff barriers. For island or coastal countries, the costs may come both from inefficiencies of the customs and other government services and from inefficiencies in ports and cargo handling. The situation is likely to be worse for land-locked countries. Their goods will incur the same costs at the point of loading, but there may also be some element of deliberate inefficiency (if the country through which goods must travel is a competitor in exporting). For Uganda, Milner *et al.*, found costs in 1993–1994 equivalent to a tariff of about 25 per cent. (This was true for all types of goods.)

There is no reason to believe that this will be worse than average in trade (and some to believe that it will be better, because of the greater contact with other countries and with traders who themselves have such contacts, and are therefore more likely to complain), and therefore it may not distort the economy away from trade. It is a cost to all activities.

Sectoral policies

The discussion of infrastructure raised a point where trade policy can come into conflict with a sectoral policy, for example, of promotion of a national airline or shipping service. Trade may be used to promote a particular sector (whether through discouraging imports or encouraging exports), but it is the existence or growth of the sector which is the objective; trade is here a tool. Such a policy is normally adopted because the virtues of improving efficiency and encouraging new exports, and perhaps also stimulating structural and behavioural changes otherwise attributed to trade are being attributed to industrialisation (Box 9.2).

Other countries have had sectoral policies for less economic reasons: to satisfy particular interest groups or to satisfy national pride by introducing a 'modern' sector. It is not always easy to distinguish between these motivations (Box 9.3).

The uncertainty about whether trade or production (or neither) has special developmental effects means that it is difficult to apply the usual argument that

Box 9.2 Industrialisation and development

> The view that industrialisation is a central characteristic (and, by extension, determinant) of development is based:
>
> - on the high income elasticity of demand for manufactures;
> - on the productivity gains through reallocation of resources and specialisation; and
> - on industry as a source of technological change.
>
> The history of development is consistent with this, although there are clear exceptions, countries which have developed on the basis of (at least initially) agricultural trade and innovation (from Denmark to Australia). The question of whether there are particular virtues of industrialisation is closely linked to the nature of the international trade environment: in the late nineteenth century, with largely open markets for agricultural goods, specialising in such production for export could offer as high potential exports as industry. In the period after the First World War, and even more after the Second, however, agricultural protection became the rule in the developed countries, so that the export potential for a country specialising in agriculture was greatly reduced. (It is possible that in the post-Uruguay Round and, potentially, post-new Agricultural Round worlds of greater opening of agricultural trade, the pattern could change again.) But even with open agricultural trade, the other arguments for manufactures, of income elasticity and of technological innovation (especially the potential to develop new sectors) remain strong, and many countries have had sectoral policies for these reasons.

it is more efficient (and less distorting) to apply a policy directly to the objective sought: that a sectoral policy should be implemented through subsidies, taxes, or other special measures directed at that sector, not indirectly through trade measures, or alternatively that a direct trade policy is better than a sectoral one. The choice of policy may also be influenced by international regulation. Restrictions on export subsidies and limits, binding or other rules, on tariffs (discussed above) limit what can be done directly. Help for a sector (or a region), if it is not directly targeted at, or proportional to, trade, is normally acceptable under international rules.

The new NICs, particularly the Asian ones, have consciously tried to combine export and industry-led and import-substitution and export-promotion strategies within a broader industrial programme. They distinguish between two stages of import substitution: light industry and then intermediate and capital goods. This provides a model in which the export promotion of the first-stage products can be occurring while the second stage is only beginning its production.

Box 9.3 Unexpected links

> Policies to encourage particular sectors and those to promote particular trade objectives may be difficult to design because the way in which businesses develop or the links which producers see may not correspond to the way in which the government sets objectives or analyses sectors. Two examples: Japanese electrical firms investing in South East Asia did so initially to supply the local market (first simple and cheap products, like rice cookers and radios, moving on to televisions, refrigerators, etc.). But this led to an interest in exports partly because that was Malaysian and Thai policy, but also because the companies considered the products suitable for these countries different (less advanced) than those for Japan, and therefore transferred all production of those products abroad. But in the 1980s, two developments, the growth in importance of electronic production, requiring large factories and large supplies of semi-skilled labour, and the rise in the value of the yen, meant that the same firms were looking for offshore production points. By then, the South East Asian countries offered not only the economic advantages of cheap labour and appropriate location, but the company advantage of familiarity, and they therefore attracted a new type of export because they had originally offered markets for a very traditional type of import.
>
> An illustration that product categories and similarities need to be rethought comes from the paper industry of Colombia: firms with experience in small paper goods (stationery, publishing) found their allied products in small leather products (also for stationery and 'gifts'), not in forward or backward linkages for the commodity 'paper'. Alternatively products may be linked not by their markets but by their type of processing: clothing and footwear are more likely to be found produced together than clothing with cotton and textiles or footwear with animals and leather processing. In both cases, the final stage requires cheap labour; the intermediate is capital- and technology-intensive, and the primary depends on natural resources.
>
> This suggests two lessons, first that countries need a sophisticated analysis of all the economic and market characteristics of a product and its mode of production to understand and exploit its potential linkages to development, and second that any contact with a foreign investor can lead to future benefits, even if these are unrelated to the initial investment (and unpredictable at the time it was made). This also suggests that we must rethink the concept of 'tariff escalation': higher tariffs on processed than on non-processed goods may not have a direct effect on the structure of production if the conditions for production at different stages are so different that they are not really competitive in the same country. (Mauritius boasts that it is a leading producer of woollen goods without a single sheep.)

Firm-level policies and instruments with trade impacts

If it is difficult to know what constitutes an economic 'sector' in a meaningful sense, and if it is uncertain whether it is trade or industrial strategy which will encourage efficiency and growth, and if we are uncertain about how to make economic actors respond to incentives, then one approach may be to look much more directly at the most basic element in production, the firm. This may provide the opportunity to make the direct impact on behaviour which seems to be necessary. The assumption behind such a policy is that the efficiency of output is being held back because of a lack of responsiveness of producers to the normal demand or price incentives which produce efficient results in a developed country, so it is necessary either to increase the strength of the incentives or stimulate a behaviour change by direct action. A successful change in behaviour may produce higher output, replacing imports or increasing exports, or it may introduce a new type of production, changing the structure of the country's economy. The result may therefore be either import reducing or export promoting, but neither is the direct objective of the policy.

What are some of the measures available which are likely to affect trade?

- exemption from import duty on the machines required for production;
- exemption from duty on inputs (tied to export of the final product incorporating them, so this is often in the form of a drawback or refund);
- profit or income tax exemptions (with or without time limits);
- special credit arrangements (at lower cost, more extended terms or with weaker security);
- promotion through marketing abroad or subsidy of firms' marketing;
- information on international markets or on competing producers;
- special schemes of technical assistance or training (often through international agencies or with donor assistance);
- differential labour laws or other exemptions from normal regulations.

As well as measures targeted at firms in their export activities, there may be government policies which favour firms with successful exports (or good intentions), through more flexible investment, location, or planning policies.

Many of these may also be available to all firms, or they may be restricted to particular types of export, or they may be restricted in other ways such as by sector or geographical region. Thus, the schemes can be more or less precisely targeted, and more or less differentiated in favour of exports. Some policies can also be grouped together, by providing export processing zones, in which firms enjoy many of the special arrangements, and additionally *de facto* the privilege of fewer administrative hurdles to obtain the special help, as they are assumed to qualify for the complete package. And of course the value of each measure will vary from country to country according to what the 'normal' regime for each is: the general reduction in tariffs has reduced the value of tariff measures, while freedom from administrative inefficiencies is frequently more useful in developing countries than it would be in developed. The problem with any of those which differentiate by sector or which require proof of export is that they can be difficult to administer, by both government and firm.

Incentives for services exports

They are also particularly difficult to apply to services: for many developing countries tourism is a major export.[5] It is, however, difficult to differentiate between purchases by tourists and by residents in many cases, for example, restaurant services, travel services, even hotels. Therefore countries with a large domestic market and/or a reluctance to offer untargeted assistance can only offer the conventional export incentives in limited cases. In contrast, those where there is less hesitation about offering incentives or a more clearly distinguished foreign tourist sector are able to offer such incentives. (Zimbabwe is an example of the first, where most services can obtain little 'export incentive' type of help, while Mauritius is an example of the second, where tourism services and hotels are treated as export industries.) This can make some types of incentives an unintentional distortion in favour of goods and against services (which may mean against labour-intensive production).

External assistance to developing country exports

For some country measures, there are parallels in measures which may be available as forms of aid by importers, although these are beyond the direct control of individual country policy. Preferences remove the distortion between production for export and production for the home market caused by tariffs in markets, but may introduce distortions between products where they differentiate (as discussed in the section 'The role of trade in development'). Except for the exclusion of clothing and textiles from the US GSP scheme, the schemes probably on balance discriminate in favour of manufactures, because of the strong protection against agricultural products (although primary minerals are also largely duty-free). Some donors (and the WTO and UNCTAD through the International Trade Centre) also offer forms of technical assistance or marketing to reinforce those provided by country governments.

Much of the assistance, by both developing and developed countries, makes the implicit assumption that production and export or consumption and import are carried out by the same firms. This ignores the role of intermediaries. Sectors with a large number of small suppliers tend to give a particularly important role to intermediaries because of the monetary and expertise costs of trade. Where there are small providers and individual consumers at either end of the chain without direct access to or experience of international transactions and payments, intermediaries are also needed for administrative reasons. They provide a means of transmitting information in the market. An important part of this information is confidence about the quality and financial security of the product. This last role is particularly important for services where their nature makes the strength of the company providing a future or ongoing service more important than for a visible and storable good. In principle, some of the tax exemption and marketing assistance could be targeted at intermediaries instead of producers (if it is trade, not production which is the target of policy), but this is rarely done. Alternatively, companies could be provided with assistance to find alternative, direct, forms of marketing where this may be efficient, and more profitable through greater opportunity for product differentiation. Again, it is difficult to find examples of this. (The only examples are on the most obvious costs, with differential freight rates offered for exports and imports or for different types of export by some national airlines or shipping services.)

An alternative way of providing marketing and other intermediating services plus awareness of technological change is through foreign investment or the various variants of this, including subcontracting, franchising and joint ventures. This will be dealt with in detail in the chapter on foreign investment, but it may be interesting for countries to look more at the opportunities for promoting this precisely in the intermediary sector where the comparative advantage of a foreign company is likely to be greatest.

Summary of policies

The uncertainty about what types of trade or industry strategy produce the most successful development path makes it difficult to judge the effectiveness of different policies. If it is a general increase in demand which will provide good results, with private decisions producing an efficient solution, then the more general measures will be the most effective; these normally require the least administrative burden (on either government or firm). But if there are special advantages to particular types of production or to trade, then more targeted measures become necessary. If these are used, however, three risks to avoid are: excessive burdens of administration and checking; the opposite: insufficient checking that they are being used effectively; and finally, lack of clarity about what types of activity meet the criteria suggested by different theories (the doubts found in the examples quoted of leather and paper or tourism). History, and, to a more limited extent, theory, suggest that there may be a sequence, in which producing for a home market, through import substitution, may come first, then production for export, but earlier history (of development based on primary production largely for export) could contradict this, and suggest that the NICs' experience was the result of particular world conditions, of unusually high protection for agriculture and unusually low for manufactures.

International policies

Multilateral policies

The scope of the WTO has increased in three significant respects in the last twelve years. More countries have joined; temperate agriculture and clothing

International trade policies 291

and textiles have been brought under WTO disciplines (although still with special regimes); and international regulation has moved into new areas, including services, intellectual property, and stronger regulation of government intervention in trade and trade-related activities.

Until the Tokyo Round, in the mid-1970s, arguably even until the Uruguay Round which began in 1986, the role of the GATT was limited in most areas of interest to developing countries. Agriculture had been effectively excluded, at the insistence of the US, while textiles and clothing had been the subject of a long-standing derogation for the MFA and its predecessors. Most primary products entered developed countries at low or zero tariffs, and some developing countries had special arrangements with the former colonial powers giving them preferred access to their principal markets, so GATT access was irrelevant. (The European arrangements with the ACP were the most important, but not the only, examples.) From the point of view of the developed countries, the developing were not important as markets or (with some exceptions) competitors. Even in 1973, the beginning of the Tokyo Round, developing countries' share in trade was only 21 per cent, and this was predominantly in primary goods. By 1986, the beginning of the Uruguay Round, their share in world trade was 26 per cent, of which 60 per cent was manufactures, a share that has now risen to more than three quarters. They now account for close to a quarter of world trade in manufactures. For both developing and developed countries, the participation of developing countries in trade is now important.

Some developing countries have always been members of GATT, and it needs to be remembered that many of the OECD countries were still at middle-income level at the time GATT was founded, so that it was never entirely restricted to developed countries and their interests. Nevertheless, the concentration of negotiations on those manufactures of interest to more developed countries, combined with a developmental strategy in the major developing countries that placed most weight on internal development and industrialisation, not on trade, meant that developing countries did not see GATT as an essential negotiating arena. And negotiations there have traditionally been between the major importers and exporters of each product, so that countries without significant roles in the goods included were left to one side.

The change in the 1960s to give differential treatment to the developing countries, under the GSP was built into the GATT agreement (as Part 4) in 1971 (see below). But the initiative and negotiations to achieve this came not from the GATT, but from UNCTAD, and it was there that most developing countries concentrated their attention. This reinforced the perception that GATT was not an effective forum for developing country interests.

The terms of the trade-off between policy freedom and participation in GATT rules were sharply altered in the 1970s and 1980s. The constraints imposed by developed countries, especially those on agriculture and clothing, were becoming more unpredictable and more damaging. As the European countries became major exporters of agricultural goods, their subsidies affected developing producers. The discretionary clauses of the MFA were used more frequently. As developing countries became competitive in new products, there was a revival of protection in the industrial countries using non-tariff barriers and trade actions like anti-dumping on goods like steel. The growth of trade's importance for the developing countries in value and in perceptions about policy, led to greater awareness of how tightly their independence of action was limited by the interventions of their trading partners, and by the *lack* of effective rules. Obstacles to exports were seen not just as barriers to static efficiency gains or extra costs, but as constraints on the most successful strategy for development. As many developing countries were lowering their own barriers, the freedom to impose or increase them ceased to be a major reason to avoid active membership in GATT. The perceived advantages of rules and predictability became major reasons for the increased interest of developing countries in the GATT system.

Developing countries' objectives in multilateral negotiations

What will be important for developing countries now is to take their own initiatives to improve trade policy between developed and developing countries. Up to now, this chapter has assumed that the external situation is given, both markets and rules. As they have become more important players, however, the developing countries no longer need to make this assumption. They have more opportunities to negotiate on improved access as their share of trade and thus their bargaining power increases. The extension to new areas will require consideration of what are the advantages and disadvantages of new rules: the advantages from certainty and from limiting the scope for others to damage them; the disadvantages from limiting their own freedom of action.

In *agriculture*, their objectives are likely to be principally liberalisation of developed country markets, while they must restrict their own liberalisation until the distortions caused in world markets by the protection and export subsidies of the developed countries have been removed. In *manufactures*, the traditional objectives remain; they still face tariff peaks in goods like clothing and shoes in some markets. Negotiating these down will reduce the distorted incentives against both trade and industrialisation of the current system. In *services*, the current regime is also biased against liberalisation of the more labour-intensive forms of those which require travel of workers from developing to developed. Developing countries were not as active as developed in the Uruguay Round services negotiations. They may have the potential to improve their position in the next round, especially as developed countries want to increase their access to developing markets. In the *new areas*, the developing countries will need to avoid any increase in the bias against those who are primarily consumers rather than producers of new technology which was introduced by having intellectual property as a part of the Uruguay Round. The *environment* could offer some developing countries arguments for trade liberalisation: protection (especially in agriculture and fishing) produces distortions towards more damaging production in less suitable (normally developed) countries; liberalisation both reduces this and provides incentives to conserve important resources as their value in world trade increases (especially important in forest products, perhaps). This could at least offset arguments using environmental standards as an argument for barriers against developing country products.

How developing countries can influence multilateral policies

Effective negotiation is itself an area of expertise, and therefore one that developing countries must acquire through experience, supplemented by assistance. How can countries, particularly those with limited resources, identify the areas in which they want to pursue (or need to defend) their interests and negotiate effectively, in the face of high actual and opportunity costs? This is not an occasional need, only when negotiations are in progress. Under current WTO and other international agencies' provisions, countries can introduce new rules implementing existing decisions unless their partners refuse, so countries must be constantly alert.

Countries must find ways of involving all the government departments with international interests (which, following the increase in the scope of the international system, means almost all departments) and also the private sector actors. Both developed and developing countries have used a variety of ways of doing this, relying on different combinations of joint committees and specialist agencies. There is insufficient research to identify 'best practice' (and this is likely to vary with the governmental and economic structure of countries and with the nature of their international interests), but some means of encouraging direct participation by the most important decision-makers of both the public and private sectors is normally required, instead of, or in addition to, indirect means. This may mean formal or informal participation by private sector representatives in international negotiations as well as in formulating countries' positions in advance.

Preferences

The Uruguay Round incorporated different requirements on the extent and timing of liberalisation for a range of the new rules on a three-tier basis: developed, developing and least developed. The agreement allowed the developing and least developed groups more time to fulfil the requirements (and, in the case of the least developed, some exemptions), but did not directly modify the existing provisions for general differential treatment of all developing countries by the rest of the world. Since then, however, there have been proposals to extend the least developed/developing distinction to GSP preferences.

In 1996, the then Director-General of the WTO proposed that developed countries give the least developed duty-free trade access, 'bound' by commitment to the WTO. A more modest plan was adopted in October 1997. It still seeks preferential access, but now with exceptions allowed. In response, most developed countries and some of the more advanced developing agreed to offer increased duty-free access to the least developed, and by March 2000, this finally seemed likely to be agreed. But this adds little to their present access to the EU, except on a few goods like leather; sensitive agricultural goods are excluded. Non-ACP countries remain excluded from the special Lomé protocols on the sensitive products, beef, sugar, rum and bananas. The US agreed to add 1,700 products to the duty-free list of its GSP, restricted to the least developed, and has also proposed additional concessions for Africa. Morocco, South Korea, Thailand, Turkey and Egypt, among others, offered limited concessions on some goods.

There is a practical conflict between preferences and the other trends in international trade. As liberalisation proceeds, every reduction in 'normal' tariffs reduces the potential size of preference margins. As new instruments and regulations appear, it becomes increasingly necessary to think of non-tariff preferences to replace tariff preferences. Differentiation among developing countries could target benefits better, but the less poor lose relative, and often absolute, preference margins.

There is also an essential internal contradiction of the GSP. While it imposes rigidities, in that it restricts the offering of preferences to that scheme, and to the group of countries accepted as developing under it, with the only differentiation permitted that which is now allowed for least developed, it is not itself 'bound' in any way under the WTO, and therefore there is no obligation to offer a particular level of preference, or to maintain the existing level. It thus does not offer guaranteed or predictable access (as WTO agreements do), and the structure (of either fixed or percentage reductions on tariffs, varying between agricultural and industrial goods, and by degree of 'sensitivity') cannot be described as transparent. The introduction by both the US and the EU of labour, environment and democracy conditions has further increased the uncertainty and reduced the transparency; these depend on assessments by each developed country of developing countries' compliance with unilaterally chosen criteria. Many of these differentiations could perhaps be challenged as contrary to the 'rules' for GSP, but its status as a recognised derogation from the WTO makes these rules and the means of enforcing them uncertain, and the fact that nothing is permanent or bound makes any challenger vulnerable to retaliation. Developing countries need to consider whether the decreasing advantages, increasing differentiation and limited certainty of the GSP have now turned the balance of negotiating advantage more to concentrating on multilateral, bound negotiations under the WTO.

Special arrangements

Developing countries have also tried to alter their external environment by forming regional groups, notably CARICOM, MERCOSUR, SADC and COMESA (Box 9.4). (There are many others, but few with more than paper results.) These offer the advantages of certainty and improved access which are attributes of multilateral negotiations, but for a more limited number of countries (and in some cases, more limited range of commodities). They, like GSP, are losing their advantages as they become less differentiated from world tariff reductions and more

Box 9.4 Border zones

Cooperation across borders need not be under the increasingly rule-based regime of trade relations. The most ancient forms involve common use of natural resources: seas and rivers. These have been surprisingly absent from policy discussion of trade, but by improving the contacts and often the infrastructure across borders, they can be among the most important engines for improving both production and trade. They also typically allow informal joint production, cross border temporary migration, and thus effectively free border regions. (Perhaps this example is one where there should be little or no mention so that it continues to increase without opposition.) For many countries they offer the advantages of a freer international regime within a controllable extent (in quantity and in geography). Developing countries need to look at examples from both developed countries (along the Rhine, the US/Canada border) and developing (Singapore/Malaysia/Brunei). Like establishing international contacts through the internal relations of foreign investors as well as through the external effort of export promotion agencies, they offer a supplement to policy-based liberalisation. It is a type of cooperation not open to island states.

bound by international trade rules. The GATT has always regulated customs unions and free trade areas by requiring them to cover 'substantially all trade' (to avoid selective liberalisation that could maximise advantages to members and costs to the excluded), but the developing countries were exempted from the full requirements and the lack of an effective enforcement mechanism for the GATT meant that all countries were effectively exempt. The Uruguay Round improved the disputes procedure, and the growth in the number of regions reduced international tolerance: one developing country region (MERCOSUR) has been brought under the full examination procedures, and mixed developing/developed regions (NAFTA, the new proposals by the EU) are also attracting opposition. As with preferences, developing countries will need to balance the advantages of greater access against the disadvantages of a more limited coverage, and ask whether the balance is changing. Developing countries have been among those most damaged by some of the new regions (the diversion of clothing trade from Caribbean countries and Sri Lanka towards Mexico, for example). But regions offer particular advantages to small developing countries, both to provide a larger 'home' market and to offer bargaining strength. In the next Round, they may want to try to improve the international rules for regions. One suggestion which has been made is to combine the WTO rules for preferences and regions, allowing regions to discriminate internally by level of development, and requiring preference regimes to be bound.

One of the oldest issues in GATT/UNCTAD discussions is special insurance or other help for commodities because of their vulnerability to price fluctuations and secular decline. The World Bank (encouraged by the EU) has recently suggested a compromise between a full market system to provide insurance (through forward buying and selling) and subsidies. Developing countries were absent from the consultations on this, but will have an interest in becoming involved if it is implemented.

For all of these measures, the increased share of trade and other foreign effects in all countries' economies makes them more of a national priority than they would have been thirty years ago.

Conclusions

If the problem is to change the productivity (i.e. the behaviour) of decision-makers, public and private, in developing countries, then little is known about how this is done. There can certainly be no presumption that one type of structural change from domestic to export demand, will work better than another, from primary to industrial output. And if governments have additional objectives, whether for a market-led economy or alternatively for particular structural characteristics, it becomes even less possible to make general prescriptions. The risk that decision-makers may be less able to respond to price or policy incentives, because of lack of experience or lack of full awareness of the alternatives suggests that all policies need to be stronger than in developed countries. In administrative terms, this may require more effort (which may be a practical difficulty in this prescription: the administrators are also in need of training). In legal terms, this may mean greater potential for conflict with the increasingly rigid international rules on discriminatory policies within countries, especially if measures are particularly directed at trade rather than at activities or sectors.

Specific problems have been identified in the types of discrimination which may result, through preference systems, tariffs, tax regimes, the special difficulties of identifying users of services. There is, however, more scope than is sometimes recognised for informal efforts.

There is probably a sequence of policies that is more effective than any single policy, as countries' potentials to trade (and to negotiate) evolve, and also as the international environment changes. The sequence of the most recent successes suggests that it is one from import substitution to exports, but there are examples of countries that have developed on the basis of successful exports of primary goods, and the international conditions may be changing now to permit this again. This is of particular interest to small countries where import substitution has always been more difficult and more limited. Mauritius, because of its special access on sugar, is an example of country that followed a 'nineteenth century strategy' in the 1980s and 1990s; other countries may be able to do so if the agricultural trade regime changes.

Although it may be inappropriate to emphasise it in a chapter on trade, trade is not an end in itself but a tool, either for raising income or for providing the technological and developmental stimulus that assists the transformation of a country.

Notes

1 The late 1980s saw a series of studies of the effects of 'openness' on growth which misused correlations and definitions as evidence for 'export led growth', most notably Papageorgiou et al. 1991. There were also attempts to simplify all the elements of a trade regime by classifying the regime as import substituting, neutral or export promoting in terms of effective exchange rates (Bhagwati, 1988).
2 Many of the 'allowed' subsidies are designed to cover measures used by the developed countries, for example, to support agriculture, and require more administrative competence than may be available.
3 The labour and environment provisions have not yet (September 1999) been used.
4 Trade liberalisation is often a component of macroeconomic adjustment programmes imposed in developing countries by the international financial institutions. Here, the trade component is intended to help in the restoration of domestic and external balance, so that this can be considered an example of using trade for other policies. The potential effects and the tools are the same as for other cases of this.
5 It is among the top five for 83 per cent of countries, and the principal export for a third of developing countries. It has the same advantage of many manufactures of being in principle exportable by any country; it is sufficiently varied not to be dependent on particular natural endowments.

References

Bhagwati, J.N. (1988), 'Export-Promoting Trade Strategy: Issues and Evidence', *World Bank Research Observer*, 3(1), 27–58.
GATT (1986), *Text of the General Agreement on Tariffs and Trade*, Geneva: GATT.
Hoekman, B. and Martin, W. (eds) (2001), *The Developing Countries and the WTO: A Pro-Active Agenda*, Oxford: Basil Blackwell.
IMF (1994, 1998), *Direction of Trade Statistics Yearbook*, Washington, DC: IMF.
IMF (2000), *World Economic Outlook*, Washington, DC: IMF.
ITC/Commonwealth Secretariat (1999), *Business Guide to the World Trading System*, Geneva: International Trade Centre.
Milner, C., Morrissey, O., Rudaheranwa, N. (1999), 'Protection, Trade Policy and Transport Costs: Effective Taxation of Uganda Exporters', Paris: EADI.
Page, Sheila (1990), *Trade, Finance and Developing Countries*, Savage, MD: Barnes and Noble Books.
Page, Sheila (1994), *How Developing Countries Trade*, London and New York: Routledge.
Page, Sheila (2000), 'Trade in the 2000s: New Rules, New Preferences, New Priorities', in Wahiduddin Mahmud (ed.), *Adjustment and Beyond: The Reform Experience in South Asia*, Basingstoke (UK): Palgrave, pp. 166–182.
Papageorgiou, D., Michaely, M. and Choksi, A.M. (eds) (1991), *Liberalizing Foreign Trade*. 7 Volumes, Cambridge: Basil Blackwell Inc.
United Nations, *Monthly Bulletin of Statistics*, New York: United Nations.
United Nations (1998), *World Investment Report 1998. Trends and Determinants*, New York and Geneva: United Nations.
Wignaraja, Ganeshan (1998), *Trade Liberalization in Sri Lanka*, Basingstoke: Macmillan.
World Bank Group, The (2000), *Global Development Finance 2000. Analysis and Summary Tables*, Washington, DC: World Bank.

Index

ACP countries 279, 291, 294
Adam, C. 255
administrative costs 285
adverse selection 107
Africa 78, 176, 178, 281; North 73; Sub-Saharan 5–6, 73, 145; *see also* ACP; Nigeria; South Africa; Uganda
agglomeration effects 185
air and sea cargo 37
Alchian, A. 252, 258
Altenburg, T. 172
Alvey programme 119
Andean drug countries 279
Anglo-Irish Free Trade Agreement (1965) 186
anti-trust 243, 247
Arrow, K. 106, 107, 108, 109
Asia 78, 207, 281; Central 73; East 5, 7, 28, 43, 248, 250; South 5; *see also under individual country names*
Astek Fruit Processing 24–5
asymmetries 107, 110, 203, 211
Audretch, D. B. 249
Australia 132, 185, 286

Bacon, R. 253
Bain, J. S. 244
balance of payments 17, 34, 276
Balasubramanyam, V. N. 168
bandwagon effect 176, 185, 187
banks: cooperative 218; development 219–23; monitoring, regulation and supervision 206, 208–11; potential for instability 203, 205; runs on 208; *see also* BIBFs
Barbados 212
Barrel, R. 182
barriers 241, 242, 277, 292, 293
Basle Committee 203, 205, 210

Baum, C. 222
Belize 73
benchmarking 38, 44, 47, 61, 158, 226
Bergsman, J. 179
Best, M. N. 127
best practice 146–62, 176, 257, 267; lack of familiarity with 285
Bhatt, V. V. 222
BIBFs (Bangkok International Banking Facilities) 227
Blomstrom, M. 170, 180, 181
Blonigen, B. A. 177
bonds 226, 227, 262
booms 210
Borensztein, E. 170
BOT (Build-Operate-Transfer) 41
bottlenecks 37
bottom-up policies 269
Brada, 256
Brazil 139–40, 142–4, 150, 154, 222
Britain *see* United Kingdom
Brunei 295
'bubbles' 207
budget deficits 34
bureaucracy 178
business angels 202, 223, 224–5
business development services 152–8
Business Excellence 120
business parks 218
business strategy perspective 18–20, 21, 30
buyers and sellers 107

Callanan, B. 188
Cameron, R. 220
Canada 64, 132, 295
capital: accumulation 253; efficient allocation of 205, 206; external 202, 203; investment 145; risk 104, 224; scarce 145; working 143

capitalism 96, 106, 110, 246, 253; knowledge generating system characteristic of 98
capital markets 199, 201, 204, 223; developing 219, 225–8, 260; imperfections 7
Caribbean 5, 64, 73, 295; *see also* Barbados; Grenada; Jamaica; St Kitts; Trinidad
Caves, R. E. 168–9, 180
CCI (Current Competitiveness Index) 63
CDTs (Centres for Technological Development) 125
CEFE (Competency-based Economies, Formation of Enterprise) 147–8, 154, 155
Central and Eastern Europe 6, 178, 181, 206, 251, 259; *see also* Czech Republic; Hungary; Poland; Russia
central banks 208, 209, 210
centrally planned economies 205, 251
CFIs (community finance initiatives) 200, 218, 224
Chamberlain, Edward 107
change 6, 106, 268; social 134; structural 276; technological 22, 45, 161, 246, 261
Chile 5, 40, 70, 259
China 34, 70–3, 140, 205, 210; gradualism 257
CIT (communications and information technology) 198, 224
Clapham, J. H. 220
Clinton administration 6
closed-door policy 186
clusters 191, 249, 268, 269; *see also* industrial clusters
Coase, R. 245, 257
code of conduct 181
collaboration 116, 120, 128; flexibility and 140; inter-firm 137
collective action 137, 140; *see also* learning
Colombia 96, 124–5, 287
communications 156
comparative advantage 19, 34, 159; major theories of 26; short- vs long-term 49
compensation packages 41, 52
competitive advantages 18, 21, 31, 261
competitiveness: assessment 35, 43–5, 47–8, 62, 112, 147; business strategy perspective 15, 18, 21, 44, 62; definition 16, 19, 21–2; indices 61–4, 67–9, 74–6, 78, 85, 88; macroeconomic perspective 15–17, 19–20; monitoring 36, 43, 50–1, 55; strategy 6, 8; technology and innovation perspective 8, 15–16, 21–2, 26, 28, 30, 53, 67–8
computer industry 144–5, 150
concentration 246; market power and 240–4
consultancy firms 6
consumption 42
coordination 50, 111, 135; systems to encourage 277
Corden, M. 16
corruption 178, 200, 205
Costa Rica 40, 70, 137, 172
Côte d'Ivoire 42
countertrade 283
Cowling, K. 243, 244
creative destruction 246
creativity 99, 106
credit 36, 135, 204, 211, 212; rationing 199–200, 208, 210, 211, 219; rural markets 205; SME 37
'cronyism' 200
current account deficit 16
Cyprus 69, 73
Czech Republic 168, 207; *see also* Prague

Davies, R. B. 177
Dahlman, C. J. 56
DBFO (Design-Build-Finance-Operate) 41
debt 200–1, 202, 205, 226, 227; bad 206–7; crises 34; external 78
demand: changing patterns 133; conditions 3, 19; consumer 4; external 51, 276
democracy 134, 269
Demsetz, H. 252, 258
Denmark 38–9, 286
deposit insurance 209
deregulation 19, 204, 247
diamond framework 20, 21
differential competences 246–7
diffusion 101, 104, 126
DINP (Denmark Industrial Network Programme) 38
disclosure 211
discretionary grants 175
disintermediation 226
distribution: equitable 248; income 277; production and 104, 127
distribution costs 107
division of labour 104, 106, 110, 139, 246

Index

domestic competition policies 35, 239–73
Donaldson, D. J. 262
Dosi, G. 17, 21
'downstream' firms 159
Dugan, A. 222
Dunning, J. 173
duty-drawback schemes 280, 282

East Asian 'tigers' 248, 250
Eastern Europe *see* Central and Eastern Europe
ECNs (Electronic Communication Networks) 198
economic growth 133, 140, 171, 200; exports and 22; stable 226
economies of scale 26, 34, 249, 276
Ecuador 149
education 142, 177, 186, 284; and investment 77; and training 31, 35, 45, 134, 285
efficiency: allocative 240, 242; capital 199; cost 211; SMEs 133–4; traditional gain 276
Eggertson, Y. 252
Egypt 294
El Salvador 149
Eltis, W. 252
employment 42, 138; full 16
EMU (Economic and Monetary Union) 227
entrepreneurship 112, 219, 221, 223, 263, 276
EPZs (export processing zones) 36, 177, 178
EU (European Union) 186, 249, 294, 295, 296
Eureka programme 120
Euromarkets 226
European Community policy 119
evolution 105–6
evolutionary theory 9, 11, 22, 26, 56, 96, 98, 105–6
exchange rates 225, 283; pegs 207; real 16, 17, 34, 63
experimentation 99, 100; markets provide framework for 105
exports: and economic growth 22; labour-intensive 48; major 289; manufactured 17, 44, 144, 187; poor performance 26; relative price of 283; services, incentives for 289; sudden fall in demand for 47; technology-intensive proportion of 74; *see also* MECI

externalities 221; spill-over 107, 108

factor conditions 18
factor markets 28
Fagerberg, J. 17, 77
FDI (foreign direct investment) 3, 21, 31, 77, 84; complementary 112; countries less attractive to 51; cumulative inflow 78; encouragement of 124; export-oriented 6, 80; favourable destination for 41; government policies towards 166–97; history of seeking 145; policy 36
Feldstein, M. 253
Fifth Framework programme 120
financial sector policies 198–235
Finland 64
firm-level policies 288–9
fiscal policy 63, 277, 283–4
Fischer, E. 10, 35, 56, 131, 227, 229–30
'fixed costs problem' 217
flexibility 153, 207–8; decision making 172
foreign investors 173–81
Foresight Programme 121–2
Forrant, R. 127
France 218, 226
fraud 199, 209
free rider problem 211, 252
Free Trade Zone 187
Friedman, M. 252
FRNs (floating rate notes) 226
Frydman, R. 259

Galicia 97
Garelli, Stephane 62
GATS (General Agreement on Trade in Services) 172, 177, 207
GATT (General Agreement on Tariffs and Trade) 277; Tokyo Round 280, 291; Uruguay Round 3, 172, 207, 278, 280–1, 282, 285, 291, 292, 293, 295
GCI (Growth Competitiveness Index) 63
geographical costs 186
geographical proximity 137, 138, 247
Georghiou, L. 121
Germany 138, 147, 217, 226; East 178; industrial banking 220
Gerschenkron, A. 220
Ghana 23, 24–5
Gibson, A. 152, 154, 157–8, 163
Glasgow 136

globalisation 45, 207, 211; pervasive 77; progressive 3, 4; rapid 7, 8
Goh Keng Swee 188
goodwill 51
Gorg, H. 186, 187
government: efficient 254; failure of 39, 261; intervention 40, 199; policies towards FDI 166–97; private sector and 30–1; procurement program 142–3; spending, sectoral allocation 284–5; strong capabilities 51–2
gradualism 257
graduates 41, 120, 179
Grameen Bank 217
grants 54, 186
Greece 69, 73, 186
Grenada 73
growth 17, 139, 205, 248, 277; catalyst for 143; constraint to 5; differential 246; export 48, 74, 276; new areas 142; productivity 168, 176; *see also* economic growth
'growth' enterprises 200
GSP (Generalised System of Preferences) 279, 289, 291, 293, 294

Haiti 70
Hamel, G. 18, 19
Hannah, L. 243
Hayek, F. von 105, 252
HDI (Human Development Index) 68–9
Herfindahl index 243
'hidden costs' 155
high/low performing countries 82
Hines, J. R. 178
'hit and run' behaviour 241
Hobday, M. 12, 25, 56, 123
Hong Kong 5, 69, 73, 188, 275
horizontal integration 139
Howes, C. 7
Hu, Y. S. 220, 221
human capital 80, 84, 168; and productivity 82
Humphrey, J. 56, 158, 160–1
Hungary 40, 70
hypotheses testing 105

ICT (information communication and telecommunications) 3, 4, 80
IDA (Industrial Development Act 1969) 185, 186, 187
IDB (Inter-American Development Bank) 148

ILO (International Labour Organisation) 152–3, 155, 157–8, 163
IMD (International Institute for Management Development) 61, 62–7, 68, 75–6
IMF (International Monetary Fund) 6, 42, 207, 210, 255, 279
imitation 146
imperfect competition 241
imperfect markets 21–2
implementation 49–50, 226, 261, 262–3, 266–70
imports 179, 186, 275, 283; *see also* GATT; substitution
incentives: financial 175; fiscal 175, 185, 224; increased 119; justified 171; policies 31, 237–97; R&D 124; tax 175, 179, 190, 199, 224, 277; widespread use of 170
income: high 5, 73, 74, 75; low 63, 74, 180, 185; middle 74, 124; per capita 6; real 22, 67, 68; rising 4
India 73, 145, 154, 222
indices *see* CCI; GCI; HDI; Herfindahl; IMD; MECI; TI; WE
Indonesia 168
industrial clusters 28, 29, 131–65, 176; promotion of 37–9
Industrial Master Plans 49
industrial policy 185
industrialisation 43–4, 219, 220, 286, 291; labour-intensive 35; new risks for 3; rapid 122
inefficiencies 252; deliberate 285
inflation: high 34; tightly controlled 34; tolerable level of 16
infrastructure: administrative 18; communications telecommunications 66; educational 112; innovation system 127; legal 224; national innovation 124; physical 18, 28, 63, 80; private sector participation in projects 41; quick provision of 172; technological 117–18, 126
innovation 18, 244, 245; differential 246; improved 48; technology and perspective 21–30; *see also* STI
Institute for Management Development 6
insurance sector 226
integration 3, 5, 180, 182, 249–50
Intel 137
intellectual property 108, 277, 278, 292; *see also* TRIPS

interest rates 135, 199, 200, 204, 207, 209
intermediaries 289
International Trade Centre 289
invention 100–1, 108
investment 23, 176, 205; education and 77; equity 202, 223; infrastructure 117; inward 19, 124; returns to 284; R&D 124; *see also* FDI
IO (industrial organisation) 240, 241, 244
IP (industrial policy) 199, 204, 239, 249, 250
IPAs (investment promotion agencies) 175, 176
IPOs (initial public offerings) 225, 226
Ireland 41, 167, 168, 172, 175, 178, 182–8, 191–2
ISO standards 4, 35, 36, 46, 54, 149
Israel 5, 69, 73
Italy 25; Northern 136; Southern 97
ITS (information technology service) 120

Jamaica 73, 127
James, J. 77, 79
Japan 96, 119, 121, 123, 132, 144, 185, 209, 210, 250, 287; cooperative banks 218; Meiji 220; MITI 248; *see also* Tokyo
Jequier, N. 221
joint ventures 24, 41, 178, 290
Jun, K. W. 177
'just-in-time' techniques 161, 180

Kalotay, K. 181
Kaufman, 209
Kay, J. A. 243, 254
Kenya 149
Keynes, J. M. 257
Kikeri, S. 259, 261, 262
Kim, L. 124
Knight, Frank 98
knitwear industry 145
'know-how' 207
Krugman, P. 6–7, 19
Kuwait 69, 73

labour 132, 171; poor standards 177; semi-skilled 287; skilled 104; unproductive 253
labour costs: cheaper 5; relative unit 16–17
labour market: flexibility 63; policies 177, 179
Lall, S. 7, 20, 67, 102, 172, 181

'lame ducks' 247
Latin America 5, 281; *see also under individual country names*
LDCs (least developed countries) 278
learning: collective 28, 29, 37, 137; interactive 28; inter-firm 138, 143
Lee Kuan Yew 188
'lender of last resort' role 208, 209
liberalisation: financial rapid 8; tariff 180; trade 19, 80, 82, 84, 125, 161, 207
liberalisation plus approach 9, 53
LINK programme 120
Littlechild, Stephen 267
loan guarantees 212, 217
Lomé convention (1975) 279, 294
London Stock Exchange 225
Lundvall, B. A. 27
Luxembourg 64

machining industry 141–2
McKinnon-Shaw model 204
Malaysia 5, 49, 70, 188, 259, 287, 295
management 43–52
Manchurian Incident (1931) 141
manufacturing 6, 17, 144, 182; export-intensive 185; local capabilities 186; new 3
margins 142, 199; fixed 204; interest rate 207; price-cost 243
market failures 7, 20, 106, 119, 160, 191; banks and 199, 200; governments intervene carefully to remedy 40; informational related 181; most likely 171–2; pervasive 254; structural 240
marketing 41, 46, 145, 146, 152; capabilities 39; effective strategies 40; opportunities 48
market power 245; concentration and 240–4
market processes 105–6, 108
market shares 44
Marx, K. 246
Mauritius 40, 46–7, 73, 96, 126, 132, 287, 289
Mazumdar, S. C. 199
MECI (manufactured export competitiveness index) 67–84
Megginson, 256, 257, 260
MERCOSUR 294, 295
mergers 35, 145; and acquisitions 172, 242, 244–5, 249
MES (minimum efficient scale) 241
metrology 35, 118

Index 303

Metcalfe, J. S. 9, 12, 35, 55–6, 95, 105, 115
Mexico 5, 70, 137, 168, 178, 259, 295
MFA (Multi-Fibre Arrangement) goods 279, 280, 281, 291
M-form organisation 253, 254
MGS (mutual guarantee scheme) 218
Michie, J. 258
microeconomics 67
Middle East 73
Milner, C. 285
mission 221
MNCs (multinational corporations) 3, 4, 22, 24, 45, 247; protection and controls on 29
Modigliani, F. 244
Mody, A. 177
monetary policy 226, 283–4
monitoring 50, 157
monopoly 35, 107, 240, 241, 242, 246, 247, 250, 254, 255; limited term right 108
Moon, H. C. 20
moral hazard 107, 199
Moran, T. H. 176, 179
Morocco 168, 294
Most Favoured Nation status 278
MSMEs (micro and small and medium enterprises) 6
Mullineux, A. W. 10, 37, 198, 206–7, 210–11, 217, 227
multilateral policies 290–3
multinationals *see* MNCs
Murinde, V. 37, 198–200, 202, 204, 206–8, 219–20, 223, 226, 231
mutuality 218
MVA (manufacturing value added) 80, 84
Mytelka, L. 56

nationalisation 247
national policies 280–90
natural resources 51, 171, 178, 277; cheaper 5
Neary, J. P. 181
Nelson, R. 27, 108
neoclassical perspective 240–7, 249, 250, 251
Netherlands 64
Netter, J. M. 256, 257, 260
networking 120
networks 151–2; regional 149; social 156
NGOs (non-governmental organisations) 30, 47
Nicaragua 70, 151

NICs (newly industrialised countries) 275, 278, 284, 285
NIEs (newly industrialising economies) 5, 43; central aspect of 7
Nigeria 45, 68
NIS (national innovation system) 27–30, 53
non-tradeables sector 16, 17
Noorbakhsh, F. 78, 79
Nordic banking crisis 206
North, D. C. 250, 268

OECD (Organisation for Economic Cooperation and Development) 22, 37, 67–8, 77, 103, 111–12, 115, 186, 251, 259, 283, 291
OEM (original equipment manufacture) 24, 123
Ohmae, K. 19
oligarchies 207
oligopoly 36, 241, 244, 245
open-door policy 172
opportunistic behaviour 107
opportunities 102, 103, 104, 134, 222; innovation 116; lending 226
opportunity costs 293
output 39, 240, 260, 267; intermediate 113
ownership advantage 173

Pacific region 5, 64, 73
Page, S. 11, 34, 193, 274
Paloni, A. 78, 79
Paraguay 148–9
Pareto principles 105, 106, 107, 109, 255
patents 108, 145
PCs (personal computers) 80, 82
Penrose, E. 244, 245, 246, 248, 252
perfect competition 240, 241, 247
Philippines 5, 70
Pioneer Industries Ordinance (1959) 190
Pitelis, C. 11, 35, 239, 243, 245, 247, 249, 254, 271
Poland 206, 257
policy framework 47–50
political stability 51
Pollitt, M. 259–60, 267
Porter, M. E. 18, 19, 20, 21, 62, 63, 247, 249
Portugal 69, 70, 73, 186
Posner, M. V. 21
PPP (purchasing power parity) 16
Prague 222
Prahalad, C. K. 18, 19

Index

preferences 295; *see also* GSP
price signals 76
prices 107, 240, 241, 245, 275; collusion over 242, 243, 244; commodity 51; consumer 16; oil 47; wholesale 16
private sector 6, 47, 50, 75, 79, 122, 199, 254; government and 30–1, 52; role of 39–43; tiny 52; uncertainty in 49
privatisation 35, 176, 225; bank 206–7; regulation and 251–70; state-owned enterprises 19
'problematic enterprises' 247
product differentiation 290
production 276; and distribution 104, 127; 'just-in-time' 161
production costs 107
profitability 48, 66, 222; crucial 105; decreased or eliminated 140; resource (re)allocation 34, 79; tradeables 17
profits 210, 221, 242, 258; above normal 241; loss of 168; maximising 18, 39, 254, 267; new export 185; tax treatment of 201
promotion 175, 222, 223; campaigns 51; export 180, 276, 277; investment 189; policies 135
property rights 106, 108, 110; *see also* intellectual property
prosperity 3, 4
protection(ism) 34, 186, 279, 285; agricultural 286; patent 100
public expenditures 40
public policies 19, 21, 31, 153; initiatives 146; narrow scope for 17
public–private partnerships 9, 12, 42
public procurement 19, 280, 282
public sector 40, 253, 254–5, 256; support from 140; private sector and 41, 42

quality 113, 187
quantitative restrictions 34
quotas 280–1

R&D (research and development): collaborative projects 126, 128; encouraged 178, 179, 181; expenditures 66, 77, 78, 79, 80, 103, 115, 123; institutions expected to invest in 149; joint 39; product-centred 48; raising the level of 186; subsidies 115–16
Rabelloti, R. 138
Raffeisen banks 218
rate of return 267

recession 190, 210
redistribution 254
regional groups 294
regulatory bodies 258
Reinert, E. S. 7, 19–20
rent-seeking behaviour 36
repeat ordering 37
resource allocation 79, 105, 134, 252; efficient 246, 268, 276; equilibrium 110; shifts in 17; profitability of 34
resources 119; bureaucratic 171; comparable 158; financial 135, 171; global 5; human 35; inefficient utilisation of 252; lack of 3; need to mobilise 275; political 171; scarce 156; technical 49; *see also* natural resources; resource allocation
restrictive practices 243–4, 268; policy 172
restructuring 5; concerns about 6; financial sector 204–8, 224; industrial 211
Reuber, R. 35, 56, 131
reward 99, 108, 227
Richardson, G. 246–7
risk 106, 200, 205, 224, 227; credit 210, 211, 212; exchange rate 225; political 178
Rodrik, D. 171
Romijn, H. 77, 79
Rosenberg, N. 109
Ross, D. 243, 244
royalty payments 175
Ruane, F. 186, 187
rural enterprises 200
Russia 207

Sachs, Jeffrey 62, 80, 82
St Kitts and Nevis 73
salaries 41, 182
SAPs (structural adjustment programmes) 54
savings 205, 207, 218
Scherer, F. M. 243, 244
Schmitz, H. 160, 162
Schumpeter, J. A. 21, 100, 246
science *see* STI
SCM (Subsidies and Countervailing Measures) 172
Second World War 141, 144, 275
sectoral policies 285–7
segmentation 140; ineffective 155
shocks 51, 276
shoemaking industry 139–40

SID (systems of innovation for development) 27
Silicon Valley 136
Singapore 5, 40, 41, 50, 64, 69, 70, 73, 75, 167, 168, 172, 175, 179, 188–92, 248, 275, 295
Singh, A. 7
Singh, H. 177
skills gaps 35
small firms 31, 120, 248
SMART programme 120
Smarzynska, B. K. 178
SMEs (small and medium-sized enterprises) 23, 24–5, 46, 131–65, 227, 251; 'bulking-up' through networks 38; exposure to credit rationing 199–200, 208, 219; financing 201, 211–19; 'growth' 224, 225, 226; innovation 119; metrology and technical services for 54; part-grants for 36; private equity investment in 223–4; production relations between large firms and 40–1; promotion agencies 29; specialist 'soft terms' funding windows for 37; support for clusters of 268; technical extension services for 35
Smith, Adam 246
social capital 137
SOEs (state-owned enterprises) 35, 210, 211, 251, 256; disappointing performance 255; inefficient 261; private vs 253–4; privatising 53; reforming 263–6; sale of 260
South Africa 73, 185, 279
South Korea 5, 69, 70, 96, 171, 176, 222, 275, 294; *Chaebol* 123, 124, 202, 205, 210; STI 122–4
Soviet Union (former) 178
specialisation 19, 111, 141, 142, 153; sectoral 136
spillovers 107, 108, 182, 191; inter-industry 168; productivity 181
Sri Lanka 49, 73, 295
SRIs (specialist local research institutions) 117
stakeholders 47, 49, 202
standard of living 18, 20, 22
standards 4, 118
state intervention 48
Stigler, G. 252
Stiglitz, J. 12, 43, 107, 199–200, 208, 211, 254
STI (science, technology and innovation) policy 95–130

Stockes, M. T. 222
stock markets 225, 226, 260, 263
subcontracting 46, 141, 160, 290; and partnership exchange 149–50
subsidiarity 269
subsidies 37, 38, 149, 180, 247, 277, 283; export-related 282; import-related 281–2; R&D 181; WTO rules for 39
substitution 276, 277, 278
success 22, 40; acquiring firm-level technological capabilities 26; industrial 76; necessary conditions for 50–2
supply-chain relationships 3, 4, 125
supply-side policy 31, 239, 249
Sweden 209
'Swiss' competitiveness indices *see* IMD; WEF
Swisscontact 150, 152, 153, 155
synergies 27, 160
systems failures 28, 37, 106–112

Taiwan 5, 24, 40, 68, 70, 73, 139, 144–5, 150, 171, 176, 185, 275
takeovers 253, 254
targeting 191, 212; strategic 172
tariffs 34, 191, 287, 288; constant-level 283; *see also* GATT
Tatung 24
taxation 202, 253, 283; concessions 171; corporate 185; deduction scheme 35; holidays 179; *see also* VAT
Taylor, A. 9, 61
TBT (technical barriers to trade) 4
TCS (Teaching Company Scheme) 120
technical assistance 279, 288
technological capability 8, 10, 17, 20–2, 25–6, 28, 31, 35, 37, 52, 54, 56
technological effort 80, 82, 84
technological progress 22, 176; rapid pace of 67
technology 222; communications 137; innovation perspective and 21–30; SME support and 35–6; *see also* STI
technology and innovation perspective 8, 15–16, 21–2, 26, 28, 30, 53, 67–8
technology transfer 44, 173, 279
Teubal, M. 35, 43, 56–7
te Velde, D. W. 10, 36, 167, 169, 178, 181–2, 194
Thailand 5, 69, 70, 227, 287, 294
Thatcher, Margaret 119, 248
Theys, T. 198
TI (technology index) 45–7

Index

TNCs (transnational corporations) 167, 172; factors affecting 180; impact on social development 181; linkages with 182, 185, 190, 192; locational advantages/decisions 173, 178; standard required by 187; strategic asset-seeking 177; tax concessions to 178, 179
Tokyo 139, 141–2
Tonga 73
total quality management 161
tourism 289
trade: international 3, 19; role in development 274–80
tradeables sector 16, 17
trade barriers 207; falling 4, 22, 45; tariff and non-tariff 177, 186, 191
trade policy 34–5, 277, 280–2, 288–9
trade unions 30
'tragedy of the commons' 252
training 147–8; on-the-job 52; voucher scheme 148–9; *see also* education
transaction costs 28, 245–6
transition economies 211, 225
transparency 294
transport costs 4, 186
treasury bills 226
Treaty of Rome (1957) 243, 247
trial and error 105
TRIMs (Trade-Related Investment Measures) 172, 177, 179, 180
Trinidad and Tobago 49, 70, 73
TRIPs (Trade-related aspects of Intellectual Property Rights) 172, 177
Turkey 70, 294

Uganda 285
Ul Haque, I. 79
uncertainty 106, 110, 208
UNCTAD (UN Conference on Trade and Development) 125, 175, 289, 291, 295
UNDP (UN Development Programme) 69, 188
unemployment 133, 260
unfair competition 180
UNIDO (UN Industrial Development Organisation) 28, 29, 149, 151, 154
United Kingdom 22, 41, 50, 123, 212, 217, 218, 226, 259; regulatory bodies 209, 245, 258, 267; STI 118–22
United States 6–7, 24, 64, 96, 123, 139, 154, 180, 186, 187, 227, 291, 294, 295; anti-trust policy 243, 247; balance of policy 111; Competitiveness Policy Council 22; FDI 177, 178; manufacturing sectors 181; National Institute of Standards and Technology 118; provisioning policy 210; retail market 140; savings and loans 206, 218; Small Business Administration 212, 217, 219
unpredictability 102
utilities 37

value chains 44, 47; globally integrated 4
VAT (value-added tax) 282
Venezuela 168
venture capitalism 37, 201, 202, 219; developing 223–5
Vernon, R. 21
vertical linkages 141, 142
Vickers, J. 254

wages 182, 277
Wagie, D. M. 262
Wagner's Law 251
Wales 97
Warner, A. 80, 82
Waterson, 243, 244
wealth: creation of 240, 246, 247, 248, 257; national 19; personal 202
WEF (World Economic Forum) 6, 20, 61, 66, 68, 75–6; core elements of indices 62–4
Wei, S. J. 178
'weightless economy' 186
Weinberg, A. M. 121
welfare 242
Wheeler, D. 177
Wignaraja, G. 35, 46, 68, 78, 126–7, 132, 154
Williamson, J. 52
Williamson, 242, 245
Winsemius, Albert 188–9
Wolf, C. 252
women 133, 212
woodworking industry 142–4
World Bank 6, 42, 69, 73, 219, 255, 257, 258, 262, 263, 264, 265, 279, 296
WTO (World Trade Organisation) 4, 34, 35, 41, 172, 177, 178, 180, 277, 278, 280, 281–2, 283, 289, 290–5 *passim*; rules for subsidies 39; stringent compliance 48

Yamamura, K. 220
Yarrow, G. 254
Yip, G. 18, 19
Young, P. 198

Zimbabwe 289